HVAC
Control Systems

AMERICAN TECHNICAL PUBLISHERS, INC.
HOMEWOOD, ILLINOIS 60430-4600

Ronnie J. Auvil

American Technical Publishers, Inc. Editorial Staff

Editor in Chief:
> Jonathan F. Gosse

Production Manager:
> Peter A. Zurlis

Technical Editors:
> Peter A. Zurlis
> Russell G. Burris

Copy Editor:
> Richard S. Stein

Illustration/Layout:
> Maria R. Aviles
> Carl R. Hansen
> William J. Sinclair
> Sarah E. Kaducak
> Aimée M. Brucks
> Tina T. Biegel
> Ellen E. Pinneo

CD-ROM Development:
> Carl R. Hansen
> Maria R. Aviles

Adobe and Acrobat are trademarks of Adobe Systems Incorporated. Arcnet is a trademark of Datapoint Corporation. BACnet is a registered trademark of American Society of Heating, Refrigerating, and Air-Conditioning Engineers (ASHRAE), Inc. Echelon is a trademark of Echelon Corporation. National Electrical Code and NEC are registered trademarks of National Fire Protection Association, Inc., Quincy, MA 02269. Netscape is a registered trademark of Netscape Communications Corporation. Microsoft, Windows, Windows NT, Windows XP, Windows Paint, and Internet Explorer are either registered trademarks or trademarks of Microsoft Corporation.

1 2 3 4 5 6 7 8 9 – 03 – 9 8 7 6 5 4 3 2

Printed in the United States of America

ISBN 0-8269-0750-4

 **Acknowledgments**

The author and publisher are grateful for the technical information and assistance provided by the following companies and organizations:

Air Conditioning Contractors of America
Alnor Instrument Company
American Society of Heating, Refrigeration and Air Conditioning Engineers, Inc.
Carrier Corporation
Cleaver-Brooks
Fluke Corporation
Gateway Community College
Greenheck Fan Corp.
HAI (Home Automation, Inc.)
International Union of Operating Engineers
Jackson Systems, LLC
JUN-AIR USA, Inc.
KMC Controls
Leslie Controls, Inc.
Novar Controls Corporation
Ranco, Inc.
Saylor-Beall Manufacturing Company
The Trane Company
Trerice, H. O., Co.
Weiss Instruments, Inc.
White-Rodgers Div., Emerson Electric Co.

Contents

CD-ROM Contents

- **Using the CD-ROM**
- **Quick Quizzes**
- **Illustrated Glossary**
- **Media Clips**
- **Reference Material**

Introduction

HVAC Control Systems covers all aspects of commercial HVAC control systems. The topics included are specifically designed to aid HVAC and building maintenance technicians. The textbook includes an introduction to HVAC fundamentals, energy sources, and control principles. The main focus of the text is on pneumatic, electrical, electronic, and building automation control systems and components. The textbook also covers the latest technology in energy efficiency practices, networking fundamentals, direct digital control, building automation system retrofitting, maintenance management, and troubleshooting.

The Appendix contains reference material pertinent to the HVAC trade. The Glossary provides definitions of HVAC control system terms introduced in the textbook. The PC-compatible CD-ROM in the back of the book includes a Quick Quiz for each chapter, Illustrated Glossary, Media Clips, and related HVAC reference material. Information about using the *HVAC Control Systems* CD-ROM is included on the last page of the book. To obtain information about related training products visit the American Tech web site at www.go2atp.com.

The Publisher

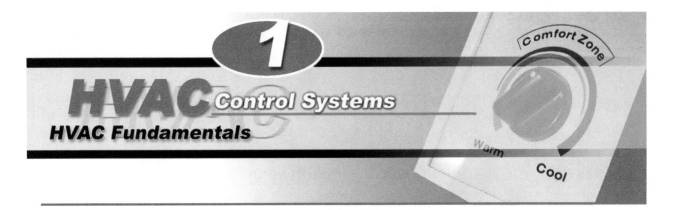

Occupants of a commercial building require that the indoor environment be comfortable. Comfort is a product of the temperature, humidity, filtration, circulation, and ventilation of the air. HVAC systems add or remove heat from building spaces. Heat is transferred to building spaces by conduction, convection, and/or radiation. Psychrometrics is the study of the properties of air and the relationships between them, which are related to comfort.

COMFORT

Heating, ventilating, and air conditioning (HVAC) systems are used in buildings to provide comfort to occupants. *Comfort* is the condition that occurs when people cannot sense a difference between themselves and the surrounding air. The five requirements for comfort are proper temperature, humidity, filtration, circulation, and ventilation. See Figure 1-1. *Discomfort* is the condition that occurs when people can sense a difference between themselves and the surrounding air. Discomfort can occur when any of the five requirements for comfort are not met.

Temperature

Temperature is the measurement of the intensity of the heat of a substance. Controlling the temperature of the human body is an important physiological function. A *physiological function* is a natural physical or chemical function of an organism. The body produces energy by digesting food. Some energy is used for normal living processes, some is stored, and some is used as thermal energy (heat). Physiological systems regulate body temperature to maintain comfort.

Normal body temperature is 98.6°F. The body has natural heating and cooling systems to maintain this temperature. These systems control heat output by responding to the conditions of the air relative to the internal temperature of the body. The body responds by controlling blood flow at the surface of the skin, radiating heat from body surfaces, or using evaporation of perspiration from skin. *Evaporation* is the process that occurs when a liquid changes to a vapor by absorbing heat. When the body is clothed, the body's temperature control system provides comfort at an air temperature of approximately 75°F. If the air temperature varies much above or below 75°F, the body begins to feel uncomfortably warm or uncomfortably cool.

Signals (electrical impulses) from different points in the body are sent through a network of nerves to the hypothalamus, a gland in the brain. The hypothalamus regulates body temperature by controlling blood flow to capillaries (tiny blood vessels) located in the skin. Capillaries regulate perspiration flow to the surface of the skin. If the body temperature rises, blood flow to the skin increases. Blood carries heat to the skin, where it is given off to the air.

While comfort can be defined and measured, comfort may also be affected by the difference between the outdoor and indoor environmental conditions. For example, an individual entering a building on a cold day may feel comfortable even though the indoor temperature is below comfort standards.

COMFORT REQUIREMENTS

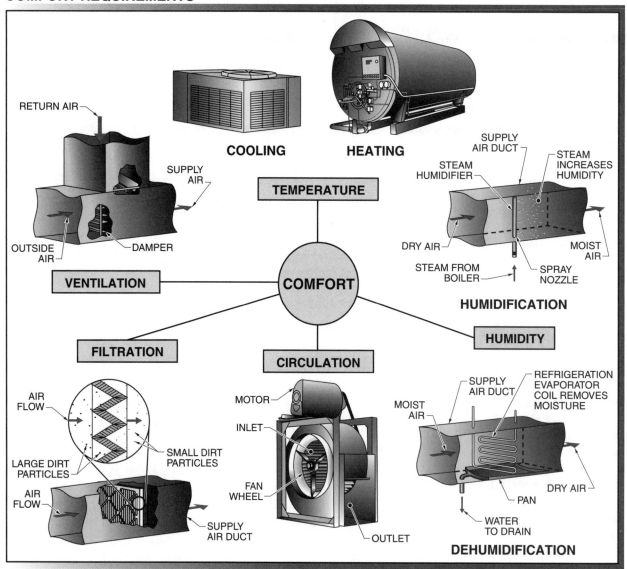

Figure 1-1. Heating, ventilating, and air conditioning equipment work together to provide the proper temperature, humidity, filtration, circulation, and ventilation required for occupant comfort.

A person who becomes overheated also becomes flushed because blood flow to the skin increases. If signals indicate the body is cooling off, the system reduces the cooling effect by allowing less blood flow to the skin. In cold weather, fingers and toes become cold before the rest of the body because blood flow to these areas is reduced. Cold areas also appear pale because of the reduced blood flow.

Technicians working in the health care field must consider the physical condition of the patients because it affects their comfort level.

Temperature control in buildings is provided by cool air from air conditioning equipment or warm air from heating equipment. In cold weather, heating equipment supplies the proper amount of heat to offset heat loss from a building.

Humidity

Humidity is the amount of moisture present in the air. Humidity is always present in air. A low humidity level indicates dry air that contains little moisture. A high humidity level indicates damp air that contains a significant amount of moisture.

Humidity affects comfort because it determines how slowly or rapidly perspiration evaporates from the body. The flow of perspiration is controlled by the cooling system of the body, which regulates body temperature. Evaporation of perspiration cools the body. The higher the humidity, the slower the evaporation rate. The lower the humidity, the faster the evaporation rate. For example, with no temperature change and an increase in humidity, a person feels warmer because of the slower evaporation rate. With no temperature change and a decrease in humidity, a person feels cooler because of the faster evaporation rate.

Comfort is usually attained at normal cooling and heating temperatures with a humidity level of about 50%. If the humidity is too low, a higher air temperature is required to feel comfortable. If the humidity is too high, a lower air temperature is required for the same feeling of comfort. Humidity level is controlled by humidifiers and dehumidifiers. See Figure 1-2.

A *humidifier* is a device that adds moisture to air by causing water to evaporate into the air. In cold climates and dry climates, the humidity in a building may be too low for comfort. Humidifiers are used in buildings to add moisture to the air to maintain a comfortable humidity level.

A *dehumidifier* is a device that removes moisture from air by causing moisture to condense. *Condensation* is the formation of liquid (condensate) as moisture or other vapor cools below its dew point. *Dew point* is the temperature of air below which moisture begins to condense from the air. In buildings with swimming pools or a large number of potted plants, the humidity level may be too high for comfort.

Filtration

Filtration is the process of removing particulate matter from air that circulates through an air distribution system. Airborne particulate matter may create an environment inside a building that is unhealthy for building occupants. *Indoor air quality (IAQ)* is a designation of the contaminants present in the air. Airborne particulate matter includes dirt particles, spores, pollen, leaves, etc. All airborne particulates possess their own unique physical properties and possible effects on health. The size of airborne particles is measured in microns. A *micron* is a unit of measure equal to .000039″.

Circulated air in a building must be clean. Circulated air in a building includes return air that is recirculated and fresh air that is used for ventilation. The air in a forced-air heating system is automatically cleaned by filters that are placed in the ductwork. The primary consideration in air filter selection is the size of particulates removed. Filters are commercially available in low-, medium-, and high-efficiency levels. In small commercial buildings, low-efficiency filters are sufficient for normal filtering applications. Medium- and high-efficiency filters are used according to the degree of filtration required. Some filters are designed for disposal after use, while others may be cleaned and reused. Special filters are also available which remove organic and biological compounds, or absorb specific types of gases and vapors, from the air stream.

HUMIDITY CONTROL

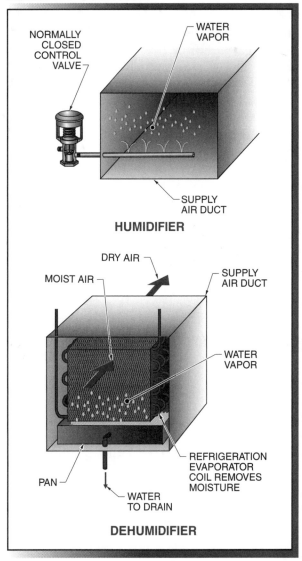

Figure 1-2. Humidifiers and dehumidifiers control the level of humidity in a building.

Circulation

Circulation is the movement of air. Air in a building must be circulated continuously to provide maximum comfort. In a building where there is improper circulation, air rapidly becomes stagnant and uncomfortable. *Stagnant air* is air that contains an excess of impurities and lacks the oxygen required for comfort. When air is improperly circulated, temperature stratification occurs. *Temperature stratification* is the variation of air temperature in a building space that occurs when warm air rises to the ceiling and cold air drops to the floor. Temperature stratification in a building space causes discomfort.

Air velocity is the speed at which air moves from one point to another. Air velocity is measured in feet per second (fps). If air is not moving, heat cannot be carried away from skin surfaces. Air circulation is required for comfort because it helps cool the body by evaporating perspiration. An increase in air velocity increases the rate of evaporation of perspiration from the skin, causing a person to feel cool. A decrease in air velocity or no air velocity reduces the evaporation rate, causing a person to feel warm. Air movement is used to supply clean and fresh air and to remove stagnant air. Air movement at 40 cubic feet per minute (cfm) is considered ideal.

Supply air ductwork and registers are used to distribute air to different building spaces. Supply air ductwork is sized and located for maximum efficiency.

The type, location, and size of registers determine the amount of supply air introduced into each building space and the distribution pattern of the air.

Return air ductwork and grilles are used to move air from building spaces to the heating/cooling equipment. Return air ductwork and grilles, like supply air ductwork and registers, are sized and located for maximum efficiency. See Figure 1-3.

> *Commercial buildings are often reconfigured by adding walls, changing the location of the controller, etc. When a building space is reconfigured, airflow and comfort may suffer. Any time a reconfiguration of the areas in a building is performed, the ductwork must be checked to ensure adequate air flow, circulation, and comfort.*

Ventilation

Ventilation is the process of introducing fresh air into a building. Ventilation and air circulation are necessary for comfort inside a building. Offensive odors and vapors are created in closed spaces. The air in building spaces is kept comfortable by diluting the contaminated air with fresh air. This is accomplished by introducing fresh air from the outdoors with recirculated air.

CIRCULATION

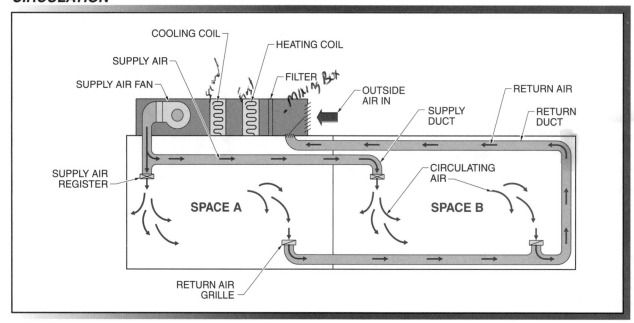

Figure 1-3. Supply and return air ductwork is sized and located to provide efficient flow of air through building spaces.

Some odors that originate in a building may be offensive and/or dangerous. Offensive odors are eliminated by bringing fresh air into an HVAC system from outdoors. See Figure 1-4. The amount of ventilation necessary for comfort is determined by the number of occupants and the activity taking place inside the building. If odors or contaminants originating in a building are potentially toxic, the air in the building may require continuous exhaust with no recirculation. The building would require 100% makeup air for ventilation. *Makeup air* is air that is used to replace air that is lost to exhaust.

VENTILATION

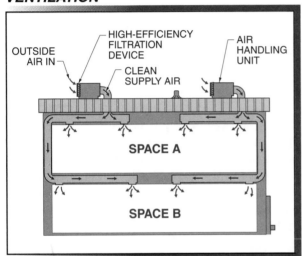

Figure 1-4. Ventilation is the process of introducing fresh air into a building.

THERMODYNAMICS

Thermodynamics is the science of thermal energy (heat) and how it transforms to and from other forms of energy. Thermodynamics relates to events that occur when air is heated or cooled in a building space. The two laws of thermodynamics that apply to heating and air conditioning systems are the first law of thermodynamics and the second law of thermodynamics.

First Law of Thermodynamics

The *first law of thermodynamics,* the law of conservation of energy, states that energy cannot be created or destroyed but may be changed from one form to another. An example of the first law of thermodynamics is the combustion process. In the combustion process, as fuel burns, the elements hydrogen and carbon com-

bine with oxygen in the air. As these elements combine, a chemical reaction occurs, and chemical energy in the elements is released in the form of thermal energy (heat). The hydrogen, carbon, and oxygen recombine to form new compounds. During the process, energy has changed from chemical energy to thermal energy, but no energy has been created or destroyed.

Another example of the first law of thermodynamics is a compressor operating in an air conditioning system. An electric motor uses electrical energy to drive the compressor. The electrical energy is converted to mechanical energy in the motor. The mechanical energy is used to compress the refrigerant in the compressor. As the compressed refrigerant expands, it produces a cooling effect. Most of the electrical energy used to drive the motor which drives the compressor results in a cooling effect from the system. Some of the mechanical energy is converted to thermal energy because of friction, but no energy is created or destroyed in the process.

Second Law of Thermodynamics

The *second law of thermodynamics* states that heat always flows from a material at a high temperature to a material at a low temperature. The flow of heat is natural and does not require energy. The second law of thermodynamics applies to all cases of heat transfer. For example, air in a furnace is heated by the products of combustion. Heat flows from the warm burner flame to the cool air.

Cleaver-Brooks

The second law of thermodynamics is applied in a boiler as heat is transferred from the hot gases of combustion inside the boiler tubes to the cool water surrounding the tubes.

Heat Measurement

Heat is the measurement of energy contained in a substance and is identified by a temperature difference or a change of state. All substances exist in either a solid, liquid, or gas state. *Change of state* is the process that occurs when enough heat is added to or removed from a substance to change it from one physical state to another, such as from ice to water or water to steam. Substances may contain sensible heat and/or latent heat.

Sensible heat is heat measured with a thermometer or sensed by a person. Sensible heat does not involve a change of state. *Latent heat* is heat identified by a change of state and no temperature change. Latent heat is heat added to ice that changes it to water, or heat added to water that changes it to steam. See Figure 1-5. Heat transferred in heating systems is usually sensible heat. Heat transfer in air conditioning systems includes both sensible heat and latent heat.

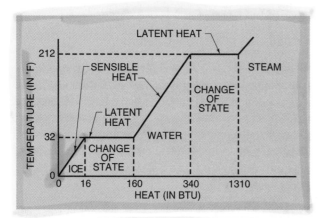

Figure 1-5. Sensible heat raises the temperature of a substance. Latent heat causes the substance to change state.

In the U.S. system of measurements, quantity of heat is measured in British thermal units (Btu). A *British thermal unit (Btu)* is the amount of heat energy required to raise the temperature of 1 lb of water 1°F. Heating equipment is rated in Btu per hour (Btu/hr). Large quantities of heat are measured in thousands of Btu per hour (Mbh). For example, 2000 Btu/hr equals 2 Mbh.

Air conditioning equipment supplies the proper amount of cooling in hot weather to offset heat gain to a building. Air conditioning equipment is rated in Btu per hour or ton of cooling. A *ton of cooling* is the amount of heat required to melt a ton of ice (2000 lb) over a 24-hour period. One ton of cooling equals 288,000 Btu per 24-hour period or 12,000 Btu/hr (12,000 Btu/hr × 24 hr = 288,000 Btu/24-hr).

Heat Transfer

Heat transfer is the movement of heat from one material to another. Heat is transferred from one substance to another when a temperature difference exists between the two substances. Heat is always transferred from a substance with a higher temperature to a substance with a lower temperature. Heat transfer rates increase with the temperature difference between two substances. Three methods of heat transfer are conduction, convection, and radiation. See Figure 1-6.

Conduction. *Conduction* is heat transfer that occurs when molecules in a material are heated and the heat is passed from molecule to molecule through the material. For example, if one end of a metal rod is heated, heat is transferred by conduction to the other end. Heating one end of a metal rod heats molecules, making them move faster. The faster-moving molecules transfer energy from molecule to molecule through the metal rod.

Convection. *Convection* is heat transfer that occurs when currents circulate between warm and cool regions of a fluid. For example, as air is warmed by a fire, the warm air rises and is replaced by cool air. The movement of air creates a current that continues as long as heat is applied. Convection transfers heat from an object to an intermediate substance which then transfers its heat to the desired object. In a hot water heating system, heat from a hot heating element is transferred to water, which then transfers its heat to air at a terminal unit. The warm air flows over the occupants, transferring heat to the occupants.

Radiation. *Radiation* is heat transfer in the form of radiant energy (electromagnetic waves). Radiation occurs from object to object directly without the objects touching. The amount of heat transferred depends on the intensity of heat (temperature) and the distance between the heat source and object. The heat transferred is inversely proportional to the distance between the heat source and object. Radiant energy waves move through space without producing heat. Heat is produced when the radiant energy waves contact an opaque object. A common application of radiation is an electric heating element. The heat rays from the element pass through the air without heating it. These heat rays strike the occupants in the building space, heating them. Heat produced on Earth by light waves from the sun is a form of radiation.

HEAT TRANSFER

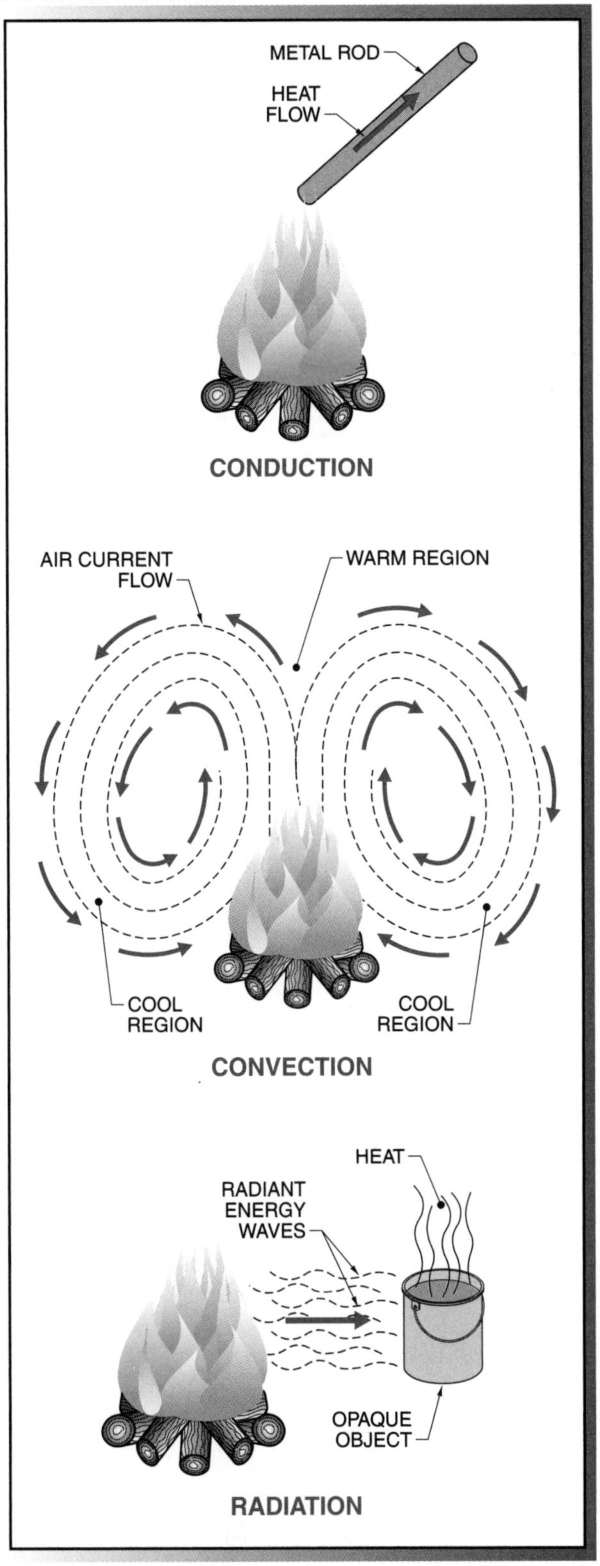

CONDUCTION

CONVECTION

RADIATION

Figure 1-6. Heat is transferred from one substance to another by conduction, convection, or radiation when a temperature difference exists between the two substances.

HVAC systems use conduction, convection, and radiation in various combinations to provide comfort in a building space. The quantity of heat involved in heat transfer is a function of weight, specific heat, and temperature difference. *Weight* is the force with which a body is pulled downward by gravity. In heat calculations, weight is used instead of volume because weight, unlike volume, remains constant despite changes in temperature. The volume of most materials changes with a change in temperature. In heat transfer calculations, weight is expressed in pounds (lb) or grams (g).

Different substances absorb and release heat energy at different rates. *Specific heat* is the ability of a material to hold heat. Specific heat is expressed as the ratio of the quantity of heat required to raise the temperature of a material 1°F compared to the quantity required to raise the temperature of an equal mass of water 1°F. Specific heat is used in calculations and is normally given in Btu/lb/°F.

The specific heat of water is 1 and is used as the standard for calculating specific heat. Since the specific heat of water is 1 Btu/lb/°F, it takes 1 Btu to change the temperature of water 1°F. The specific heat of most materials is less than the specific heat of water because water holds a large quantity of heat. All substances have a constant value for specific heat that can be found on tables and charts. See Figure 1-7. The specific heat of air is .24 Btu/lb/°F. This means that 1 lb of water holds as much heat energy as approximately 4 lb of air.

SPECIFIC HEAT*					
Solid		**Liquid**		**Gas**	
Aluminum	.214	Alcohol	.615	Air	.24
Brass	.09	Ammonia	1.099	Butane	.377
Coal	.3	Kerosene	.5	CO_2	.20
Concrete	.156	Mineral oil	.5	Chlorine	.117
Glass	.18	Petroleum	.4	Helium	1.241
Gold	.031	R-22	.26	Methane	.520
Ice	.487	R-502	.255	Neon	.246
Iron	.12	Saltbrine	.745	Oxygen	.218
Rubber	.48	Turpentine	.42	Propane	.375
Wood	.45	Water	1.00	Steam	.48

* in Btu/lb/°F

Figure 1-7. Specific heat is the ability of material to hold heat. Values of specific heat are constants that are given in tables and charts.

PSYCHROMETRICS

Psychrometrics is the scientific study of the properties of air and the relationships between them. *Atmospheric air* is a mixture of dry air, moisture, and particles. *Dry air* is the elements that make up atmospheric air with the moisture and particles removed. *Moist air* is the mixture of dry air and moisture. The properties of air are the characteristics of air, which are temperature, humidity, enthalpy, and volume. The properties of air determine the condition of the air, which is related to comfort.

Temperature Measurement

Temperature is the most important variable that is measured and controlled in a commercial HVAC system. The two substances that are most often measured are air and water. A measurement of 70°F indicates that there is enough heat in the air to register a 70°F reading on a thermometer. The temperature of 70°F indicates intensity of heat, not quantity of heat. Quantity of heat is expressed in Btu.

All thermometers, including electronic sensors, may need to be checked for accuracy.

Temperature is commonly expressed using the Fahrenheit or Celsius scale. The scale most often used to measure temperature in the U.S. is the Fahrenheit scale. On the Fahrenheit scale, 32°F is the freezing point and 212°F is the boiling point of water at normal atmospheric pressure (14.7 psia). The Celsius scale is used in Canada, Europe, Japan, and other areas of the world. On the Celsius scale, 0°C is the freezing point and 100°C is the boiling point of water at normal atmospheric pressure.

Although many modern HVAC systems are designed for worldwide use and often have temperatures listed in degrees Fahrenheit and degrees Celsius, it is sometimes necessary to convert temperature readings from one scale to the other. See Figure 1-8. The relationships between the freezing and boiling points on the Fahrenheit and Celsius scales make it possible to convert from one scale to the other.

TEMPERATURE CONVERSION

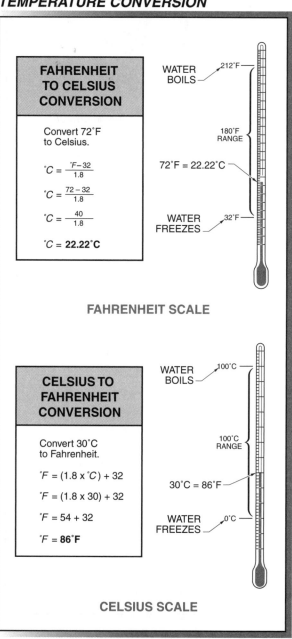

FAHRENHEIT TO CELSIUS CONVERSION

Convert 72°F to Celsius.

$$°C = \frac{°F - 32}{1.8}$$

$$°C = \frac{72 - 32}{1.8}$$

$$°C = \frac{40}{1.8}$$

$$°C = \mathbf{22.22°C}$$

WATER BOILS — 212°F

180°F RANGE

72°F = 22.22°C

WATER FREEZES — 32°F

FAHRENHEIT SCALE

CELSIUS TO FAHRENHEIT CONVERSION

Convert 30°C to Fahrenheit.

$$°F = (1.8 \times °C) + 32$$

$$°F = (1.8 \times 30) + 32$$

$$°F = 54 + 32$$

$$°F = \mathbf{86°F}$$

WATER BOILS — 100°C

100°C RANGE

30°C = 86°F

WATER FREEZES — 0°C

CELSIUS SCALE

Figure 1-8. Temperature is commonly expressed by using the Fahrenheit or Celsius scale.

To convert between the two scales, the difference between the bases and the ratio of the difference between the bases are used. On the Fahrenheit scale, 32°F is the base. On the Celsius scale, 0°C is the base. Therefore, 32° is the difference between the bases. To find the difference between the bases, apply the formula:

base difference = base F – base C
where
base difference = difference between bases
base F = base of Fahrenheit scale (32)
base C = base of Celsius scale (0)

For example, what is the difference between the bases of the Fahrenheit and Celsius scales?

base difference = base F – base C
base difference = 32 – 0
base difference = **32°**

The ratio of the difference between the two scales is determined by the difference between the freezing point and the boiling point of water on each scale. There is a range of 180°F between 32°F and 212°F on the Fahrenheit scale and 100°C between 0°C and 100°C on the Celsius scale. The ratio for this conversion is found by dividing 180 by 100. To find the ratio between the bases, apply the formula:

$$ratio = \frac{range\,F}{range\,C}$$

where
ratio = ratio of the ranges of the temperature scales
range F = difference between Fahrenheit freezing and boiling points
range C = difference between Celsius freezing and boiling points

For example, what is the ratio of the ranges of the Fahrenheit and Celsius scales?

$$ratio = \frac{range\,F}{range\,C}$$
$$ratio = \frac{212 - 32}{100 - 0}$$
$$ratio = \frac{180}{100}$$
$$ratio = \mathbf{1.8}$$

There is 1.8°F for every 1.0°C. Both of these numbers are used in the formulas for converting from one scale to the other. To convert a Fahrenheit reading to Celsius, subtract 32 from the Fahrenheit reading and divide by 1.8. A Fahrenheit temperature is converted to Celsius by applying the formula:

$$°C = \frac{°F - 32}{1.8}$$

where
°C = degrees Celsius
°F = degrees Fahrenheit
32 = difference between bases
1.8 = ratio between bases

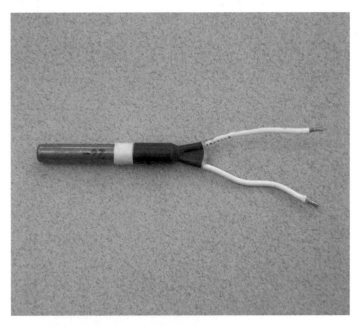

Most temperature measurement uses an electronic sensing element that indicates temperature on Fahrenheit and Celsius scales.

For example, what is 72°F converted to the Celsius scale?

$$^\circ C = \frac{^\circ F - 32}{1.8}$$

$$^\circ C = \frac{72 - 32}{1.8}$$

$$^\circ C = \frac{40}{1.8}$$

$$^\circ C = \mathbf{22.22^\circ C}$$

To convert a Celsius reading to Fahrenheit, multiply the Celsius reading by 1.8 and add 32. A Celsius temperature is converted to Fahrenheit by applying the formula:

$$^\circ F = (1.8 \times {^\circ C}) + 32$$

where
$^\circ F$ = degrees Fahrenheit
1.8 = ratio between bases
$^\circ C$ = degrees Celsius
32 = difference between bases

For example, what is 30°C converted to the Fahrenheit scale?

$^\circ F = (1.8 \times {^\circ C}) + 32$
$^\circ F = (1.8 \times 30) + 32$
$^\circ F = 54 + 32$
$^\circ F = \mathbf{86^\circ F}$

Infrared thermometers may be used to measure the temperature of high-temperature devices such as steam traps and steam lines.

A thermometer should be chosen based on the application requirements. For example, if accuracy is critical, an electronic thermometer may be used. If certified readings must be provided to a governing authority, a master thermometer, calibrated at regular intervals to another master thermometer, may be used. These are referred to as traceable instruments.

Temperature is most often measured using a thermometer. Thermometers are available in many configurations. Some thermometers use a bimetallic element to sense the temperature, while many modern thermometers use an electronic sensing element. Electronic thermometers display the temperature in digits. A thermometer should be of good quality and should be checked against a known accurate thermometer regularly to prevent inaccurate readings. See Figure 1-9. Many good quality thermometers allow calibration. *Dry bulb temperature (db)* is the temperature of the air without reference to the humidity level. Dry bulb temperature is measured with a thermometer.

THERMOMETERS

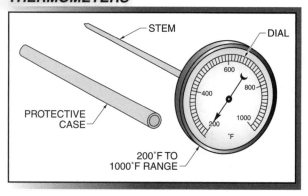

Figure 1-9. Thermometers are used to measure temperature.

Humidity Measurement

Humidity is produced from water that has evaporated into the air. The volume of moisture in the air compared to the total volume of the air is small. An inadequate or excessive amount of humidity can cause discomfort. For example, very dry air can cause throat and lung irritation and very humid air can make people feel uncomfortably hot or cold. Humidity represents latent heat. Latent heat is identified by a change of state and no temperature change. Therefore, latent heat cannot be measured with a thermometer.

Humidity is measured using a hygrometer. A *hygrometer* is any instrument used for measuring humidity. A *dimensional change hygrometer* is a hygrometer that operates on the principle that some materials absorb moisture and change size and shape depending on the amount of moisture in the air. An *electrical impedance hygrometer* is a hygrometer based on the principle that the electrical conductivity of a material changes as the amount of moisture in the air changes. Most hygrometers give direct percentage readings of humidity (relative humidity). See Figure 1-10. Humidity is expressed as either relative humidity or specific humidity.

HYGROMETERS

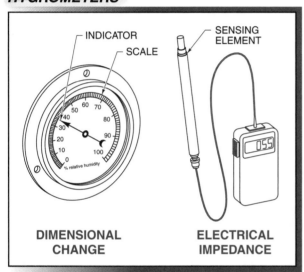

Figure 1-10. Hygrometers measure and provide readings of relative humidity.

Relative Humidity. *Relative humidity* is the amount of moisture in the air compared to the amount of moisture that it could hold if it were saturated (full of water). Saturated air carries as much moisture as possible before the moisture forms into water droplets. The temperature of the air is the main factor that determines how much moisture the air could hold. See Figure 1-11. For example, air at 70°F with a relative humidity of 100% holds 100% of the maximum moisture it can hold at 70°F. Likewise, air at 70°F with a relative humidity of 50% holds 50% of the maximum moisture it can hold at 70°F. The unit of measure used for relative humidity is percent relative humidity (% rh). The amount of moisture required to saturate the air changes as the dry bulb temperature changes. In addition, the relative humidity and the capacity to hold moisture change as the dry bulb temperature changes.

> *The advent of electronic sensing element materials has provided greater accuracy and reliability. This has dramatically changed the humidity control industry.*

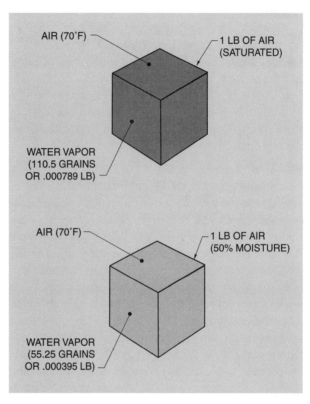

Figure 1-11. Relative humidity is the amount of moisture in the air compared to the amount of moisture that it could hold if it were saturated.

Humidity sensor packages may be mounted on the wall in a building space to provide accurate humidity level control.

Relative humidity is a humidity measurement that is always dependent on the temperature of the air. In some cases, this causes the measurement to be misleading. A measurement of 50% relative humidity indicates a certain quantity of moisture at an air temperature of 40°F and a different quantity of moisture at an air temperature of 65°F.

Specific Humidity. *Specific humidity* is a measurement of the exact amount or weight of the moisture in the air. Specific humidity is not dependent on the temperature of the air. Specific humidity is measured in grains of moisture per pound of air. A *grain* is a unit of measure equal to ¹⁄₇₀₀₀ lb. Specific humidity is not usually measured directly by an instrument but is derived by referring to a psychrometric chart. A *psychrometric chart* is a graph that defines the properties of the air at various conditions.

Wet Bulb Temperature. *Wet bulb temperature* is the temperature of the air taking into account the amount of humidity in the air. The dry bulb temperature measures only the temperature of the air without reference to humidity. Thus, dry bulb and wet bulb temperatures are different. Wet bulb temperature is measured using a sling psychrometer. See Figure 1-12. A sling psychrometer consists of two thermometers. One thermometer has a wick (sock) covering it which is saturated with water. The sling psychrometer is twirled around in the air to be measured. The water evaporates from the wick at a rate determined by the amount of humidity in the air. This evaporation cools the thermometer to a lower reading than the thermometer without a wick. The psychrometer often has a chart for determining relative humidity from the wet and dry bulb temperatures.

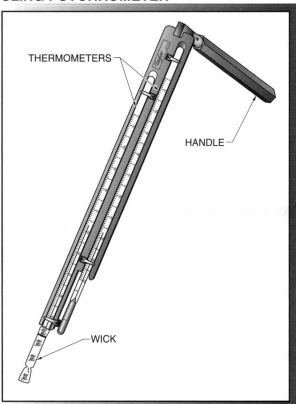

THERMOMETERS

HANDLE

WICK

Figure 1-12. A sling psychrometer is used to measure the wet bulb temperature of the air.

Dew Point Temperature. *Dew point temperature* is the dry bulb temperature of the air at which the moisture in the air condenses and falls out as dew, rain, sleet, ice, or snow. Dew point temperature is a function of the dry bulb temperature and the relative humidity of the air. Dew point temperature of the air is found by referencing the temperature and relative humidity of the air on a psychrometric chart.

Wet bulb and dew point temperatures are used in various HVAC applications. For example, cooling towers release heat to the ambient air through evaporation. The ability of a cooling tower to reject heat is directly related to the ambient wet bulb temperature. Cooling towers are often rated by the amount of heat they can reject at specific wet bulb temperatures. In addition, some advanced humidity sensors cool a mirror and optically detect the temperature at which condensation forms on the mirror (dew point temperature).

Cooling towers are rated using typical ambient wet bulb temperatures.

Enthalpy

Enthalpy (h) is the total heat contained in a material. Enthalpy is the sum of sensible heat and latent heat. Enthalpy is expressed in Btu/lb of moist air. Enthalpy is a better indicator of the condition of the air than dry bulb temperature. Outside air enthalpy is often measured and used to determine the suitability of outside air for use in place of mechanical cooling.

Pressure

Pressure is the force created by a substance per unit of area. See Figure 1-13. For example, if a weight of 1 lb is placed on a surface of 1 sq in., the pressure exerted is 1 lb per sq in. (1 psi). In addition, if a weight of 100 lb is placed on a surface of 1 sq in., the pressure exerted is 100 lb per sq in. (100 psi). Pressure measurements may be atmospheric pressure, gauge pressure, absolute pressure, or inches of water column.

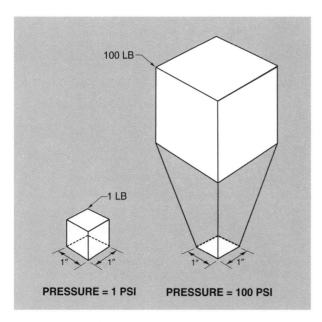

Figure 1-13. Pressure is the force created by a substance per unit of area.

Atmospheric Pressure. *Atmospheric pressure* is the pressure due to the weight of the Earth's atmosphere pressing against an object on the Earth's surface. At sea level, there is 14.7 lb of air pushing against every square inch of an object on the Earth's surface. This pressure is 14.7 lb per sq in. (14.7 psi). Atmospheric pressure is relatively constant and varies only a small amount due to elevation and humidity.

Small duct pressures, in inches of water column (in. wc), can be measured accurately by using high-quality sensors.

Gauge Pressure. *Gauge pressure* is the pressure above atmospheric pressure that is used to express pressures inside a closed system. Gauge pressure assumes that atmospheric pressure is zero (0 psi). Gauge pressure is expressed in pounds per square inch gauge (psig). For example, 50 lb of pressure per square inch inside a tank equals a gauge pressure of 50 lb per square inch gauge (50 psig). Gauge pressure is the most common pressure measurement used in HVAC systems.

Absolute Pressure. *Absolute pressure* is pressure above a perfect vacuum. Absolute pressure is the sum of gauge pressure and atmospheric pressure. Absolute pressure is expressed in pounds per square inch absolute (psia). The relationship of absolute, gauge, and atmospheric pressure is summarized as psia = psig + 14.7 psi.

Inches of water column (in. wc) is a unit of measure used when indicating very small pressures due to air movement in ducts or gas pressures in a pipe. An inch of water column equals .036 psi. It takes 27.7″ wc to equal 1 psi. Inches of water column (in. wc) is also expressed as inches of water gauge (in. wg).

Psychrometric Charts

A psychrometric chart defines the properties of the air at various conditions. A psychrometric chart provides a technician with a method of correlating dry bulb temperature, wet bulb temperature, relative humidity, specific humidity, and dew point temperature of air at various conditions. See Figure 1-14. If two properties of air are known, the remaining properties can be found using a psychrometric chart. Comfort is defined as an area on the psychrometric chart. In general, a dry bulb temperature of 73°F to 75°F and a relative humidity of approximately 40% to 50% is considered comfortable.

PSYCHROMETRIC CHART

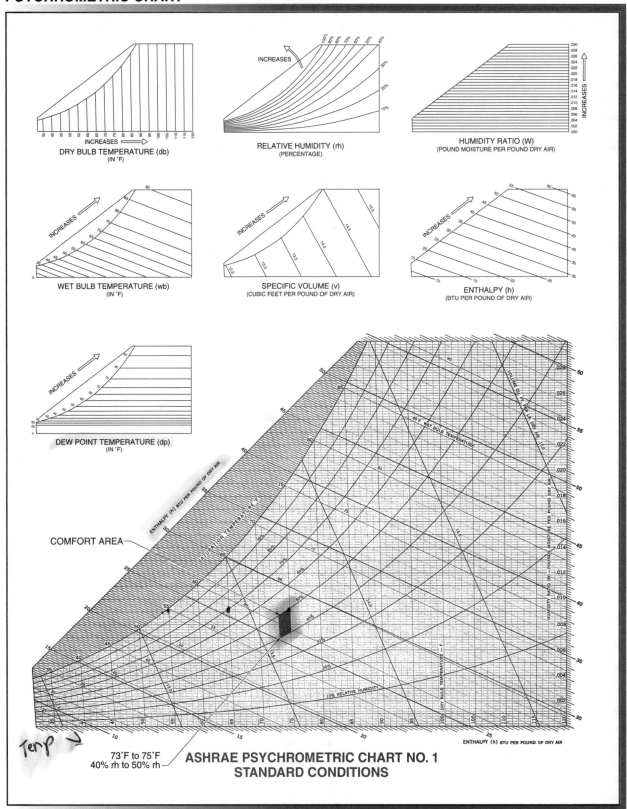

Figure 1-14. A psychrometric chart is a graph that defines the properties of the air at various conditions.

In commercial HVAC systems, heat is produced by hot water, steam, electricity, heat pumps, natural gas, or fuel oil-fired systems. Cooling is produced by outside air economizers, direct expansion cooling, and water chillers. Commercial HVAC systems normally consist of an air handling unit that conditions the air sent to the building space. Air handling units may be constant-volume or variable air volume air handling units.

COMMERCIAL HVAC SYSTEMS

A *commercial HVAC system* is a heating, ventilating, and air conditioning system used in office buildings, strip malls, stores, restaurants, and other commercial buildings. Commercial HVAC systems must provide comfort to building occupants during all times of the year. A commercial HVAC system may contain a heating system, ventilation system, cooling system, humidification system, dehumidification system, and/or air filtration system.

Heating Systems

A *heating system* is a system that increases the temperature of a building space. Heating systems consist of a heating device, fan or pump, heat exchangers, ductwork or pipes, and controls. Heating systems are classified based on the medium or heat source used. Common heating systems used to heat commercial buildings include hot water, steam, electric heat, heat pump, natural gas, or oil-fired systems.

Hot Water Heating Systems. A *hot water heating system* is a heating system that uses hot water as a medium to heat building spaces. Hot water heating systems are the most common heating systems used in commercial HVAC systems. Hot water heating systems consist of a hot water source, circulating pump,

piping, and controls to ensure the water is at the correct temperature and location to provide heat for building spaces. Hot water sources include hot water boilers or steam-to-hot-water heat exchangers. See Figure 2-1. A *heat exchanger* is a device that transfers heat from one substance to another substance without allowing the substances to mix. In a tube-and-shell heat exchanger, steam surrounds tubes through which water flows. Heat is transferred from the steam to the water in the tubes. The hot water is then used to heat building spaces.

Hot water heating systems circulate hot water throughout a building using a circulating pump. The hot water passes through heating units in the building spaces or through coils located in ductwork. Ductwork coils require a fan to force air over the coils to transfer heat to the air. The warm air flows through the ductwork to warm building spaces.

Steam Heating Systems. A *steam heating system* is a heating system that uses steam as a medium to heat building spaces. Steam heating systems are similar to hot water heating systems in that a boiler is used to provide heat. Due to the high heat content of steam, steam is an excellent medium for providing heat for building spaces. Steam is generated in a boiler and flows through heating units or coils located in ductwork that transfer heat to the building spaces.

HOT WATER HEATING SYSTEMS

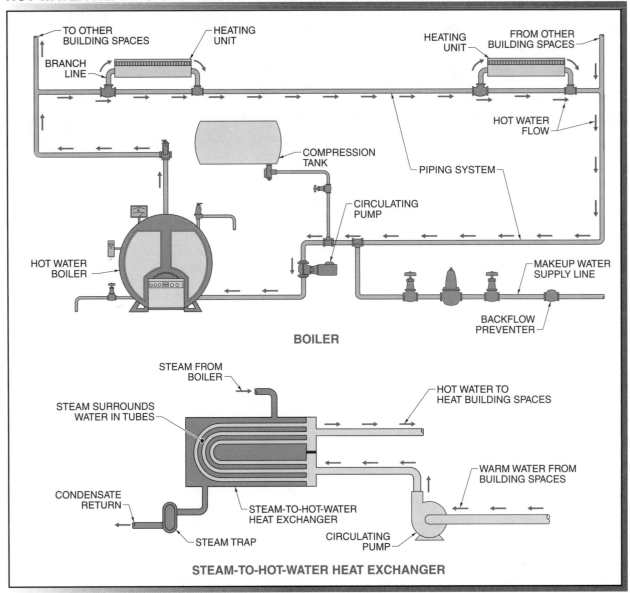

Figure 2-1. Hot water heating systems use hot water boilers or steam-to-hot-water heat exchangers to provide hot water for heating commercial buildings.

Some boilers can be converted from a hot water heating system to a steam heating system. Steam heating systems do not require circulating pumps and use steam traps attached to the discharge side of heating units to remove the condensation created as the steam releases heat. See Figure 2-2. The high amount of heat energy contained in steam causes the air temperature in building spaces to rise quickly. Steam heating system control is more difficult than hot water heating system control due to the rapid rise in air temperature.

Electric resistance heating elements are widely used in commercial HVAC systems and may include a wire guard to prevent accidental contact and injury to maintenance personnel.

STEAM HEATING SYSTEMS

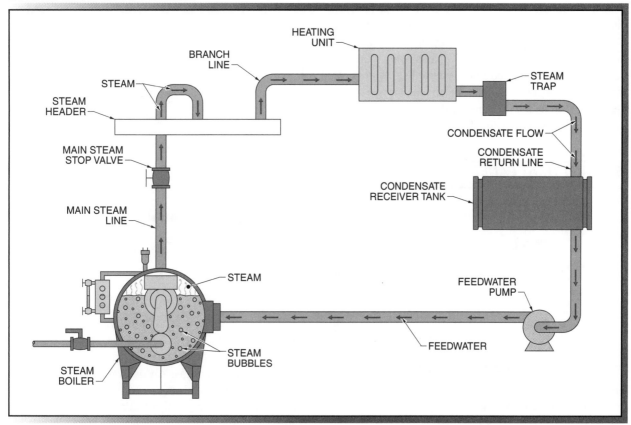

Figure 2-2. Steam heating systems use steam, which provides a large amount of heat energy to heat building spaces.

Electric Heating Systems. An *electric heating system* is a heating system that consists of electric resistance heating elements and a fan that circulates air across the heating elements. See Figure 2-3. Resistance to the flow of electric current causes the resistance heating elements to become hot. The fan circulates air over the elements, transferring the heat to the air. The advantages of electric heating systems include simple control, ease of installation, and low installation cost due to the lack of piping required. The disadvantages of electric heating systems include the high cost of electricity and the precautions that must be taken to ensure that the heating elements and building are not damaged due to excessive heat.

Electric heat is widely used in moderate climates because the cost of the electricity and necessary precautions are offset by the simple control schemes and low installation and maintenance costs.

ELECTRIC RESISTANCE HEATING ELEMENTS

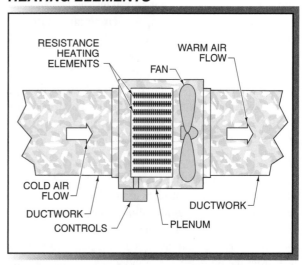

Figure 2-3. Electric heating systems use resistance heating elements located in building ductwork to provide heat for building spaces.

Electric heating systems may also contain radiant heat panels. A *radiant heat panel* is a device that uses an electric resistance heating element embedded in a ceiling panel. The ceiling panel is heated by electricity and the heat is radiated directly to the occupants below. Radiant heat panels are used in building spaces that require heat directed to a localized area. The advantage of radiant heat panels is that only a minimal amount of space is required for installation. See Figure 2-4.

RADIANT HEAT PANELS

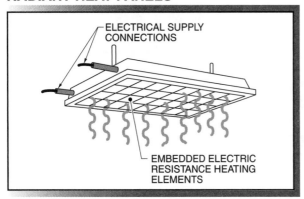

Figure 2-4. Radiant heat panels are heated by electricity and radiate the heat directly to the area below.

Electric heating systems may also use electric baseboard heaters to provide heat. An *electric baseboard heater* is a device that uses electric resistance heating elements and is located along the base (bottom) of outside walls of a building. An electric baseboard heater may have a thermostat that energizes the heating elements when the building space temperature falls below a setpoint. Electric baseboard heaters use natural convection for air circulation and do not require a fan for air flow. Electric baseboard heaters supplement the heating system of a building due to heat losses through the exterior walls of the building or falling outside temperatures. See Figure 2-5. Electric wall-mounted units are also used with or without a circulating fan.

Heat Pump Systems. A *heat pump* is a direct expansion refrigeration system that contains devices and controls that reverse the flow of refrigerant. A direct expansion refrigeration system has a cooling coil containing refrigerant located in the air stream of the unit. The reversal of the flow of refrigerant enables a heat pump to transfer heat from the outdoors to the indoors to produce a heating effect and to transfer heat from the indoors to the outdoors to produce a cooling effect. Some commercial buildings use heat pumps to provide heating and cooling.

ELECTRIC BASEBOARD HEATERS

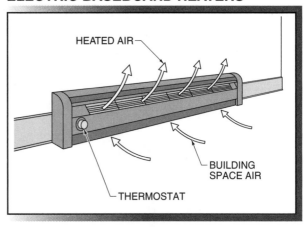

Figure 2-5. Electric baseboard heaters include a thermostat and use natural convection for air circulation.

While heat pumps are commonly used in residential systems, they are also used in commercial buildings. Residential heat pumps are commonly air source (air-to-air) heat pumps. An *air-to-air heat pump* is a heat pump that uses air as the heat source and heat sink. Air-to-air heat pumps transfer heat from the air outside a building to the air inside a building. Commercial heat pumps are commonly water source (water-to-air) heat pumps. A *water-to-air heat pump* is a heat pump that uses water as the heat source and heat sink. Water-to-air heat pumps have a coil heat exchanger with water as the heat transfer medium. Heat is transferred from the water in the coil to the air in the building spaces. See Figure 2-6.

Natural Gas or Fuel Oil-Fired Systems. Natural gas or fuel oil-fired systems are also common heating systems used in commercial buildings. In natural gas or fuel oil-fired systems, the fuel is burned in a heat exchanger. The heat exchanger is commonly located inside ductwork. The heat exchanger contains the hot products of combustion. The heat is transferred to air flowing over the heat exchanger. A fan circulates air over the heat exchanger to heat the air and move it into the building spaces. The products of combustion are vented outside the building. The most common type of unit that uses a fired heating system is a rooftop packaged unit. See Figure 2-7. Rooftop packaged units are common in small commercial buildings such as fast food restaurants and strip malls. Natural gas and fuel oil-fired heating systems in rooftop packaged units operate on the same principles as fired residential heating equipment.

HEAT PUMP SYSTEMS

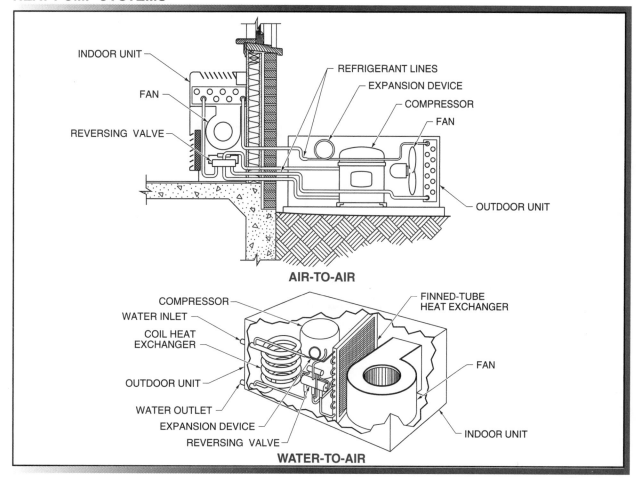

Figure 2-6. Air-to-air heat pumps and water-to-air heat pumps are used in mild climates for heating and cooling commercial building spaces.

ROOFTOP PACKAGED UNITS

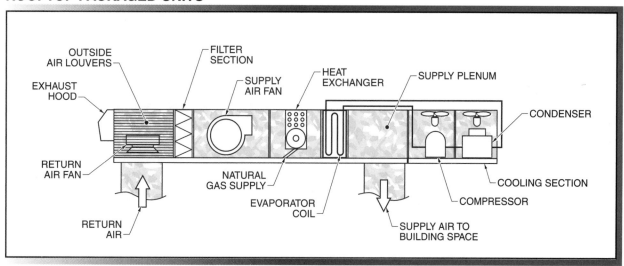

Figure 2-7. Rooftop packaged units commonly provide heat using natural gas as fuel.

Ventilation Systems

Ventilation is the process of introducing fresh air into a building. *Stale indoor air* is air that contains odors or contaminants. The ventilation requirements of a building include the number of times per hour that air must be changed in a building space. Outside air is commonly mixed with air from inside a building (return air) during ventilation. *Mixed air* is the combination of return air and outside air. See Figure 2-8. Mixed air must have enough outside air to minimize odors or contaminants. The amount of outside air admitted into a building must remain above a certain amount. Common percentages of outside air used for ventilation are 5%, 10%, and 30%. The minimum ventilation air percentage varies according to the type of activity in a building space and local code requirements. *Minimum ventilation air percentage* is the minimum amount of outside air that must be mixed with return air before the air is allowed to enter a building space. Some industrial, research, and commercial areas cannot use any return air because of dangerous contaminants and must use 100% outside air for ventilation.

> *An indoor air quality challenge is the prevention of mold growth inside buildings. Mold growth may be caused by poor airflow, excessive heat and humidity conditions, or water leaks. HVAC technicians often work with medical and health department officials to prevent mold growth.*

VENTILATION SYSTEMS

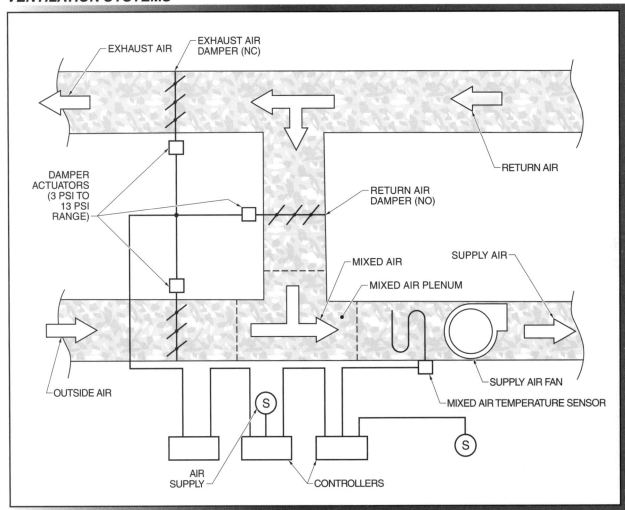

Figure 2-8. Air handling units have a mixed air plenum where outside air is combined with return air.

Indoor Air Quality. *Indoor air quality* is a designation of the contaminants present in the air. The contaminants consist of thousands of different chemical compounds of different sources and effects. Common building materials such as carpet fibers, adhesives, and ceiling tiles are sources of contaminants. Indoor air quality is a concern in all buildings. Good indoor air quality means the air is free of harmful particles or chemicals. Poor indoor air quality means that air is contaminated. Poor indoor air quality can affect human wellness and lead to serious health problems for building occupants. The Environmental Protection Agency (EPA) ranks indoor air quality as one of the top five environmental threats to human health.

As indoor air contaminants build up to unacceptable levels, outside air is introduced in greater quantities to dilute the contaminants. Currently, no universal standard exists for measuring the amounts of contaminating pollutants in a building space. However, carbon dioxide levels in a building space are commonly used as an indicator of indoor air quality.

The level of air contaminants is reduced by introducing outside air. The increased use of outside air for ventilation may lead to increased building energy costs because cold outside air requires heating in the winter, and hot, humid outside air requires cooling and dehumidification in the summer. Extra care is required in northern climates to prevent the freeze-up of water pipes and coils when cold outside air is introduced to mechanical systems. Freeze-ups may also indicate problems in the design or installation of the mechanical system. Alternately, when an outside air duct is excessively small, insufficient outside air is brought into the building, which can lead to indoor air quality problems. In this case, the mechanical system requires corrective action to alleviate the poor air quality problem.

Cooling Systems

Commercial building cooling systems may be required to provide year-round comfort. In the southern areas of the United States, many commercial buildings require only cooling. Methods used to provide cooling for commercial buildings include outside air economizers, direct expansion cooling, and water chillers.

Outside Air Economizers. An *outside air economizer* is a unit that uses outside air to cool building spaces. The use of cool outside air for cooling reduces the energy expenditures of a building. An *economizer cycle* is an HVAC system cycle in which building spaces are cooled using only outside air. The use of outside air for cooling is allowed when the air is at an appropriate temperature and humidity. See Figure 2-9. For example, an outside air dry bulb temperature of 45°F to 65°F is common for outside air economizer use. When the outside air is warmer than 65°F, building spaces are not effectively cooled. This requires that the outside air source be closed and the mechanical cooling system be turned ON. When the outside air is colder than 45°F, building spaces may experience excessive cooling or the water coils in the air handling unit may freeze.

Direct Expansion Cooling. Some commercial air conditioning systems, such as rooftop units and heat pumps, use direct expansion cooling. *Direct expansion cooling* is cooling produced by the vaporization of refrigerant in a closed system. Direct expansion cooling uses an evaporator coil placed in the ductwork of the air handling unit for the transfer of heat. See Figure 2-10. Warm air flowing over the evaporator coil causes the liquid refrigerant in the coil to vaporize, absorbing heat. The absorption of heat by the refrigerant cools the air. Direct expansion cooling is commonly used in residential cooling systems and in small- to medium-capacity commercial cooling systems.

Water Chillers. A *water chiller* is a device that cools water. Large commercial buildings that require large quantities of cooling use water chillers. The cool water is pumped throughout a building for cooling purposes. See Figure 2-11. Cool air is transferred to building spaces by terminal devices located in the building spaces or coils located in an air handling unit. Automatic valves at the terminal devices or cooling coils provide air temperature control. In large commercial buildings, the heat absorbed by the water may be rejected to the outside air through a cooling tower.

Water chillers include reciprocating, centrifugal, and absorption chillers. Reciprocating water chillers use piston-type, positive-displacement compressors. Reciprocating water chillers are found in small- and medium-capacity systems. Centrifugal chillers use an electric motor or steam turbine to rotate an impeller. The rotating impeller increases the velocity of the refrigerant, which increases its pressure. Centrifugal chillers are the most common commercial water chillers and are found in medium- and large-capacity systems. Absorption chillers use water as a refrigerant and steam to cause a water and lithium bromide solution to separate. Absorption water chillers are found in large-capacity systems.

OUTSIDE AIR ECONOMIZERS

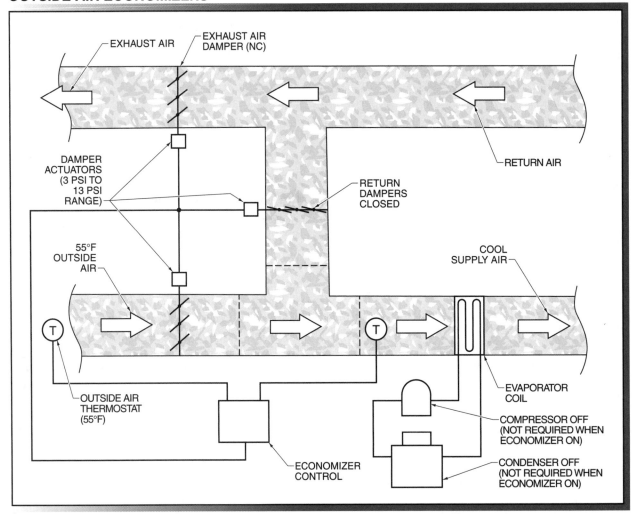

Figure 2-9. Outside air economizers allow outside air at the correct temperature and humidity conditions to be used to cool building spaces.

DIRECT EXPANSION COOLING

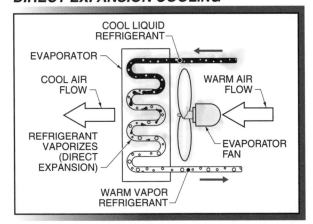

Figure 2-10. Direct expansion cooling uses the vaporization of refrigerant in a closed system to produce a cooling effect.

Humidification

A requirement of all commercial HVAC systems is to add moisture to air. *Humidification* is the process of adding moisture to air. Humidification is required in cold climates where continual heating of indoor air causes humidity levels to drop, and the low humidity levels cause building occupant discomfort. Industrial facilities require specific humidity levels to ensure product quality or because of requirements of the manufacturing process.

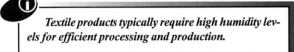

Textile products typically require high humidity levels for efficient processing and production.

WATER CHILLERS

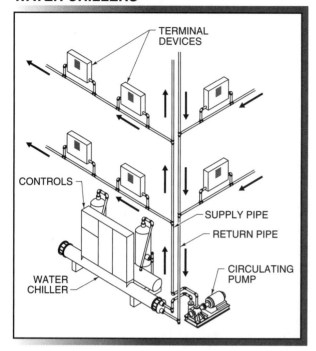

Figure 2-11. Water chillers cool water, which is pumped through terminal devices or air handling unit coils to provide cooling.

Humidification for commercial building spaces is commonly provided by introducing steam or water into the supply air ductwork. See Figure 2-12. Humidification systems contain a number of safety controls to ensure that humidity is not introduced unless it is required. Humidifiers are commonly shut OFF when the outside air is excessively warm or if the fan is shut OFF.

Dehumidification

Some commercial buildings require that the humidity in the air be reduced. *Dehumidification* is the process of removing moisture from air. Dehumidification may be required for occupant comfort or industrial process purposes. Common methods used to dehumidify air are passive and desiccant dehumidification.

Passive Dehumidification. *Passive dehumidification* is the process of removing moisture from air by using the existing cooling coils of a system. In a passive dehumidification system, no additional equipment is required. Passive dehumidification uses the condensation process to remove moisture from the air. Moisture condenses out of the air as the air is cooled.

HUMIDIFICATION SYSTEMS

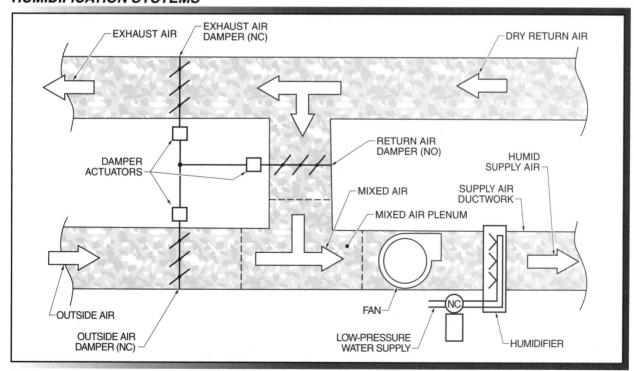

Figure 2-12. Humidification for commercial building spaces is provided by introducing steam or water into the supply air ductwork.

Passive dehumidification removes enough moisture from the air to satisfy the humidity requirements for human comfort but may not remove enough moisture for industrial applications.

Desiccant Dehumidification. *Desiccant dehumidification* is the process in which air contacts a chemical substance (desiccant) that adsorbs moisture. Desiccant dehumidifiers are used when extremely low humidity is required in a building space. See Figure 2-13. The desiccant may be in a solid or liquid form, with the solid (silica gel) being the most common. When the desiccant has adsorbed moisture, heat is used to dry it for reuse. The desiccant is returned to the air flow, ready to adsorb additional moisture. Desiccant systems are commonly used in commercial HVAC systems, in industrial systems that require extremely low humidity to preserve products, and in pneumatic systems to reduce rust and corrosion.

AIR HANDLING UNITS

An *air handling unit* is a device consisting of a fan, ductwork, filters, dampers, heating coils, cooling coils, humidifiers, dehumidifiers, sensors, and controls to condition and distribute air throughout a building. Air handling units are available in a variety of shapes, sizes, and configurations. However, the general mechanical layout of air handling units is the same. See Figure 2-14.

In an air handling unit, outside air enters through an inlet and damper arrangement. The damper provides a method of controlling the amount of outside air entering the air handling unit. Return air from the building spaces is introduced in another duct and damper arrangement. The outside air and return air ducts are joined in a mixed air plenum. The outside air and return air combine to create mixed air. Mixed air is filtered and travels through the fan, which causes airflow in the ductwork. The air passes across heating coils, cooling coils, and humidifiers, and/or dehumidifiers inside the air handling unit. The air leaves the air handling unit and passes through the ductwork and into the building spaces. Packaged air handling units such as rooftop units and heat pumps are preassembled, shipped, and installed as a unit. Large air handling units are assembled at the site from customized components.

DESICCANT DEHUMIDIFICATION

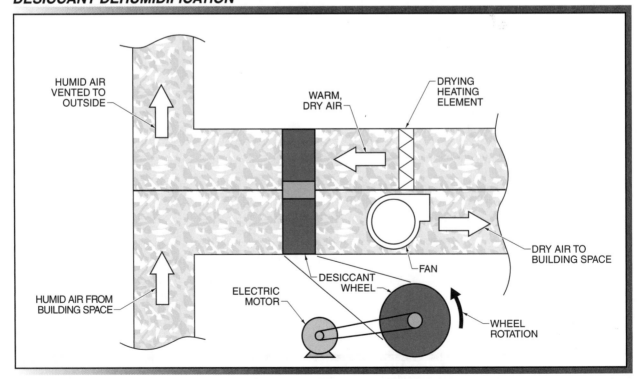

Figure 2-13. Dehumidification of air can be performed by rotating a desiccant wheel through the air stream of a duct.

AIR HANDLING UNITS

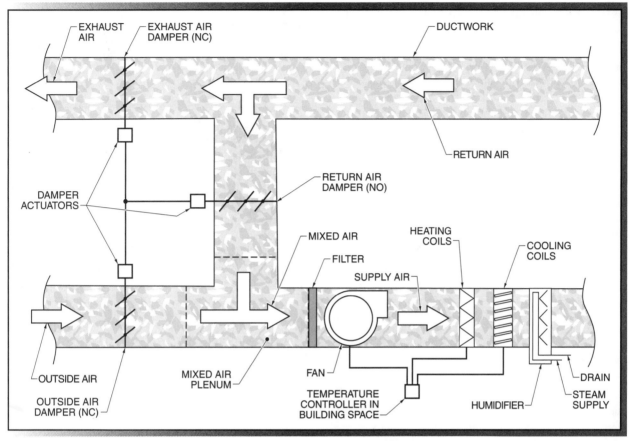

Figure 2-14. Air handling units consist of fan(s), coils, dampers, ductwork, humidifiers, filters, and controls to condition and distribute air throughout a building.

Fans

A *fan* is a device with rotating blades or vanes that move air. The fan of an air handling unit creates air flow that causes air to move throughout a building. Most air handling units have only one fan, which is referred to as the supply fan. Large air handling units often use two fans, a supply fan and a return or exhaust fan. The return or exhaust fan is located in the return air ductwork. The return fan ensures that the volume of return air flowing to the air handling unit is sufficient to provide the required air circulation. The return fan is also used to exhaust smoke from a building when a fire has been detected. During normal operation, supply and return fans are electrically interlocked to start and stop together.

The most common fan used in commercial building air handling units is a centrifugal fan. See Figure 2-15. A centrifugal fan has a housing and blades (vanes) that force air into the ductwork. A fan must be sized to deliver the correct volume of air at a reasonable efficiency, without creating excessive noise in the building.

CENTRIFUGAL FANS

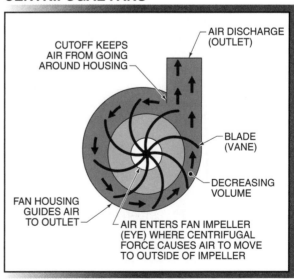

Figure 2-15. Most air handling units contain a centrifugal fan to create air flow.

Ductwork

Ductwork is the distribution system for a forced-air heating or cooling system. Ductwork directs conditioned air to the building spaces. Ductwork consists of square, rectangular, or round sheet metal that is coated for energy efficiency with a combination of insulation and soundproofing material. Ductwork is sized and located to allow the required amount and direction of air flow throughout a building for occupant comfort. See Figure 2-16.

Large air handling units are constructed of heavy-gauge steel, with the insulation and soundproofing material located between different layers. A rubber transition is commonly used between the air handling unit and the ductwork to provide a sound barrier and reduce noise transmitted to the building spaces. Ductwork is normally terminated with registers and grilles. Registers are used to disperse the air entering a building space. Grilles are used as a decorative device on return ductwork. The ductwork must be properly designed and installed to achieve the correct air flow level and comfort inside a building.

Filters

A *filter* is a porous material that removes particles from a moving fluid. Particulate matter can cause allergic reactions in building occupants or contaminate manufacturing and printing processes, and lab experiments. Particulate matter can also coat heating and cooling coils of an air handling unit, causing reduced air flow and poor heat transfer. Filters are used to reduce particulate matter in the air flow to building spaces. See Figure 2-17.

FILTERS

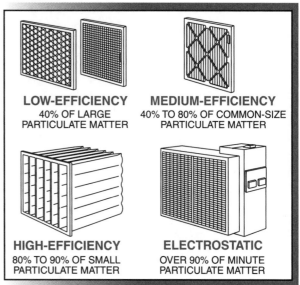

Figure 2-17. Filters are available in a variety of types and are used in air handling units to remove particulate matter from the air.

DUCTWORK

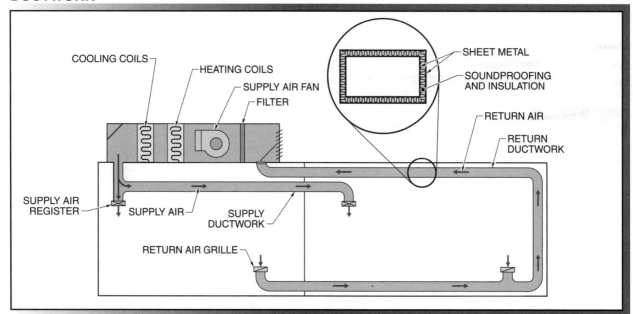

Figure 2-16. Ductwork consists of sheet metal that is coated with a combination of soundproofing material and insulation for energy efficiency.

Small air handling units such as rooftop units use residential-type disposable filters. For efficient removal of small particles, three rows (banks) of filters with progressively higher ratings are used. In some cases, electrostatic filters are used. An *electrostatic filter* is a device that cleans air by passing the air through electrically charged plates and collector cells. High-efficiency filters provide a greater resistance to air flow, require more frequent changing, and are more expensive than low-efficiency filters. All filters must be changed at intervals recommended by the manufacturer. Filters are changed as part of mechanical equipment preventive maintenance procedures.

Dampers

A *damper* is an adjustable metal blade or set of blades used to control the flow of air. Dampers are used to control the flow of air into and out of air handling units and into building spaces. See Figure 2-18. Dampers are commonly used to control outside air, return air, and exhaust air flow. Depending on the type and application of the air handling unit, dampers may also be used to control air flow across heating and/or cooling coils.

DAMPERS

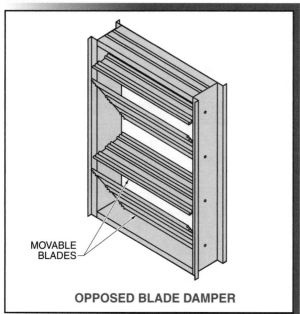

MOVABLE BLADES

OPPOSED BLADE DAMPER

Figure 2-18. Dampers are used to control the amount of air flow into and out of air handling units and into building spaces.

Heating Coils

A *heating coil* is a finned heat exchanger that adds heat to the air flowing over it. Steam, hot water, and electric heating coils are used in air handling units to provide heat for building spaces. Air handling units may contain more than one coil to provide extra heating capacity or preheat outside (ventilation) air. The heating coils must be protected against freezing or loss of air flow.

Cooling Coils

A *cooling coil* is a finned heat exchanger that removes heat from the air flowing over it. Direct expansion cooling coils or chilled water coils are located in the ductwork similar to heating coils. Cooling coils are used to precool hot, humid, outside air or provide primary cooling for building spaces. Cooling coils also provide passive dehumidification of the air as moisture condenses out of the air as it is cooled.

Humidifiers

A *humidifier* is a device that adds moisture to air by causing water to evaporate into the air. High humidity limit controls in the ductwork downstream of the humidifier cause the humidifier valve to close when excessive humidity is present in the building spaces. Normally closed humidifier valves are used in case of system failure to prevent excessive water leakage. A humidifier valve that is open and left unchecked causes water to collect in air handling units and ductwork.

Heat Recovery Devices

Building designers are constantly working to reduce the energy usage of air handling units. Reduction in air handling unit energy usage is often limited because of the required amount of ventilation air. A *heat recovery device* is a heat exchanger that transfers heat between a medium at two different temperatures. In a heat recovery device, a heat wheel (exchanger) located between the exhaust air stream and the incoming outside air is used to improve system efficiency when a high volume of ventilation air is required. See Figure 2-19. The heat wheel is attached to both air streams and transfers heat from one to the other without allowing the air streams to mix.

Heat wheels may be treated to prevent the growth and transfer of dangerous organisms.

HEAT RECOVERY DEVICES

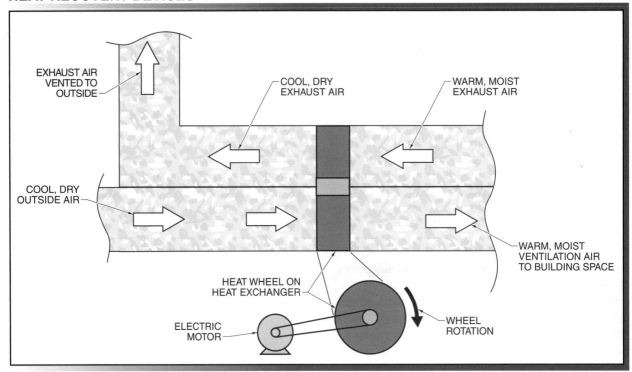

Figure 2-19. Heat recovery devices conserve energy by transferring heat from warm exhaust air to cool ventilation air.

Constant-Volume Air Handling Units

A *constant-volume air handling unit* is an air handling unit that moves a constant volume of air. Constant-volume air handling units operate at their rated capacity (in cfm) at all times. Constant-volume air handling units control building space temperature by changing the temperature of the air, not the volume. Most air handling units found in commercial buildings are constant-volume air handling units. The disadvantage of constant-volume air handling units is that the constant volume of air provides no savings in fan energy because the fan operates at rated power 100% of the time. Constant-volume air handling units used in commercial buildings include single-zone, multizone, dual-duct, terminal reheat, and induction air handling units.

Single-Zone Air Handling Units. A *single-zone air handling unit* is an air handling unit that provides heating, ventilation, and air conditioning to only one building zone or area. See Figure 2-20. A building zone or area may be a 10′ × 13′ room, a 50′ × 50′ conference area, or a whole floor of a building. The size of an individual building zone (space) is limited because as a building space becomes large, the temperature in one area can be different from another area. Single-zone air handling units are identified by the heating and cooling coils that are in series and controlled directly from the building space.

Multizone Air Handling Units. A *multizone air handling unit* is an air handling unit that is designed to provide heating, ventilation, and air conditioning for more than one building zone or area. Multizone air handling units are distinguished from single-zone air handling units by the heating and cooling coils located side by side in different ducts. The ducts are referred to as the hot duct (deck) and cold duct (deck). See Figure 2-21. A multizone air handling unit has dampers for each zone that mix the hot air and cold air and send it through separate ducts to the appropriate zone. The dampers are controlled from thermostats or controllers in each zone. Multizone air handling units are designed to work with as many as 50 zones. Multizone air handling units require labeling and documentation to indicate which zone is controlled by each thermostat or controller.

SINGLE-ZONE AIR HANDLING UNITS

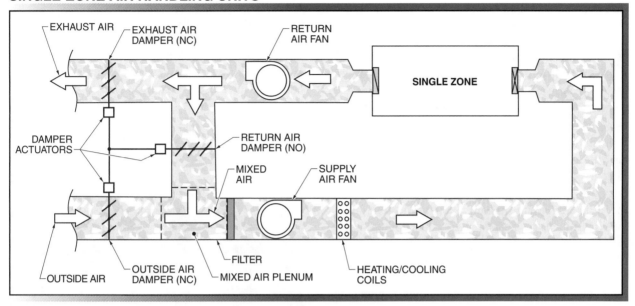

Figure 2-20. Single-zone air handling units serve only one building zone or area.

MULTIZONE AIR HANDLING UNITS

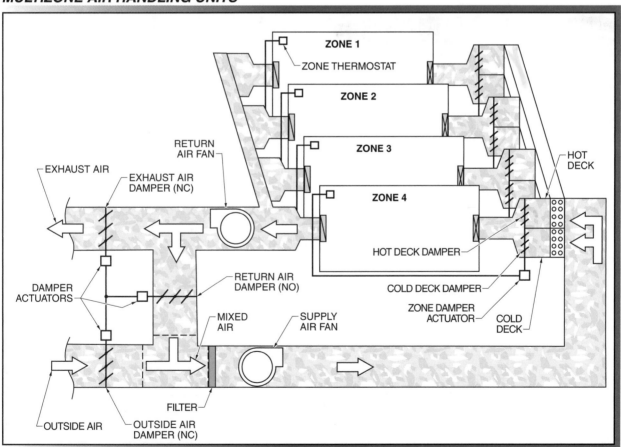

Figure 2-21. Multizone air handling units serve multiple rooms, each with its own individual building space temperature control.

Dual-Duct Air Handling Units. A *dual-duct air handling unit* is an air handling unit that has hot and cold air ducts connected to mixing boxes at each building space. A *mixing box* is a sheet metal box with inlets for hot and cold air. A damper in the mixing box opens either the cold air or hot air duct to provide HVAC capability to the building space. The damper opens and closes in response to a thermostat signal. The mixing box is typically located above the ceiling in the room.

Dual-duct air handling units are similar to multizone air handling units. The difference is in the location of the dampers that mix the air in the hot duct and cold duct. In a multizone air handling unit, the air is mixed in the air handling unit. In a dual-duct air handling unit, the air is mixed at or near the individual building space. See Figure 2-22. The mixing boxes have a mixing damper that is connected to a thermostat or controller within the building space. The similarity between multizone and dual-duct air handling units enabled the development of hybrid air handling units. A hybrid air handling unit is an air handling unit that is part multizone and part dual-duct.

Terminal Reheat Air Handling Units. A *terminal reheat air handling unit* is an air handling unit that delivers air at a constant 55°F temperature to building spaces. Each building space has a steam, electric, or hot water reheat coil in the ductwork. The reheat coil valve is attached to a thermostat or controller in the building space. When heat is required, the thermostat opens the reheat coil valve, causing the air temperature to rise from 55°F to the required temperature. See Figure 2-23. When cooling is required, the reheat coil is shut OFF and 55°F air flows into the building space. If the reheat coil contains electric heating elements, a thermostat controls a relay which turns the electric reheat coil ON and OFF as required. While terminal reheat air handling units have a simple layout, the reheat coil piping required for hot water or steam increases system installation costs.

Three-way valves are commonly used to control hot water flow through reheat coils.

DUAL-DUCT AIR HANDLING UNITS

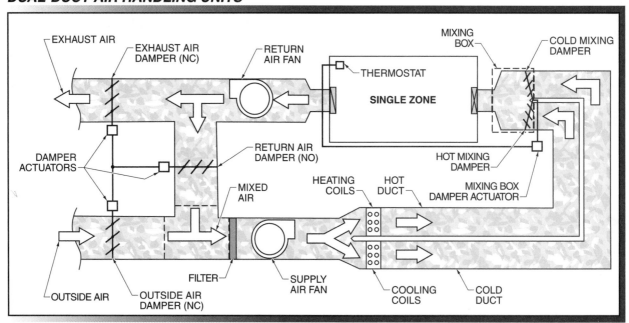

Figure 2-22. Dual-duct air handling units use mixing boxes located at or near each building space.

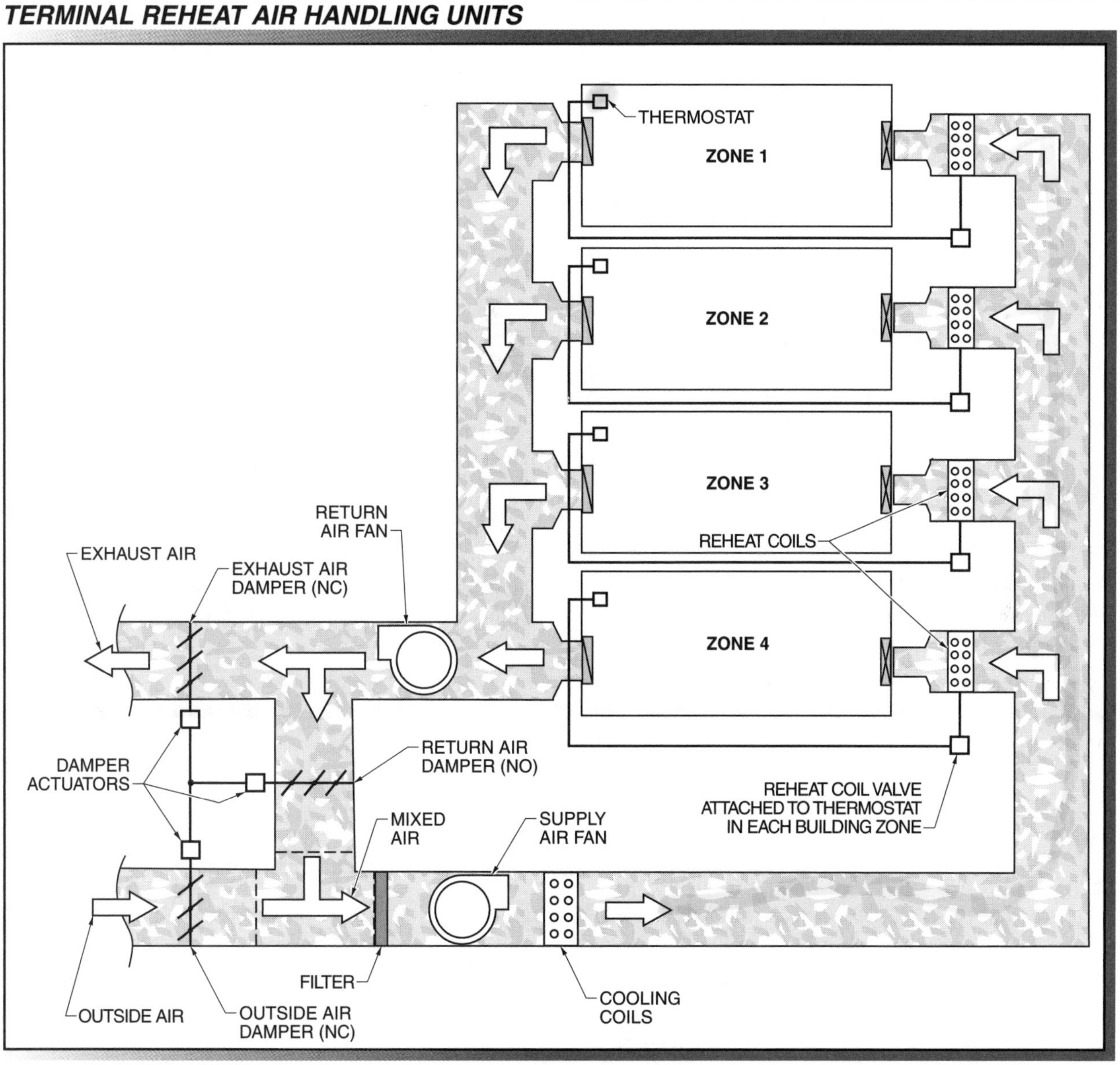

TERMINAL REHEAT AIR HANDLING UNITS

Figure 2-23. Terminal reheat air handling units heat or cool the air in individual building spaces by modulating the reheat coil of the building space.

Induction Air Handling Units. An *induction air handling unit* is an air handling unit that maintains a constant 55°F air temperature and delivers the air to the building spaces at a high duct pressure. Induction air handling units are an early constant-volume air handling unit design. In the building space, the high-pressure air is delivered to a slotted wall-mounted unit. See Figure 2-24. The high-pressure air is forced out through an induction nozzle into the building space. The high-velocity air flow causes building space air (return air) to flow into the unit. The induced air flow is directed across heating coils, which are controlled by a thermostat or room controller. Induction air

handling units are often noisy and energy inefficient due to the high-pressure air output.

Variable Air Volume Air Handling Units

A *variable air volume air handling unit* is an air handling unit that moves a variable volume of air. Variable air volume air handling units vary the amount of air delivered to a building space instead of varying the temperature of the air. Variable air volume air handling units were developed and became widely used during the 1970s as the energy crisis forced HVAC system designers to consider system energy efficiency.

INDUCTION AIR HANDLING UNITS

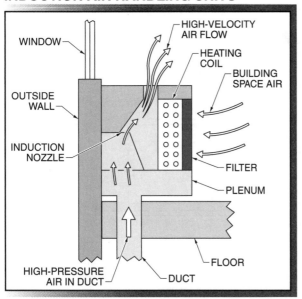

Figure 2-24. Induction air handling units use high-pressure air to force building space air across filters and heating coils.

Variable air volume air handling units are available in a variety of configurations. The purpose of a variable air volume air handling unit is to produce a constant 55°F air temperature year-round, with a varying volume. Most variable air volume air handling units have a mixed air plenum section with outside air, return air, and exhaust air dampers. Many variable air volume air handling units are used for cooling only, and any heat required is provided using return air. See Figure 2-25.

Variable air volume systems reduce the volume of air to a building space, causing the supply fan to perform less work and use less energy. Variable air volume air handling units are the most common air handling units installed in new commercial buildings today. Disadvantages of variable air volume air handling units include excessive noise and possible inability to heat a building space effectively. Variable air volume air handling units may also be unable to deliver the required volume of outside air for ventilation to building spaces, contributing to indoor air quality problems. Variable air

VARIABLE AIR VOLUME AIR HANDLING UNITS

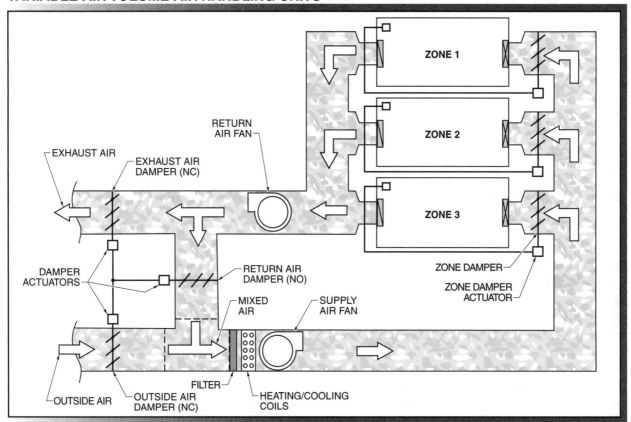

Figure 2-25. Variable air volume air handling units have standard components such as dampers, cooling coils, filters, fans, and zone dampers for volume control.

volume air handling units are also difficult to control. In some northern climates, a heating coil at the variable air volume air handling unit is required. Another disadvantage of variable air volume systems may be that the unit is sized too small for the air flow needed at full cooling. This may cause the building to be too hot during extremely hot ambient conditions.

The most important function of a variable air volume air handling unit is the ability to change the volume of air produced by the supply fan. Supply fan air volume control can be accomplished by using bypass dampers (common for small rooftop units) to circulate the supply air back to the return duct. In the past, discharge dampers in the supply fan outlet were also used to control fan volume. Today, vortex dampers and electric motor variable-frequency drives are used to control the volume of a fan. See Figure 2-26. A *vortex damper* is a pie-shaped damper located at the inlet of a centrifugal fan. Vortex

dampers, when closed, reduce the ability of a fan to grip and move air. An *electric motor variable-frequency drive* is an electronic device that controls the direction, speed, and torque of an electric motor. Vortex dampers and variable-frequency electric motor drives provide excellent speed control and save the maximum amount of fan motor electrical energy. In all variable air volume systems, the object is to maintain a specific static pressure in the ductwork. Static pressures of .5″ water column (wc) to 1″ wc are common.

VAV terminal box electric heating elements are sized based on a required airflow volume. The airflow volume is normally provided by the manufacturer of the VAV terminal box.

ELECTRIC MOTOR DRIVE VARIABLE AIR VOLUME AIR HANDLING UNITS

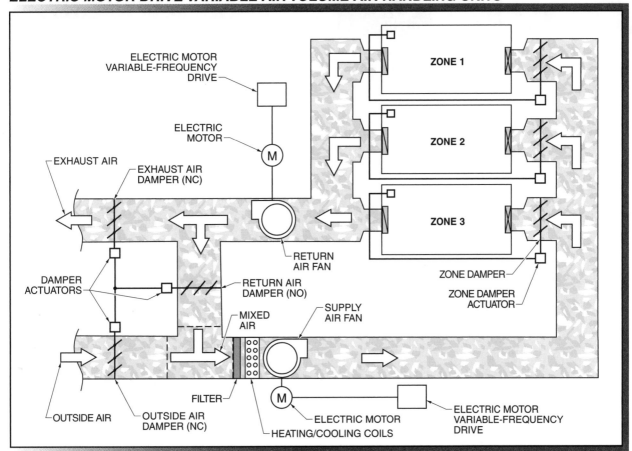

Figure 2-26. Electric motor variable-frequency drives are used in modern variable air volume systems to vary fan speed.

Variable Air Volume Terminal Boxes

A *variable air volume (VAV) terminal box* is a device that controls the air flow to a building space, matching the building space requirements for comfort. As building space temperature drops, heat is provided by an energized electric heating element within the VAV terminal box. When the building space temperature reaches setpoint, the heating element is de-energized. A hazard may occur if air flow is lost across the heating element, which may cause it to overheat and fail. Most variable air volume terminal boxes have a differential pressure switch that de-energizes the heating element if air flow falls below a minimum setting. See Figure 2-27.

Electric motor variable-frequency drives have pushbutton interfaces and displays which allow technicians to view and possibly change drive parameters.

VARIABLE AIR VOLUME TERMINAL BOXES

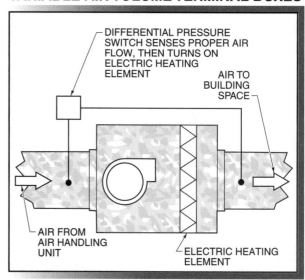

Figure 2-27. Variable air volume terminal boxes contain differential pressure switches for safety and electric heating elements.

HVAC Control Systems
HVAC System Energy Sources

HVAC systems require energy to provide heating, cooling, and air circulation in building spaces. The most common heating system energy sources are electricity, natural gas, fuel oil, solar energy, and heat pumps. The most common cooling system energy sources are outside air, electricity, chilled water, steam or hot water, and heat pumps. The energy source used for an HVAC system depends on cost and availability.

HVAC SYSTEM ENERGY SOURCES

The purpose of an HVAC system is to provide comfort to the occupants of a building space. Comfort is provided to building occupants by controlling humidity throughout the year, heating in the winter, and cooling in the summer. Energy is consumed when controlling the environmental conditions in a building.

Heating System Energy Sources

Commercial HVAC systems must provide adequate heat to building occupants for comfort in the winter months. Commercial building heating system failure results in occupant discomfort and sometimes building damage such as frozen pipes.

Commercial building heating system energy sources include electricity, natural gas, fuel oil, solar energy, and mechanical system heat transfer. Factors considered when choosing a heating system energy source include installation cost, energy cost per unit of energy used, and local climate.

Electricity. *Electricity* is the energy released by the flow of electrons in a conductor (wire). Electricity is a common energy source used to heat commercial buildings. Electricity is normally an existing energy source within a building, which minimizes the installation cost of piping distribution systems required for natural gas and fuel oil energy sources. The high cost of electricity must be considered when selecting it as an energy source for a heating system. In warm climates, the use of electricity as a heat source is economical because heat requirements are minimal. In most HVAC applications, it is more expensive to heat a building space using electricity than with other energy sources.

Electric heating systems commonly contain electric heating elements. An *electric heating element* is a device that consists of wire coils that become hot when energized. When placed in building ductwork, an electric heating element heats the air flowing through the ductwork. The heated air is then delivered to the building space. See Figure 3-1. Electric heating elements are used in electric baseboard heaters, radiant heat panels, air handling units (AHUs), and variable air volume (VAV) terminal boxes.

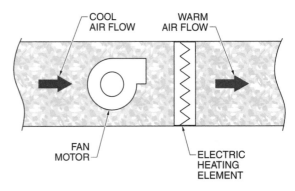

Figure 3-1. Electric heating elements heat the air flowing through ductwork and delivered to the building spaces.

Electric heat is widely used even though it is generally more expensive to heat a building with electricity than with other energy sources. The greatest advantage of electric heat is that no extensive piping or plumbing system is required. In addition, radiant heat panels can be designed for easy movement to new locations in a building. These advantages may outweigh the disadvantages for a specific application.

Natural Gas. *Natural gas* is a colorless, odorless fossil fuel. Natural gas is commonly used as an energy source for heating commercial buildings because it is plentiful and relatively inexpensive. Natural gas is also clean burning, which aids in meeting air pollution standards.

All natural gas heating applications generate heat through combustion. *Combustion* is the chemical reaction that occurs when oxygen reacts with the hydrogen (H) and carbon (C) present in a fuel at ignition temperature. *Ignition temperature* is the intensity of heat required to start a chemical reaction. Fuel, oxygen, and ignition temperature (heat) are the three requirements for combustion. See Figure 3-2. Natural gas provides chemical energy, oxygen reacts with the fuel, and the ignition temperature starts the reaction. Ignition temperature may be provided by a pilot light, electric spark, or some other means. When combustion has started, the temperature must remain at or above the ignition temperature or combustion stops. The heat produced by combustion maintains the ignition temperature. Common gas heat applications include gas-fired rooftop units, gas-fired boilers, and gas-fired radiant heaters.

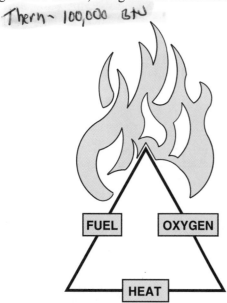

Thern~ 100,000 BTU

Figure 3-2. Fuel, oxygen, and ignition temperature (heat) are the three requirements for combustion.

Fuel oil burners require periodic inspection and cleaning by trained personnel.

In many areas of the country, natural gas is cheaper than electricity for heating applications. If large amounts of heat are needed in a cold climate, burning natural gas to create steam is efficient because steam has a large amount of heat energy (over 1000 Btu/lb). A gas-fired boiler may be set up to provide either hot water or steam heating. See Figure 3-3. In buildings such as hospitals, steam may also be needed for sterilization and laundry purposes. In facilities that use large amounts of natural gas, long-term contracts may be negotiated with the local utility, further reducing the cost.

A disadvantage of using natural gas as a heating source is the cost of installing natural gas piping in a building. Piping system integrity must be maintained to prevent contamination of the building spaces. Safety controls must be included to stop the flow of gas if improper conditions occur. In addition, the hazardous by-products of the combustion process, such as carbon monoxide, must be vented outside the building to prevent exposure to the occupants.

Fuel Oil. *Fuel oil* is a petroleum-based product made from crude oil. Different grades of fuel oil are available. The grade of a fuel oil is based on the weight and viscosity of the oil. *Viscosity* is the ability of a liquid to resist flow. The viscosity of fuel oil is lowered by raising its temperature. For example, for pumping some fuel oil, the temperature must be raised by heating in a fuel oil heater. *Heating value* is the amount of British thermal units (Btu) per pound or gallon of fuel. A *British thermal unit (Btu)* is the amount of heat energy required to raise the temperature of 1 lb of water 1°F. The Btu rating of a fuel indicates how much heat it can produce. Fuel oil is a common energy source for heating in the northeast U.S.

GAS-FIRED BOILERS

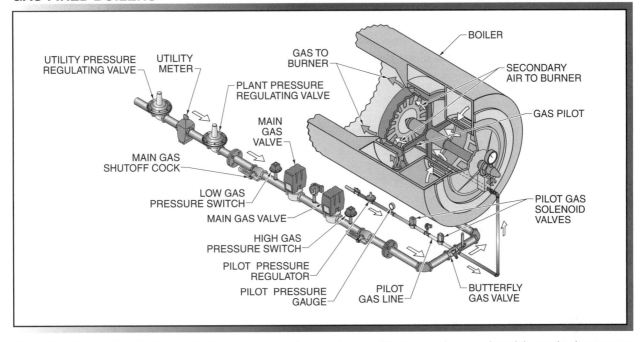

Figure 3-3. In a gas-fired boiler application, natural gas is piped from a utility to a gas burner where it is used to heat water.

The American Society for Testing and Materials (ASTM) has established standards for grading fuel oil. Each grade of oil has properties required for specific applications. The four grades of fuel oil used in boilers are No. 2 fuel oil, No. 4 fuel oil, No. 5 fuel oil, and No. 6 fuel oil. See Figure 3-4. A No. 2 fuel oil has a heating value of approximately 141,000 Btu/gal. No. 2 fuel oil does not have to be preheated. No. 4 fuel oil is heavier than No. 2 fuel oil and has a heating value of approximately 146,000 Btu/gal. No. 4 fuel oil is most commonly a blend of No. 2 fuel oil and heavier fuel oils. In colder climates, No. 4 fuel oil may require preheating to lower its viscosity for pumping.

No. 5 fuel oil has a heating value of approximately 148,000 Btu/gal. Preheating may be required in cold climates to lower the viscosity to facilitate pumping. No. 6 fuel oil (bunker C) contains heavy elements from the distillation process. No. 6 fuel oil has a Btu content of approximately 150,000 Btu/gal. Fuel oil tank heaters and line heaters must be used to heat No. 6 fuel oil to the required temperature for transport and combustion. The fuel oil temperature required depends on the type of burner and whether a straight distillate fuel oil or a blend of fuel oils is used.

Fuel oil systems include the accessories required to safely and efficiently operate the fuel oil burner. Fuel oil accessories clean, control the temperature, and regulate the pressure of fuel oil. In a fuel oil system, fuel oil is pumped from a storage tank, through pipes, to a furnace or boiler. Fuel oil may require filtering to remove impurities. See Figure 3-5.

FUEL OIL GRADES				
Characteristics	**No. 2**	**No. 4**	**No. 5**	**No. 6**
Type	light distillate	light distillate or blend	light residual	residual
Color	amber	black	black	black
Specific Gravity	.8654	.9279	.9529	.9861
Btu/gal.	141,000	146,000	148,000	150,000
Btu/lb	19,500	19,100	18,950	18,750

Figure 3-4. The American Society for Testing and Materials has established standards for grading fuel oils and their characteristics.

The cost/benefit relationship between natural gas and fuel oil is constantly changing. New natural gas power plants are relatively inexpensive to construct and are used during times of high electric demand. This causes fluctuations in the price of natural gas that affect other users.

FUEL OIL SYSTEM

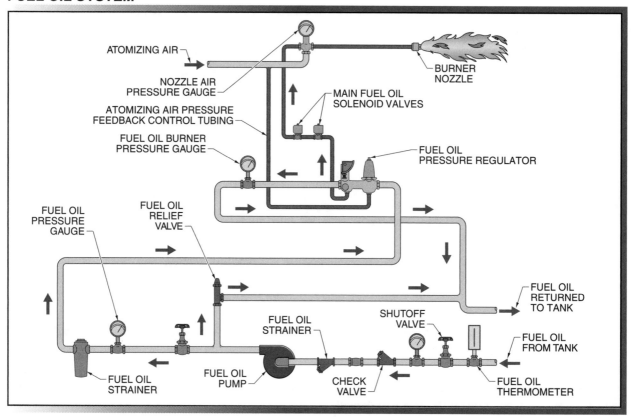

Figure 3-5. A fuel oil system includes the accessories required to safely and efficiently operate the fuel oil burner.

In HVAC applications, fuel oil and natural gas are interchangeable energy sources for heating. This allows flexibility in choosing which source to use. Common fuel oil applications include fuel oil-fired rooftop units and fuel oil-fired boilers. Advantages of using fuel oil include possible lower cost and easier availability than other energy sources. For example, fuel oil may be cost-effective in areas with exorbitant electrical costs and little availability of natural gas. In addition, because fuel oil is stored in large tanks close to the facility, fuel oil may be purchased in large quantities when costs are low and stored until needed. Disadvantages of using fuel oil include increased environmental pollution (smoke) compared to natural gas and costs associated with leak prevention and cleanup requirements. In addition, storage tanks, piping, and winter heating of the fuel oil also increase the cost of using fuel oil.

Solar Energy. *Solar energy* is energy transmitted from the sun by radiation. *Solar heat* is the heat created by the visible (light) and invisible (infrared) energy rays of the sun. These rays travel through space and the Earth's atmosphere. The rays provide energy in the form of radiant waves.

Some HVAC systems use solar energy as the primary energy source and others use it as a backup source to complement an existing system. Solar energy is plentiful and can provide substantial amounts of energy that can be used to replace more expensive or less available fuels. Most solar systems heat water for heating small commercial and residential applications. The water is heated by the solar rays and stored in tanks until the heat is needed.

The application of solar power systems is limited by the climate, the amount of available sunlight, and the size of the installed equipment. Solar energy collection and storage systems are built to collect and store the maximum amount of solar energy. The amount of energy received at the surface of the Earth can exceed 200 Btu/hr per sq ft of surface, depending on the angle of the sun's rays and the position of the solar collector. See Figure 3-6.

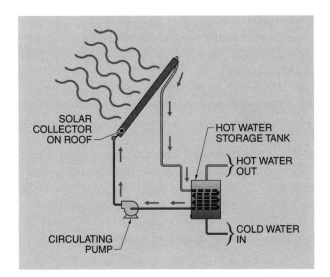

Figure 3-6. Solar collection systems may be used to heat water which is stored and then used to heat building spaces.

Mechanical System Heat Transfer. All air, even air at a moderate temperature, contains some heat. For example, when the outside air temperature is 45°F, it can still be used in a heat pump system to vaporize refrigerant that is at a temperature of 35°F. A *heat pump* is a direct expansion refrigeration system that contains devices and controls that reverse the flow of refrigerant. The reversal of the flow of refrigerant enables a heat pump to transfer heat from the outdoors to the indoors to produce a heating effect. Heat pumps are used in residential and commercial applications.

During a heat pump heating cycle, the balance point of a heat pump is reached when the heat losses through a building are the same as the amount of heat provided by the heat pump. This normally occurs at a temperature of approximately 40°F. For this reason, residential heat pumps are not widely used in northern climates. Heat pumps reduce or eliminate the need for natural gas or fuel oil piping and controls, while still being relatively efficient. The disadvantage of heat pumps is that they become less efficient as the outside air temperature drops. See Figure 3-7.

Emergency heat is heat provided if the outside air temperature drops below a set temperature or if a heat pump fails. Emergency heat is normally provided by electric heating elements. The use of electric heating elements is normally more expensive than the operation of the normal heating system. The cost of emergency heat from electric heating elements may be prohibitive in areas where a large number of hours of emer-

gency heat are required. While the use of emergency heat is discouraged, it provides heat if a heat pump fails. In some cases, an alarm light is attached to the thermostat to indicate heat pump failure. Normally, a technician is called to diagnose and repair the heat pump when the light is on.

HEAT PUMP HEATING CYCLE

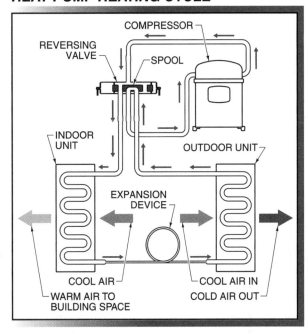

Figure 3-7. In a heat pump heating cycle, heat is collected from the outdoor air at the outdoor coil and released to the building space at the indoor coil.

Cooling System Energy Sources

Air conditioning is the process of cooling the air in building spaces to provide a comfortable temperature. An *air conditioning system* is a system that produces a refrigeration effect and distributes the cool air or water to building spaces. Air conditioning systems are rated in terms of capacity. Air conditioning system capacity is the available cooling output of an air conditioning system. Air conditioning system capacity is determined by the size and requirements of the building space. Large building spaces require a system with a greater capacity than small building spaces. Energy sources for HVAC cooling systems include outside air, electricity, cold water, steam or hot water, and mechanical system heat transfer.

Outside Air. The basic energy source used to cool a building is outside air. *Outside air* is air brought into a building space from outside the building. The use of outside air for cooling is referred to as free cooling. Outside air used for cooling purposes is limited by the temperature and humidity of the outside (atmospheric) air. When the outside air temperature or humidity is excessively high, the outside air is unsuitable as a cooling system energy source. See Figure 3-8.

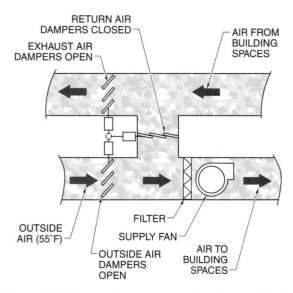

RETURN AIR DAMPERS CLOSED

EXHAUST AIR DAMPERS OPEN

AIR FROM BUILDING SPACES

OUTSIDE AIR (55°F)

FILTER

SUPPLY FAN

OUTSIDE AIR DAMPERS OPEN

AIR TO BUILDING SPACES

Figure 3-8. Free cooling uses outside air to cool building spaces.

Electricity. Electricity is a common energy source used to cool commercial buildings. Electricity is used in air conditioning systems to provide power for electric motors. The electric motors are used to operate fans and compressors. A *compressor* is a mechanical device that compresses refrigerant or other fluid. *Refrigerant* is fluid used for transferring heat (energy) in a refrigeration system. The refrigerant changes from a liquid to gas and back to liquid in the refrigeration cycle. See Figure 3-9.

When using electricity as a cooling energy source, its cost effectiveness depends on the cost per unit of electricity. For example, electrical usage costs in some parts of the country are $ .10 per kilowatt hour or less. This low cost may be due to proximity to electrical power sources such as hydroelectric dams. In other parts of the country, the cost of electricity may be $ .20 per kilowatt hour or more.

Cold Water. Cold water is a common energy source used to cool commercial buildings. Cold water is supplied from holding ponds, cooling towers, or liquid chillers. A *liquid chiller* is a system that uses a liquid (normally water) to cool building spaces. Liquid chillers contain a compressor, expansion device, condenser, and evaporator, similar to a refrigeration system. In a liquid chiller, however, the evaporator and condenser consist of tube-in-shell heat exchangers. These heat exchangers transfer heat to and from water that contacts the refrigerant-filled tubes. The warm water is pumped to a cooling tower to give up heat to the atmosphere. The cool water is pumped to building spaces to provide cooling. See Figure 3-10.

Steam and Hot Water. Steam and hot water are other common energy sources used to cool commercial buildings. While the majority of air conditioning systems are driven by electric motors, steam and hot water are energy sources that do not require the use of electric motor-compressors. Costs are lower when using steam or hot water because these energy sources are generated within the building space by a boiler, and are not totally dependent on electricity.

Steam or hot water is used to provide cooling in absorption refrigeration systems. An *absorption refrigeration system* is a nonmechanical refrigeration system that uses a fluid with the ability to absorb a vapor when it is cool and release a vapor when heated. Absorption refrigeration systems have a generator and absorber in place of the compressor to raise system pressure. A *generator* is an absorption refrigeration system component that vaporizes and separates the refrigerant from the absorbent. An *absorber* is an absorption refrigeration system component in which refrigerant is absorbed by the absorbent. An *absorbent* is a fluid that has a strong attraction for another fluid.

Absorption refrigeration systems are commonly used for large commercial or industrial applications where mechanical compression systems are not as efficient. A refrigerant and absorbent are required in absorption refrigeration systems. A common refrigerant used is ammonia. Some absorption refrigeration systems may use combinations of absorbents and refrigerants such as lithium bromide and water or lithium chloride and water.

Lithium bromide-water absorption refrigeration systems operate using a generator, condenser, chiller, and absorber. A strong lithium bromide solution absorbs water vapor, making it weaker. When heated, the weak solution releases water. See Figure 3-11.

REFRIGERATION SYSTEM

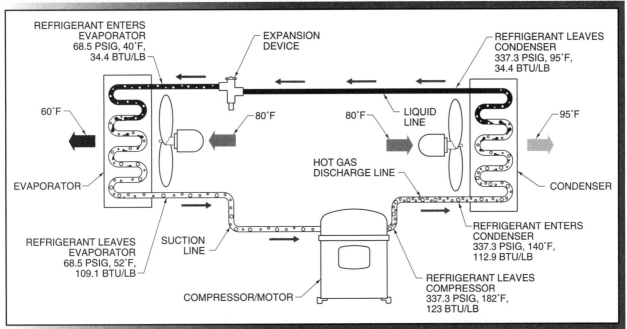

Figure 3-9. Electricity is used in a refrigeration system to power an electric motor in a compressor, which produces a refrigeration effect used to cool building spaces.

LIQUID CHILLER

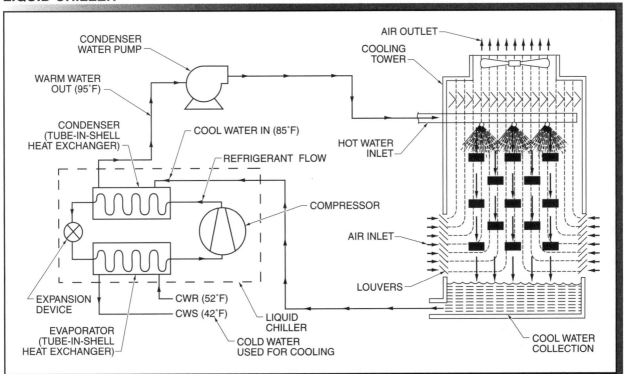

Figure 3-10. A liquid chiller contains two tube-in-shell heat exchangers that transfer heat to and from water.

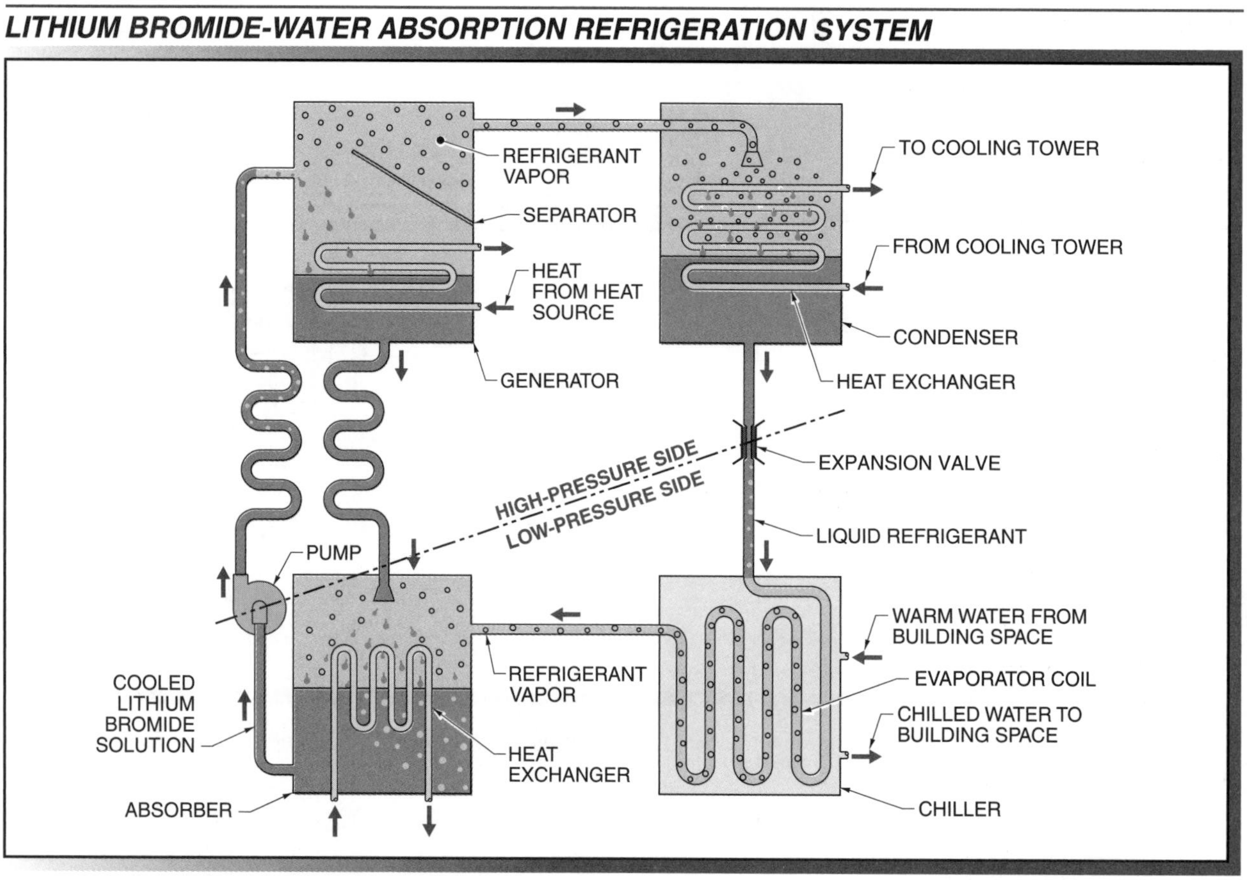

LITHIUM BROMIDE-WATER ABSORPTION REFRIGERATION SYSTEM

Figure 3-11. Absorption refrigeration systems are commonly used for large commercial or industrial applications where mechanical compression systems are not as efficient.

Heat is applied to the lithium bromide and water solution in the generator using heat from a source such as a steam coil. At the separator, the water vapor is separated from the lithium bromide solution. The heated water vapor is directed to the condenser. In the condenser, the heated water vapor is condensed into water. The liquid refrigerant passes through an expansion valve into the chiller where it flashes (absorbs heat) because of the low pressure in the system after the expansion valve. An evaporator coil inside the chiller absorbs heat and chills water to be sent to the building space. Water vapor coming from the chiller is absorbed by the lithium bromide solution in the absorber. Cooling water is directed through a heat exchanger in the absorber and condenser to remove heat. The cooled lithium bromide solution in the absorber cools the solution received from the generator.

In many commercial buildings, substantial investments are made in boiler equipment. In many cases,

the boiler equipment is operated year-round due to demands for hot water and steam. Air conditioning systems that use steam or hot water for cooling avoid problems with systems that use electric motors. For example, no large compressor motor with complicated electrical controls is required. Also, steam or hot water cooling systems can be operated off of a backup generator if power is lost. This feature is often not possible with large electrically-driven motor-compressors. These systems do not use electricity to turn compressors, so changes in the price or availability of electricity have less economic effect.

Most motor-driven systems in the past used CFC refrigerants that are believed to damage the ozone layer when inadvertently released into the atmosphere. The change out of these refrigerants has proven to be expensive. Steam and hot water cooling systems normally do not use CFC refrigerants, avoiding the problem of refrigerant replacement.

Mechanical System Heat Transfer. The mechanical equipment of a heat pump system can be used to transfer heat from the air inside a building to the air outside a building, producing a cooling effect. The heat from the building space is absorbed into the refrigerant inside the indoor coil (evaporator). The heat causes the cool liquid refrigerant to vaporize. The vaporized refrigerant flows to the compressor, where it is compressed. The hot, high-pressure refrigerant vapor is pumped to an outdoor coil (condenser) where it releases heat to the outside air. See Figure 3-12.

HEAT PUMP COOLING CYCLE

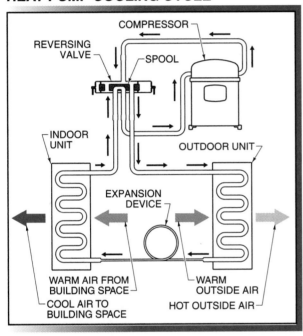

Figure 3-12. The mechanical equipment of a heat pump system can be used to transfer heat from the air inside a building to the air outside a building, producing a cooling effect.

ALTERNATIVE HVAC SYSTEM ENERGY SOURCES

A variety of HVAC system energy sources are in use today. Ponds or lakes located near buildings may be used as sources of cold water. In heavily forested areas, wood by-products may be burned to generate steam. In some plants, coal is used as a heating source. Geothermal energy is also available underground for use in some HVAC systems. The applications of these alternative energy sources depend largely on local availability of fuel, construction codes, and energy efficiency standards.

Cooling towers often use bypass valves to allow the condensing water to bypass the tower when cooling water is not required.

An HVAC control system is designed to control temperature, humidity, and pressure in building spaces. Control system components work together to maintain comfort in the building spaces. A sensor measures a variable and reports to a controller. A controller receives the measurement and sends a signal to a controlled device. A controlled device regulates the flow of a fluid to provide heating, cooling, and ventilation requirements.

CONTROL SYSTEM COMPONENTS

A *control system* is an arrangement of a sensor, controller, and controlled device to maintain a specific controlled variable value in a building space, pipe, or duct. Different HVAC control systems have been developed to provide this control. All HVAC control systems consist of the same basic components. Some variations exist in power supplies, in features such as types of adjustments, in nomenclature, and in wire/piping terminations. Control system components include sensors, controllers, controlled devices, and control agents.

Sensors

A *sensor* is a device that measures a controlled variable such as temperature, pressure, or humidity and sends a signal to a controller. The sensor output signal may be air pressure in a pneumatic control system, or resistance, voltage, or current in an electrical control system. Sensors may be mounted in a duct, pipe, or room remote from the controller. Sensors may also be integral with the controller, as in a room thermostat. Regardless of the type, the sensor must be selected for the variable to be sensed and located where it can properly sense the variable. See Figure 4-1.

ROOM SENSOR

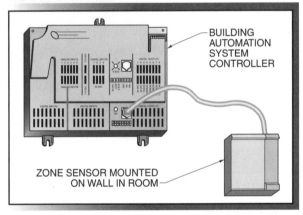

BUILDING AUTOMATION SYSTEM CONTROLLER

ZONE SENSOR MOUNTED ON WALL IN ROOM

Figure 4-1. A sensor must be chosen to suit the application and located where it can properly sense the controlled variable.

Controllers

A *controller* is a device that receives a signal from the sensor, compares it to a setpoint value, and sends an appropriate output signal to a controlled device. The controller setpoint is adjustable, whether using a knob, slider, or laptop computer connected to the controller. Along with the setpoint adjustment, there is normally an adjustment for the desired accuracy of the controller. Depending on the control system, the accuracy may be referred to as proportional band, gain, or throttling range. See Figure 4-2.

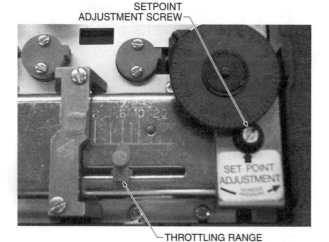

Figure 4-2. A controller receives a signal from the sensor, compares it to a setpoint value, and sends an appropriate output signal to a controlled device.

Controlled Devices

A *controlled device* is the object that regulates the flow of fluid in a system to provide the heating, air conditioning, or ventilation effect. Common controlled devices include dampers for regulating air flow, valves for regulating water or steam flow, refrigeration compressors for delivering cooling, and gas valves and electric heating elements for regulating heating.

Control system components must be compatible to provide safe, accurate, and energy-efficient control of the controlled device. For example, if electric heating elements are used to heat a building space, safety controls must be provided to shut them off in case air flow across them is lost. If not, they may overheat and possibly start a fire. See Figure 4-3.

Figure 4-3. Controlled devices regulate the flow of fluid in a system to provide the heating, air conditioning, and ventilation effect.

Commercial HVAC control systems are commonly upgraded over time by the installation of different controllers. Control panels may contain controllers from multiple manufacturers and several generations of controllers from the same manufacturer. A controls technician must be familiar with the different combinations of controls and hardware that may be present in an HVC system.

Control Agents

A *control agent* is fluid that flows through controlled devices to produce a heating or cooling effect in the system or building spaces. The most common control agents are hot water, chilled water, steam, hot air, and cold air. Control agents pass through controlled devices to condition the building spaces. Control agents are distributed in different ways in a building. For example, hot or chilled water is pumped through coils where heat is transferred to or from the air. Steam is distributed in a main steam header from a boiler and returned by steam traps and condensate pumps to the boiler. Steam or water may also be introduced directly into the air stream in a duct to provide humidification.

Heating Systems. Heating systems use hot water or steam in heat exchangers mounted in the ductwork of an air handling unit. Hot water or steam heat exchangers may also be mounted directly in the building space to provide heat by convection. Valves regulate the flow of the hot water or steam. Electric heating elements may also be mounted in the ductwork of an air handling unit. Electric heating elements heat the air as the air passes over them. Combustion is also commonly used to provide heat. In a combustion system, natural gas is burned in a heat exchanger, which heats the air passing over it. The toxic by-products of the combustion process are vented safely outside. See Figure 4-4.

Cooling Systems. Cooling systems use chilled water, direct expansion cooling, or outside air as control agents. Chilled water from a central cooling plant is pumped through valves into coils mounted in an air handling unit. In direct expansion cooling, a compressor is turned ON and OFF, which causes liquid refrigerant to flow through a heat exchanger (evaporator). The evaporator absorbs heat from the air flowing over it. Direct expansion cooling is commonly used in rooftop and packaged units.

Figure 4-4. Heating systems may use electronic combustion control devices to safely control combustion heating equipment.

Cooling systems may also use outside air for cooling. If the outside air conditions such as temperature and humidity are low enough, a damper in a duct may be opened to admit outside air into the ductwork of the HVAC system. The admission of outside air for cooling is referred to as the economizer cycle. Normally, mechanical cooling equipment is shut off when economizer cooling is available. See Figure 4-5.

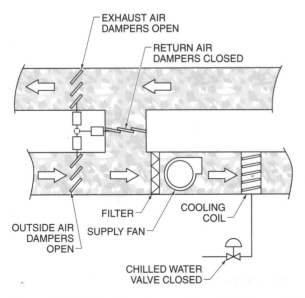

Figure 4-5. Outside air may be used to cool building spaces.

Humidification Systems. In some environments, the air inside a building becomes dry, which can lead to health problems and/or building damage. This is especially true of northern climates in the winter. Humidity may have to be added to the air in building spaces to prevent health problems and/or building damage. Humidity is added to the air in building spaces by introducing a control agent such as low-pressure steam or hot water vapor into the air stream of an air handling unit. Some humidification systems use an open sump of hot water over which air passes. Safety controls are always used with humidification systems to prevent excessively high humidity levels in a duct or building space.

Energy can be saved by using outside air for cooling. However, outside air dampers must be closed to minimum position if the outside air is excessively hot or humid.

In many areas of the U.S., excessive humidity is a problem. This is especially true in the mid and southern states. In these areas, humidity must be removed from the air (dehumidification). Excess humidity may cause discomfort and mold growth inside the building structure. Dehumidification is normally accomplished by the same method that cools the air. When air is cooled below its dew point, the extra humidity condenses and can be drained away. This condensation forms on the cooling coil of an air handling unit. The control agent in a dehumidification process is the same chilled water or direct expansion cooling that provides cooling for a building space. See Figure 4-6.

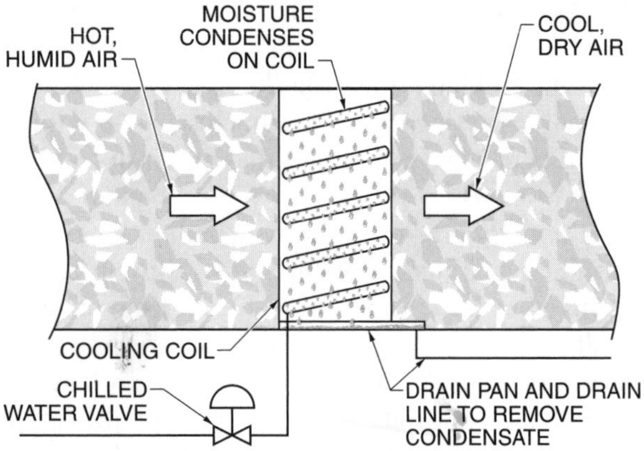

HOT, HUMID AIR

MOISTURE CONDENSES ON COIL

COOL, DRY AIR

COOLING COIL

CHILLED WATER VALVE

DRAIN PAN AND DRAIN LINE TO REMOVE CONDENSATE

Figure 4-6. Excess humidity is removed from the air in a building by passing the air across a cooling coil to condense the moisture.

Ventilation Systems. Toxins and pollutants may build up to unacceptable levels, causing health problems to occupants, if enough fresh air is not circulated in a building space. To control the quality of indoor air, dampers open to bring in fresh air to flush out pollutants. The control agent for ventilation systems is fresh outside air. A minimum amount of outside air must be introduced even if it causes extra heating or cooling to occur.

Control Functions

HVAC control systems must ensure the comfort of building occupants. The function of the control system is based on the mechanical equipment used to condition the air. HVAC system control functions include temperature control, humidity control, and pressure control.

Temperature Control. Temperature control is one of the most important components of comfort. Whether in heating or cooling mode, temperature control is accomplished by controlling either the building space temperature, return air temperature, or air volume.

In building space temperature control, a room controller or thermostat is connected directly to a heating or cooling controlled device. As the temperature in the building space changes, the controller or thermostat allows more or less heating or cooling medium to flow, affecting the building space temperature. The controlled device may be a heating or cooling coil located in a duct or mounted directly in the building space. See Figure 4-7.

BUILDING SPACE TEMPERATURE CONTROL

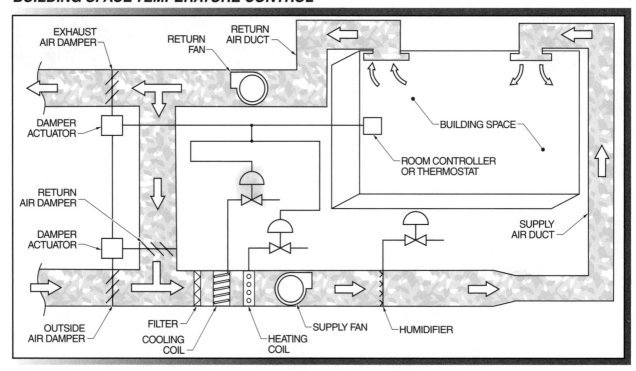

Figure 4-7. In building space temperature control, a room controller or thermostat controls a heating or cooling controlled device.

Problems may exist with building space temperature control because the temperature in a large room can vary significantly. There may be a lack of comfort in one part of a room because the room controller or thermostat only senses the temperature where it is located. Also, the room controller or thermostat may be mounted in a location that is influenced by conditions such as direct sunlight or drafts.

In return air temperature control, the return air sensor or controller is mounted in the common return duct at an air handling unit. Return air temperature control is more accurate than building space temperature control because the return air is at a temperature which is an average of all the temperatures in the building space. Return air temperature control also prevents tampering of the room controller or thermostat by building space occupants. See Figure 4-8.

Building space and return air temperature control the temperature at a controlled device such as a heating or cooling coil. In an air handling unit, the volume of air passing over a heating or cooling coil and into the building space is constant. In effect, the system is a constant-volume, variable-temperature system. However, instead of passing a constant volume of air across a variable-temperature coil, the amount of air can be varied while the air temperature remains constant. In effect, the system becomes a constant-temperature, variable-volume system. This concept is the principle behind variable air volume air handling systems. Variable air volume air handling systems reduce fan volume and therefore horsepower, saving a significant amount of energy and money compared to constant-volume systems. A mechanism such as a variable-speed drive or variable inlet vanes permits volume control of the supply fan or pump. See Figure 4-9.

RETURN AIR TEMPERATURE CONTROL

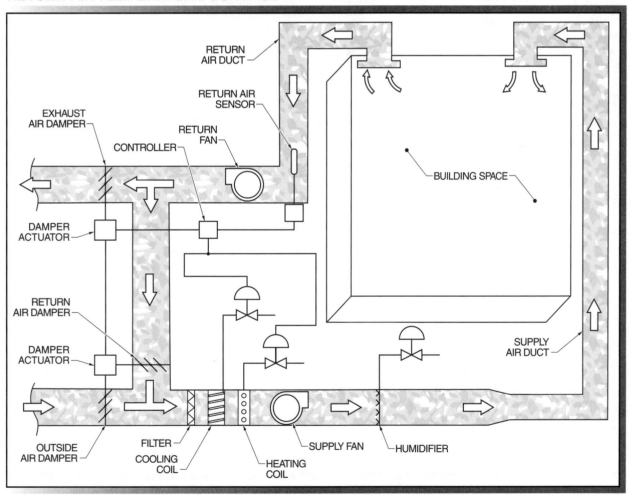

Figure 4-8. In return air temperature control, a sensor or controller is mounted in the common return duct in an air handling unit.

VOLUME CONTROL

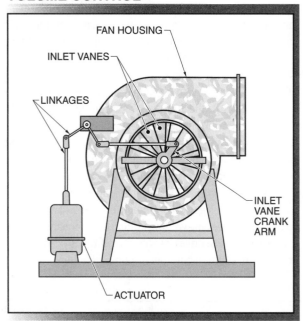

Figure 4-9. In volume control systems, supply fan air volume and required horsepower are reduced, saving a significant amount of energy and money over constant-volume systems.

Safety controls must be provided to ensure that excessively low or high temperatures do not damage the mechanical system or endanger the safety of the building occupants. Safety controls include low-temperature limits that shut down the supply fan to reduce the possibility of a freeze-up, and high-temperature limits that are used to sense and respond to high temperatures in a duct. High temperatures in a duct may cause or be caused by a fire in the building.

Humidity Control. The proper humidity level also helps ensure comfort in a building space. The most common method of humidity control is to include a humidity sensor or controller in the building space. The controller opens or closes a humidification device such as a steam valve in the air stream to keep the humidity close to setpoint. For dehumidification, the controller may cause the cooling system to operate longer than normal, drying out the air. Multiple levels of humidification safety controls are used to prevent a humidification device from staying open. For example, a discharge controller may be used to shut down the humidifier if the discharge humidity approaches 90% rh. The humidifier may also be interlocked with the air handling unit supply fan so that the humidifier is off if the fan is not operating. See Figure 4-10.

HUMIDITY CONTROL

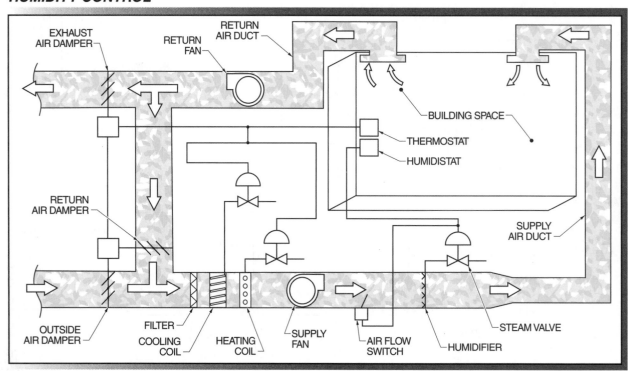

Figure 4-10. In humidity control, a humidity controller or sensor opens or closes a humidifier in the air stream to keep the humidity close to setpoint.

Pressure Control. Air handling units, pumping systems, and specialized areas such as labs require pressure control. Pressure control includes system pressure control and differential pressure control.

In system pressure control, a pressure sensor is normally mounted in a pipe or duct. The pressure sensor is connected to a controller which opens or closes a controlled device such as a damper or valve.

In differential pressure control, a specific pressure in the system is maintained based on a difference in pressure between two points in the system. Differential pressure control between supply and return pressure is common in piping systems. Differential pressure control is also used occasionally between room air pressure and pressure outside the room or even outside the building. This is done to maintain a specific difference in air pressure. This pressure difference may be used to isolate a room and prevent the escape of fumes. In addition to controlling the pressure, safety controls must be provided that shut down the mechanical equipment if the pressure becomes excessive.

> *As building design and construction techniques evolve, HVAC control systems must also evolve to provide greater accuracy, more sophisticated control sequences, and higher energy savings.*

Control System Characteristics

All HVAC control systems operate using the same set of control principles. Common control system characteristics include setpoint, control point, offset, and feedback.

Setpoint. *Setpoint* is the desired value to be maintained by a system. For example, it may be desired to maintain a temperature of 72°F in a building space. The 72°F temperature is referred to as the setpoint of the control system. Setpoints can be stated in different variables including temperature, pressure, humidity, light level, dewpoint, enthalpy, etc.

Control Point. A *control point* is the actual value that a control system experiences at any given time. For example, a temperature sensor in a building space measures a temperature of 74°F. The 74°F temperature value is referred to as the control point. At any given time, the setpoint may be different from the control point.

Offset. *Offset* is the difference between a control point and a setpoint. Depending on the accuracy of the controller, the actual value in the system may be different from the value the controller is able to maintain. See Figure 4-11.

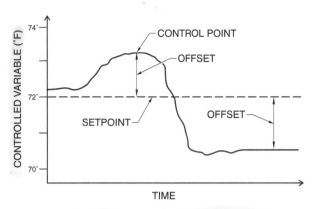

Figure 4-11. Offset is the difference between the actual variable (control point) and setpoint.

Feedback. *Feedback* is the measurement of the results of a controller action by a sensor or switch. For example, a controller measures the temperature in a building space. The controller operates a controlled device to maintain a setpoint in the building space. The controller measures its effect by comparing the result of its action against its setpoint. The measurement of the results is the feedback of the system. See Figure 4-12.

Closed Loop Control

A *control loop* is the arrangement of a controller, sensor, and controlled device in a system. *Closed loop control* is control in which feedback occurs between the controller, sensor, and controlled device. For example, a hot water terminal device is controlled by a thermostat. The controller (thermostat) positions a hot water valve to maintain the setpoint in the building space. The thermostat in the building space provides the feedback which controls the hot water valve and temperature in the building space. See Figure 4-13. In a closed loop control system, if one part of the control system malfunctions, it causes the other parts to have the wrong value or position.

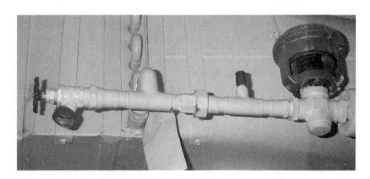

In a closed loop control system, a reheat valve is opened or closed by a room thermostat. The warm air from the reheat coil is carried to a building space and sensed by the room thermostat, completing the loop.

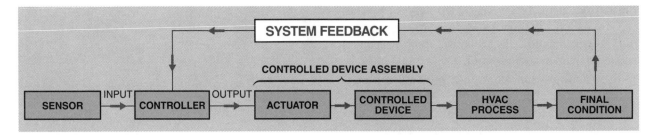

Figure 4-12. Feedback is used to measure the results of the control system output.

CLOSED LOOP CONTROL

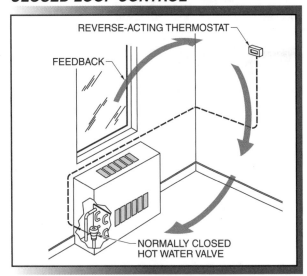

Figure 4-13. Closed loop control systems provide feedback to the controller.

Open Loop Control

Open loop control is control in which no feedback occurs between the controller, sensor, and controlled device. For example, a controller cycles a chilled water pump ON when the outside air temperature is above 65°F. In this application, the controller has no feedback regarding the status of the pump. See Figure 4-14.

ON/OFF Control

ON/OFF (digital) control is control in which a controller produces only a 0% or 100% output signal. ON/OFF control is used in open loop and closed loop control systems. The most common ON/OFF control is a set of electrical contacts that are closed (100% electrical flow) or open (0% electrical flow). ON/OFF control is often referred to as two-position control. ON/OFF control uses controllers and controlled devices

that have only two positions. These positions are normally listed as ON and OFF. ON/OFF controllers include residential electromechanical thermostats. ON/OFF controlled devices include air conditioning compressors, electric heating stages, gas valves, refrigeration compressors, and constant-speed fans.

OPEN LOOP CONTROL

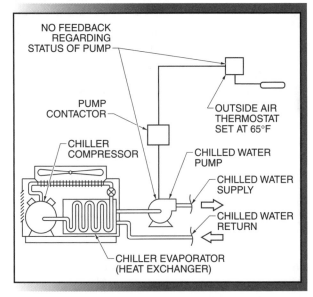

Figure 4-14. Open loop control systems do not have feedback between the controller, sensor, and controlled device.

ON/OFF control systems energize the controlled device when needed above or below setpoint. See Figure 4-15. The controlled device operates until the setpoint is reached. In an ON/OFF control system, there is a tendency for the controlled variable to go above (overshoot) or below (undershoot) the setpoint. This is because the heating or cooling equipment is either 0% (OFF) or 100% (ON).

ON/OFF CONTROL

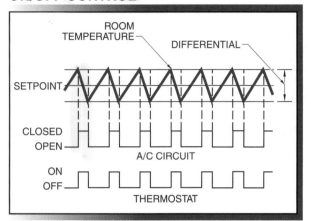

Figure 4-15. In an ON/OFF control system, the controlled device is either fully open or fully closed.

Overshooting is the increasing of a controlled variable above the controller setpoint. *Undershooting* is decreasing of a controlled variable below the controller setpoint. If left uncorrected, overshooting and undershooting can lead to occupant discomfort. For example, occupants may be comfortable at a setpoint of 74°F. If overshooting occurs, the temperature may increase to 76°F or more, which may make the occupants uncomfortable. In ON/OFF control systems, overshooting and undershooting are compensated for with anticipators. An *anticipator* is a device that turns heating or cooling equipment ON or OFF before it normally would. This reduces the possibility of overshooting and undershooting. ON/OFF control systems are used primarily in residential and small commercial systems such as heat pumps, rooftop units, and residential split systems.

Proportional Control

Proportional (analog) control is control in which the controlled device is positioned in direct response to the amount of offset in the system. For example, in proportional control a 10% increase in room temperature results in a cooling control valve opening 10%. Proportional control is also used in open loop and closed loop control systems. Proportional control systems use a response that is a number between two values such as 0 VDC to 10 VDC. Proportional control systems use controllers and controlled devices that respond to this variable signal. See Figure 4-16. Most commercial control applications today use proportional control. Proportional control systems have a lower tendency toward undershooting or overshooting. While proportional control is used successfully in many applications, it may be inaccurate if not set up properly.

Control System Requirements

Comfort is required in modern commercial buildings. Temperature, humidity, pressure, and indoor air quality must be maintained within acceptable levels. In addition, potentially harmful indoor air pollutants must be kept to acceptable levels. In facilities dedicated to manufacturing processes, the tolerances of building variables may be excessively tight because products may be ruined by improper environmental conditions. HVAC control systems are designed to provide safe, automatic, and accurate regulation of the environmental conditions in a building space.

PROPORTIONAL CONTROL

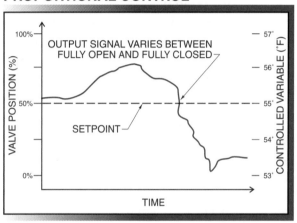

Figure 4-16. In a proportional control system, the output signal varies between fully open and fully closed.

Control System Safety. Safe operation of the control system is essential. Improper operation of a control system should pose no immediate risk to the health of the building occupants. See Figure 4-17. The control system power supply should not pose a safety risk to the occupants. For this reason, electrical controls often use 24 VAC. Likewise, the low-pressure compressed air used in pneumatic control systems and the low-voltage DC power used for electrical control systems does not pose significant risk to occupants. Any possibly dangerous systems should be securely locked in a panel which is only accessible to trained personnel.

HVAC control technicians must follow national and local codes when designing, installing, testing, and operating a control system. Good safety habits should be developed and followed routinely to prevent injury.

HVAC HAZARDS

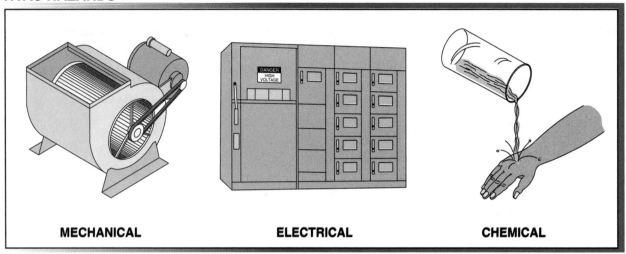

Figure 4-17. HVAC control systems may include mechanical, electrical, and chemical hazards.

The safety of a building structure is also important. The fire risks of a control system must be eliminated or minimized. Low-limit temperature controls should be provided to prevent damage due to water leakage from frozen coils and pipes. National and local codes must be followed when designing, installing, testing, and operating a control system. These codes are enforced by local inspectors who check the control system installation for compliance.

Automatic Control System Operation. Automatic operation is essential in a control system. Most commercial buildings today are not staffed to enable the continual adjustment of environmental control devices. Once adjusted, a control system should operate automatically. Any conditions that may affect the indoor environment must be compensated for without intervention by an operator. Any control system that requires continuous attention has a flaw in the design, installation, or testing process.

Control System Accuracy. Accuracy of a control system is required for customer satisfaction. The difference between the desired condition (setpoint) and the actual condition (control point) should be minimized as much as possible. See Figure 4-18. To avoid problems with expectations, the accuracy of the control system should be measured and provided to the customer. Expectations of control system accuracy have increased as the available technology has become more sophisticated. In the past, an accuracy of ±2°F was considered acceptable. With the widespread use of automated control systems, much tighter accuracies (±.5°F) are possible. It is not uncommon today for customers to expect control system accuracies of less than 1°F. In addition, demanding control applications such as pharmaceutical and manufacturing facilities expect this routinely. In these demanding applications, care must be taken to select a sensor with the correct tolerances to deliver the close control required.

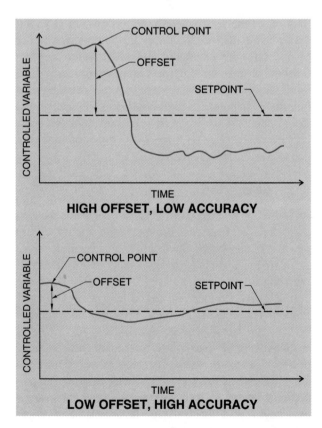

Figure 4-18. The smaller the difference between setpoint and control point (offset), the more accurate the control system.

HAI (Home Automation, Inc.)

Commercial building HVAC control systems are designed to provide comfort to the occupants of building spaces. Commercial HVAC control systems include self-contained, electric, pneumatic, electronic, automated, system-powered, and hybrid control systems. These systems differ in the technology used to provide control.

CONTROL SYSTEM HISTORY

Individuals have attempted to improve their living environment for thousands of years—whether by gathering fuel for a fire, experimenting with new fuels, or improving their shelter from the elements. Over time, innovations in technology have increased the level of comfort in living environments. For example, simple huts were followed by permanent structures made of wood or brick. Similar improvements have occurred in the area of heating, ventilation, and air conditioning (HVAC) control.

Early control of the level of comfort in a living environment was done by hand. For example, an individual may have controlled the temperature in a living environment by adding fuel to a fire or allowing the fire to die down. As structures grew in size and contained a greater number of rooms, hot air from a central fire was regulated manually by opening or closing a diffuser (damper) through the use of an adjusting pulley.

In the early 1900s, electricity came into wide distribution in many cities. This new power source was adapted to provide automatic building environment control. Electricity allowed increased control and required less time and attention than a manual control system. Early electric controls were large, bulky, unreliable, inaccurate by modern standards, and dangerous, partly because early electrical power quality and distribution were poor by today's standards.

In the first quarter of the 20th century, pneumatic control systems came into use in commercial buildings. A *pneumatic control system* is a control system in which compressed air is used to provide power for the control system. In a pneumatic control system, an electrical power supply is used to turn the motor of an air compressor. The compressed air supply is piped through the building to power the HVAC control system.

In the 1960s, solid-state, low-voltage direct current (DC) devices came into common use. This technology was adapted for use in commercial HVAC control systems. Early solid-state systems were known as electronic control systems. More recently, with the advent of microchips that can hold thousands of solid-state devices on a ¼″ square chip, automated control systems were created that enabled improved control of the environment in commercial buildings.

The control of the environment in commercial buildings is based on energy efficiency and comfort. During the energy crisis of the 1970s, the cost of energy escalated dramatically. This spurred developments in advanced control systems and techniques. Many commercial buildings were not designed with energy-saving control systems. As a result, there has been a tremendous effort to equip old buildings with systems that can significantly reduce energy expenditures.

COMMERCIAL BUILDING CONTROL SYSTEMS

Comfort is a requirement of a commercial building. *Comfort* is a condition that occurs when people cannot sense a difference between themselves and the surrounding air. Occupants of modern commercial buildings demand precise control of the environment as a condition of leasing space. See Figure 5-1. Accurate control of the environment in a building space leads to higher occupancy rates and higher profits.

Figure 5-1. Comfort is a requirement of occupants of modern commercial buildings as a condition of leasing space.

A recent HVAC systems control issue is that of indoor air quality. *Indoor air quality (IAQ)* is a designation of the contaminants present in the air. Modern HVAC systems are expected to provide enough outside air (OA) for ventilation to flush out any contaminants inside a building that may cause health concerns for the occupants.

The demands for energy savings, comfort, and occupant health have placed a tremendous burden on modern control systems and increased their complexity and cost. These costly control systems have created great opportunities for technicians and maintenance personnel who are trained in their installation, operation, service, and use.

The purpose of an HVAC control system is to maintain the proper environment inside a building. In most cases, the environment is based on human comfort. In other cases, the environment is based on maintaining a product or industrial process condition.

In either case, the indoor environment of a building must be maintained by providing the proper levels of variables such as temperature, humidity, pressure, etc.

Commercial buildings require HVAC control systems that provide a quality indoor environment for individuals and products. As time and technology have advanced, the capability and complexity of the control systems have changed. Commercial building control systems include self-contained, electric, pneumatic, electronic, automated, system-powered, and hybrid control systems.

Self-Contained Control Systems

A *self-contained control system* is a control system that does not require an external power supply. Self-contained control systems are widely used, basic control systems. Self-contained control systems have been successfully applied in certain applications for almost 100 years.

The power to a self-contained control system is supplied by a sealed, fluid-filled element (bulb). The fluid may be a gas, liquid, or mixture of both. The fluid-filled element is known as a power head or power element. The element may be attached to a pipe containing water or refrigerant. The element may also be attached to a wall and used as a building space thermostat.

As the temperature of the fluid in a pipe or air in a building space changes, heat is transferred to or from the fluid-filled element. The heat transfer changes the pressure of the fluid in the element. The pressure acts against a diaphragm that moves a valve body to regulate the flow of refrigerant, steam, hot water, or chilled water through the valve. Adjustments are often provided to change the temperature setpoint of the system. See Figure 5-2.

Self-Contained Control System Applications. Self-contained control systems are used to control basic closed loop systems that require relatively low accuracy. A *closed loop system* is a control system in which feedback occurs between the controller, sensor, and controlled device. The two most common applications of self-contained control systems are thermostatic expansion valves and building space temperature control in steam or hot water heating systems.

The advantages of self-contained control systems are that no external power supply is needed, and the operation of the control system is unaffected by fluctuations or failure of a power supply. In addition, installation cost is low because of the inexpensive control elements and lack of external wiring or piping.

SELF-CONTAINED CONTROL SYSTEMS

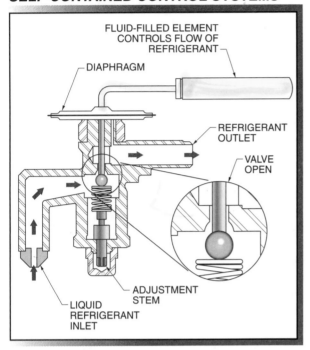

Figure 5-2. Self-contained control systems commonly use fluid-filled elements that do not require an external power supply for operation.

The disadvantages of self-contained control systems are that they cannot be expanded to provide sophisticated control sequences because only one setpoint adjustment is provided. Most self-contained control systems have relatively poor accuracy. In addition, self-contained control systems do not have diagnostic means available to troubleshoot or diagnose system failure, or remote indication to point out control system failure. Replacement is normally the only option.

The sensing element of a self-contained control system must be located properly (away from localized hot or cold areas) to prevent inaccurate control.

Electric Control Systems

An *electric control system* is a control system in which the power supply is low voltage (24 VAC) or line voltage (120 VAC or 220 VAC) from a step-down transformer that is wired into the building power supply. See Figure 5-3. Line-voltage control systems are rarely used in commercial building spaces because of the danger to

the occupants from electrical shock, fire, and possible explosion due to arcing of the circuit in an area that may contain flammable substances.

Electric control systems commonly use a mechanical device to control the flow of electricity in a circuit. The mechanical device varies from system to system. For example, thermostats may use a bimetallic element to make or break contacts to control building space temperature. A *bimetallic element* is a sensing device that consists of two different metals jointed together. The different metals bend at different rates when heated or cooled. See Figure 5-4. On a change in temperature, the bimetallic coil changes position, causing contacts to open or close or a mercury switch to tilt. The contacts or mercury switch makes or breaks a circuit which energizes or de-energizes the heating equipment. Pressurestats use a bellows for pressure control. Humidistats use a synthetic moisture-sensing element for humidity control.

ELECTRIC CONTROL SYSTEMS

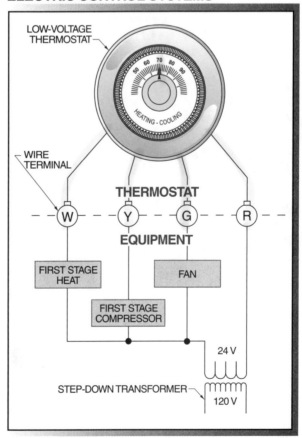

Figure 5-3. Electric control systems use low voltage (24 VAC) or line voltage (120 VAC or 220 VAC) as the power supply.

THERMOSTAT SWITCHES

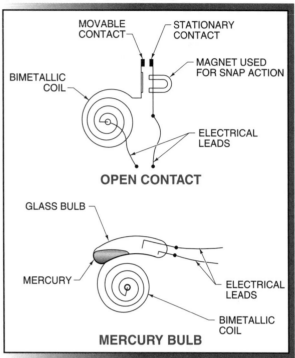

OPEN CONTACT

MOVABLE CONTACT

STATIONARY CONTACT

MAGNET USED FOR SNAP ACTION

BIMETALLIC COIL

ELECTRICAL LEADS

MERCURY BULB

GLASS BULB

MERCURY

ELECTRICAL LEADS

BIMETALLIC COIL

5-4. A bimetallic element (coil) is commonly used to switch power in an electric control system.

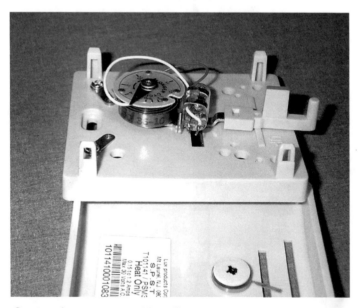

Common thermostats use a bimetallic coil to sense room temperature and switch the power circuit.

Electric Control System Applications. Electric control systems are the most common control systems used today. Electric control systems are used to control building space temperature through the use of room thermostats. Electric control systems are also used to control packaged rooftop units and split system air conditioning units. Relays can be used to enable low-voltage control circuits to switch line-voltage devices. Line-voltage controls are often used in industrial and commercial systems where access by occupants is not possible. Line-voltage control devices are wired directly to the controlled devices such as compressors, fans, and pumps because both are at the same voltage level.

The advantages of electric control systems are that they provide adequate temperature control of the air in building spaces and use small, thin wires that are easily installed in building walls. Electric control systems are relatively safe because of their low voltage levels (24 VAC), thus reducing the danger of shock to building occupants. In addition, individual controls are relatively inexpensive, easy to install and service, and easily wired in different combinations to add control features.

The disadvantages of electric control systems are that they cannot be used in areas that contain explosive products without the proper precautions. Electric control systems may be difficult to modify and are often unable to perform complex control sequences required in commercial buildings. Electric control systems are not often used for analog (proportional) control because only digital (ON/OFF) control devices are used. In addition, electric control systems are not easily designed to allow central reporting of failures and alarms.

Pneumatic Control Systems

A *pneumatic control system* is a control system in which compressed air is used to provide power for the control system. Pneumatic control systems were developed in the early 1900s as a primary method of controlling the environment in commercial buildings. Pneumatic control systems were developed because the existing electric control systems were dangerous or could not provide the flexibility and control necessary in commercial buildings.

While the popularity of building automation systems has increased dramatically during the past 10 years, pneumatic controls continue to be widely used. Many buildings use a building automation system to control the central chiller, boiler, or air handling unit, and use pneumatic controls to maintain the room temperature.

Pneumatic control systems can be separated into four main groups of components based on their function. These groups are the air compressor station, transmitters and controllers, auxiliary devices, and controlled devices. See Figure 5-5.

Air Compressor Station. The power source for a pneumatic control system is compressed air. The air compressor station consists of the air compressor and other devices that ensure that the air has the qualities needed by the control system. The pneumatic control system functions improperly if the air compressor does not deliver clean, dry, oil-free air at the correct pressure. The air compressor station dryers, filters, and pressure-reducing valves help ensure these qualities of the air.

Transmitters and Controllers. Transmitters and controllers sense and control the temperature, pressure, or humidity in a building space or system. Transmitters and controllers are connected to the air supply from the air compressor station. Transmitters and controllers change the pressure to the controlled devices, which causes the controlled devices to regulate the flow of the heating, cooling, or other medium into the building spaces.

> *A controls technician must be familiar with the different groups and components of a pneumatic control system as well as the different generations of controllers from different manufacturers. It is not uncommon for a building control system to have been upgraded with different controllers many times over the years.*

Auxiliary Devices. Auxiliary devices are devices that are normally located between the transmitters and controllers and the controlled devices. Auxiliary devices change or reroute the air supply from the transmitter or controller before it reaches the controlled devices.

Controlled Devices. Controlled devices include dampers, valves, actuators, and switches that deliver the heating or cooling into the building spaces or system. Controlled devices are driven by the compressed air that is supplied to a controller and that causes the controlled devices to open or close properly.

PNEUMATIC CONTROL SYSTEMS

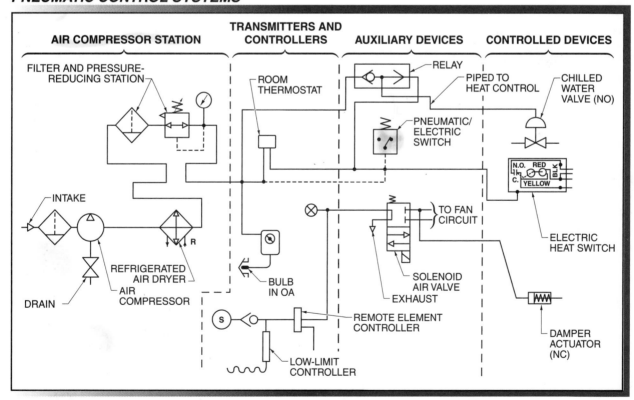

Figure 5-5. Pneumatic control systems include the air compressor station, transmitters and controllers, auxiliary devices, and controlled devices.

Many pneumatic controllers use bimetallic elements to cover or uncover a bleedport. A *bleedport* is an orifice that allows a small volume of air to be expelled to the atmosphere. The air pressure delivered by the controller to the valve or damper changes as the bimetallic element covers or uncovers the bleedport. The air pressure in the line changes proportionally to the temperature change at the bimetallic element. The air pressure signal is piped to an actuator, which causes a valve or damper to open or close to control the temperature of the air in a building space.

Pneumatic Control System Applications. Pneumatic control systems are used primarily in large commercial buildings such as hospitals and schools. Pneumatic control systems can be used in most control sequences because of their flexibility. See Figure 5-6. Pneumatic control systems can be used in central air handling units, boilers, chillers, and cooling tower systems. Zone temperature, pressure, and humidity control schemes can also be provided. Pneumatic control systems are rarely used in residential systems or packaged (rooftop or heat pump) units.

SINGLE-ZONE AIR HANDLING UNIT

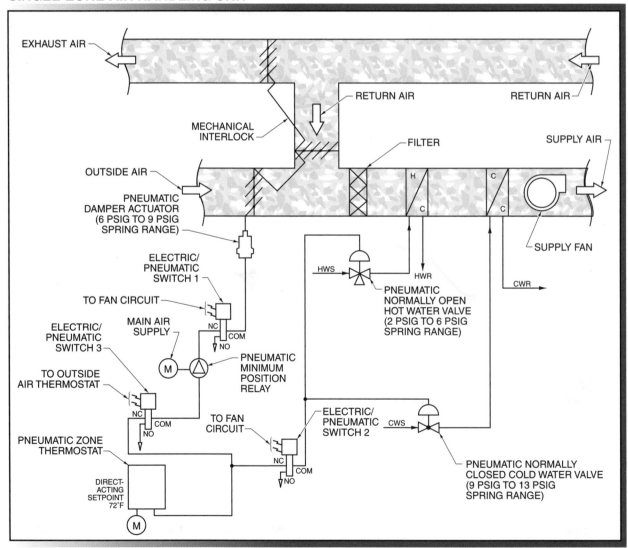

Figure 5-6. Pneumatic controls are flexible enough to be used with many different HVAC systems.

The advantages of pneumatic control systems are that, due to their principle of operation, pneumatic controls easily provide analog (variable) control. Pneumatic control systems do not produce a shock hazard to building occupants and are suitable for buildings that may contain flammable or explosive vapors because they use compressed air as the power supply. Pneumatic control systems are relatively inexpensive and rugged, and provide years of service if properly set up. Most pneumatic control systems are accurate, can be easily expanded, and can provide many different control sequences.

The disadvantages of pneumatic control systems are that pneumatic air compressor stations require regular maintenance. In addition, pneumatic control systems may require specialized tools, calibration, and setup procedures.

Electronic Control Systems

An *electronic control system* is a control system in which the power supply is 24 VDC or less. Electronic control systems are analog control systems. An *analog control system* is a control system that uses a variable signal. Electronic control systems were originally developed to replace pneumatic control systems. Electronic control systems use solid-state components and are often confused with automated control systems.

Electronic control systems commonly have supply voltages of 10 VDC, 12 VDC, or 18 VDC. See Figure 5-7. The power supply of an electronic control system consists of a transformer which drops the supply voltage to between 10 VAC to 18 VAC and rectifiers which convert the voltage from AC to DC. Filters may also be included to prevent voltage spikes from passing through the power supply and damaging the control system components.

In an electronic control system, a signal from a sensor is wired to a resistive bridge circuit. A *resistive bridge circuit* is a circuit containing four arms and four resistors. Two of the resistors are fixed and two are variable resistors. The variable resistors normally consist of a temperature sensor and setpoint potentiometer. The bridge circuit output value is zero when the system is at setpoint. See Figure 5-8. When the sensor signal changes, the resistance bridge circuit becomes unbalanced, generating a response voltage in relation to the sensor input. The voltage signal is sent to an actuator, which opens or closes a valve or damper in response to the signal change. The signal may also be changed by adjusting the setpoint potentiometer.

ELECTRONIC CONTROL SYSTEMS

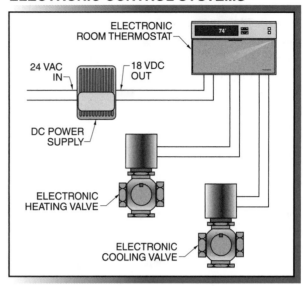

Figure 5-7. Electronic control systems commonly have supply voltages of 10 VDC, 12 VDC, or 18 VDC.

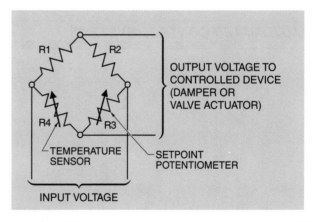

Figure 5-8. Resistive bridge circuits use fixed resistors and variable resistors to vary the output voltage to an actuator.

Electronic Control System Applications. Electronic control systems can be used in any HVAC application. Electronic control systems are used in packaged units such as heat pumps and split system air conditioning units. Electronic control systems may also be used in large commercial building variable air volume and multizone systems.

The advantages of electronic control systems are that they are reliable, accurate, and relatively inexpensive. In addition, electronic control systems can be easily expanded to provide many different control sequences.

The disadvantages of electronic control systems are that they may require special diagnostic tools and procedures. Electronic actuators are normally expensive and complex, especially when compared to pneumatic control actuators. Electronic control system components may become obsolete rapidly and may be difficult to replace. In addition, electronic control systems are not as powerful and flexible as automated control systems.

Automated Control Systems

An *automated control system (building automation system)* is a control system that uses digital solid-state components. A digital control system uses ON/OFF (1/0) signals which represent numbers, setpoints, control sequences, etc. Automated control systems have become popular in commercial buildings because of their increased reliability and the increase in the power and capacity of personal computers (PCs).

An automated control system power supply is the same as that of an electronic control system. Automated control systems use a step-down transformer and rectifiers that change the AC power to DC. The DC signal is filtered to eliminate any interference or problems with the incoming power supply. See Figure 5-9.

Automated control systems commonly consist of intelligent local controllers that are connected to HVAC equipment such as VAV terminal boxes, heat pumps, or rooftop units. Each local controller has a central processing unit that can perform complex control functions. Each controller also has an individual program in its on-board memory. The program determines setpoints, control response times, and any additional calculations needed. The program parameters may be modified and new control sequences and options added by reprogramming the controller with a desktop PC, notebook PC, portable operator terminal, or keypad display.

AUTOMATED CONTROL SYSTEMS

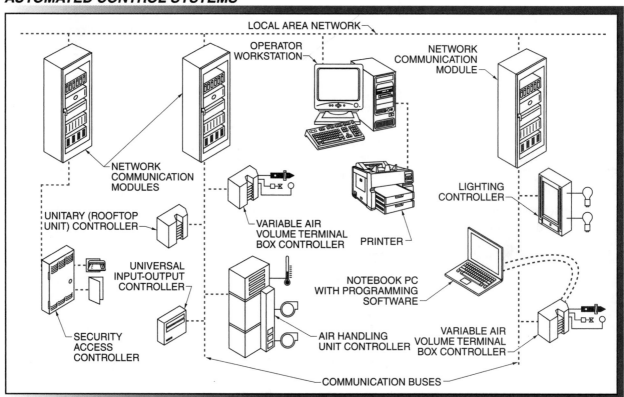

Figure 5-9. Automated control systems (building automation systems) have become more popular in commercial buildings due to their increased reliability and the increase in the power and capacity of PCs.

Automated Control System Applications. Automated control systems are used in large commercial buildings such as hospitals, schools, military bases, and colleges. In addition, they are also becoming more common in relatively small applications such as retail stores, restaurants, and strip malls.

The advantages of automated control systems are that they are extremely accurate if set up properly, can be set up for remote data acquisition and monitoring, and can integrate fire and security services. Automated control systems are versatile and can be reprogrammed easily. In addition, they can use many common components which are identical to consumer electronics such as printers, PCs, and monitors.

The disadvantages of automated control systems are that the programming may be complex and require specialized software knowledge, they may need to be upgraded regularly to obtain new software features, and they can become obsolete rapidly. In addition, service for automated control systems can be expensive.

System-Powered Control Systems

A *system-powered control system* is a control system in which the duct pressure developed by the fan system is used as the power supply. System-powered control systems are used sparingly. System-powered control systems were developed to avoid the installation costs of pneumatic piping runs from an air compressor station.

The power supply of a system-powered control system is the HVAC system itself. In a system-powered control system, the control system is connected to a duct that supplies air to a building space. This small amount of pressure, measured at about 1″ wc, powers the control system. The air duct has a pressure tap which directs the duct system pressure to a filter. The filter supplies the low-pressure air to a system-powered thermostat. See Figure 5-10. The thermostat has a bimetallic element that senses building space temperature. The bimetallic element moves in response to changes in the building space temperature. As the bimetallic element changes position, the air pressure to an inflatable bellows changes. As the bellows inflates or deflates, it controls the primary air flow from the duct into the building space, causing a change in the building space temperature. This eliminates the need for pneumatic piping or wires from a transformer to power the control system.

System-Powered Control System Applications. System-powered control systems are generally used in VAV terminal box applications. Variable air volume terminal box applications are suitable for system-powered control systems because of the proximity of the ducts that deliver air to the VAV terminal boxes. The use of duct air makes system-powered controls unusable for refrigeration or large commercial control systems because air may not be available or the pressure generated in the duct may be excessively low.

SYSTEM-POWERED CONTROL SYSTEMS

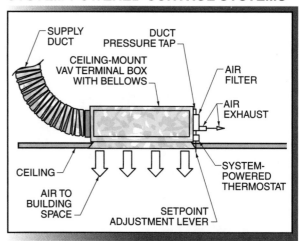

Figure 5-10. System-powered control systems are used in limited applications such as VAV terminal boxes.

One advantage of system-powered control systems is reduced installation time because controls are normally factory mounted and calibrated on the VAV terminal boxes. System-powered control systems also allow flexibility of zoning in a building because zones can be easily installed or moved as needed without changing piping or wiring. In addition, system-powered control systems are relatively inexpensive and require no external power supply that may fail or cause problems.

The disadvantages of system-powered control systems are that the control system fails if duct air is dirty or at the wrong pressure, air leaks cause the control system to operate improperly, and there is little flexibility in application. In addition, system-powered control systems cannot be adapted to other types of control; no remote monitoring, adjustment, or alarm or fault indication is possible; and the systems are relatively inaccurate.

Hybrid Control Systems

A *hybrid control system* is a control system that uses multiple control technologies. As control systems evolve, it is often necessary to use multiple control system technologies together in the same HVAC unit. See Figure 5-11. In hybrid control systems, transducers are often used as an interface between different control system technologies. Hybrid control systems are common in commercial building control.

HYBRID CONTROL SYSTEMS

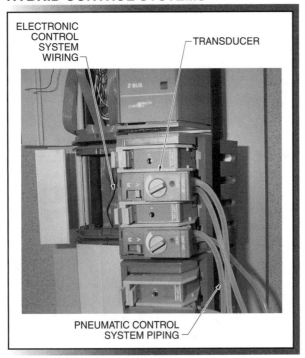

Figure 5-11. Hybrid control systems include different HVAC control technologies to control a single HVAC unit.

Hybrid control systems contain multiple control technologies requiring different power supplies. Care must be taken that each power supply is operating properly for the entire system to work correctly. For example, a hybrid control system that uses an automated control system to control pneumatic damper and valve actuators requires that the air compressor station and the electronic power supply are working to deliver the proper building space control.

Hybrid Control System Applications. The most common application of hybrid control systems is in the retrofit of an automated control system into a pneumatic control system. In retrofit applications, pneumatic controllers are removed, but the actuators remain. Pneumatic actuators are often kept because they are trouble-free and powerful. The automated control system is con-

nected to pneumatic actuators through transducers. The transducer has an input signal of 0 VDC to 10 VDC and a corresponding output of 0 psig to 20 psig. See Figure 5-12. Other common hybrid control system applications include the use of an automated control system to control the central plant equipment while pneumatic controls are used for individual building space control.

Figure 5-12. Retrofitting requires the use of electric/pneumatic transducers to enable one control technology to interact with another.

The advantages of hybrid control systems are that they take advantage of the best characteristics of each system. For example, pneumatic actuators are trouble-free and powerful, and automated control systems provide accurate control. Hybrid control systems minimize cost by reusing old components and installed materials such as control cabinets, wiring, and piping. Accuracy and dependability may be increased by using modern control system components. New features such as alarms, dial-up, and remote indication are provided.

The disadvantages of hybrid control systems are that additional knowledge is needed to service the control system. Troubleshooting difficulty may be increased due to system complexity. In addition, failure of one of the control systems may cause the entire system to fail.

An air compressor station provides clean, dry, oil-free air that is at the correct pressure and volume to pneumatic control systems. An outside air intake or intake air filter is used to keep compressed air clean. A manual drain, automatic drain, outside air intake, refrigerated air drier, desiccant drier, air line filter, and oil removal filter are used to keep compressed air dry. A pressure switch, pressure regulator, and safety relief valve are used to keep compressed air at the correct pressure.

AIR COMPRESSOR STATION

An air compressor station consists of the air compressor and auxiliary components used to provide pressurized air to the pneumatic controls. Air provided by an air compressor station must have the proper qualities or the HVAC control system may operate improperly or fail. An air compressor station includes the air compressor and auxiliary components, which remove particulate matter (outside air intake, intake air filter), remove moisture (drains and driers), remove oil (oil removal filters), and control the air supply pressure and volume (pressure switches, regulators, and safety relief valves). See Figure 6-1.

Air Compressors

An *air compressor* is a component that takes air from the atmosphere and compresses it to increase its pressure. Air compressors convert the mechanical energy provided by an electric motor into the potential energy of compressed air. Air compressors may be positive-displacement or dynamic compressors.

Positive-Displacement Compressors. A *positive-displacement compressor* is a compressor that compresses a fixed quantity of air with each cycle. Positive-displacement compressors have lower flow rates than dynamic compressors, but system pressure does not normally affect the operation of a positive-displacement compressor. Positive-displacement compressors may be reciprocating or rotary. See Figure 6-2.

A *reciprocating compressor* is a compressor that uses reciprocating pistons to compress air. Reciprocating compressors are the most common compressors used for pneumatic HVAC control systems. The pistons are connected to a crankshaft that is rotated by a motor. The rotation of the crankshaft causes the pistons to reciprocate inside cylinders. On the suction stroke, the suction valve (inlet valve) opens, and air from the atmosphere is drawn into the cylinder. At the same time, the discharge valve is pulled closed. On the compression stroke, the suction valve is pushed closed, compressing the air and discharging it at high pressure through the discharge valve to the receiver. The compressor valves are made from thin pieces of flexible metal that cover and uncover the inlet and discharge ports. Reciprocating compressors are designed to operate efficiently over long periods while preventing problems such as oil carryover. *Oil carryover* is lubricating oil that leaks by the piston rings and is carried into the compressed air system. Reciprocating compressors range in size from ½ horsepower (HP) to multiple installations of compressors over 25 HP.

AIR COMPRESSOR STATIONS

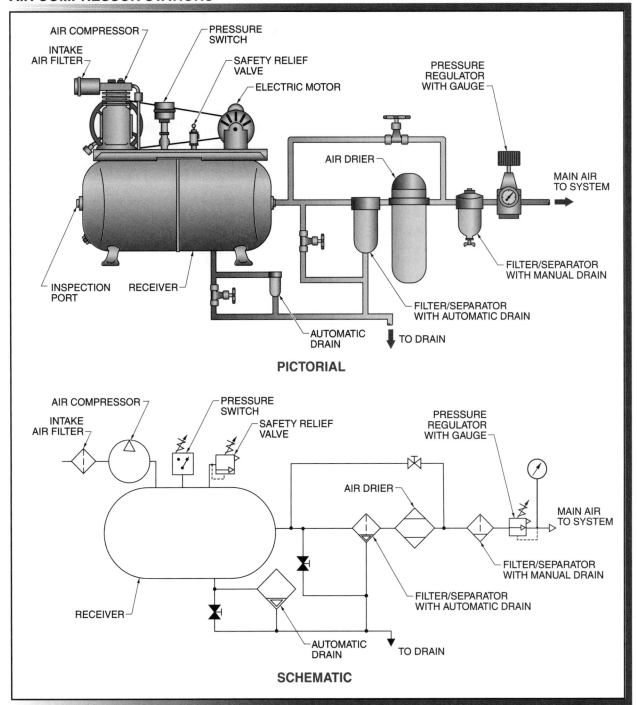

Figure 6-1. An air compressor station includes the air compressor, air drier, drains, filters, and pressure-regulating valve stations.

A *rotary compressor* is a compressor that uses a rotating motion to compress air. Rotary compressors may be screw or vane compressors. A *screw compressor* is a compressor that contains a pair of screw-like rotors that

interlock as they rotate. The rotors are located in a tight-fitting housing and are rotated by an electric motor. As the rotors rotate, air is drawn into the inlet port and then forced through the housing by the interlocking lobes of the rotors. The air is compressed as the opening between the rotors and the housing becomes smaller. The compressed air is then discharged through the outlet port.

> *The air compressor used in an HVAC pneumatic control system must be replaced with a compressor specifically designed for this application. A general-purpose air compressor operates at a higher speed than an HVAC pneumatic control system air compressor and has a greater amount of oil carryover.*

A *vane compressor* is a positive-displacement compressor that has multiple vanes located in an offset rotor. The vanes form a seal as they are forced against the cam ring. At different points inside the compressor, the offset of the rotor in the cam ring produces different distances between the rotor and cam ring. As the rotor rotates, its offset position allows the vanes to slide out and draw air from the inlet port. As the rotor continues to rotate, the volume between the vanes and the cam ring decreases, pushing the vanes into their slots in the rotor. The decreasing volume compresses the air, forcing it from the outlet port.

POSITIVE-DISPLACEMENT COMPRESSORS

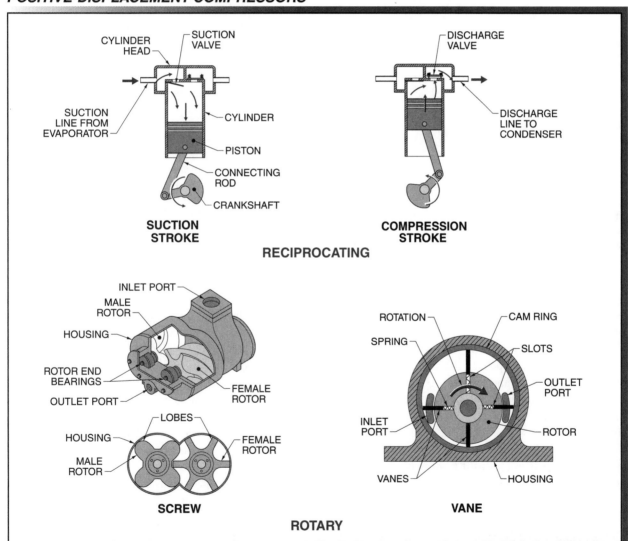

Figure 6-2. Reciprocating and rotary compressors are positive-displacement compressors that decrease the volume of the air to increase its pressure.

Dynamic Compressors. A *dynamic compressor* is a compressor that adds kinetic energy to accelerate air and convert the velocity energy to pressure energy with a diffuser. Dynamic compressors are used for flow rates of hundreds of thousands of standard cubic feet per minute, but have limited pressure ratings (under 100 psi). Dynamic compressor operations are affected by changes in system pressure. Dynamic compressors include centrifugal compressors. See Figure 6-3.

CENTRIFUGAL COMPRESSORS

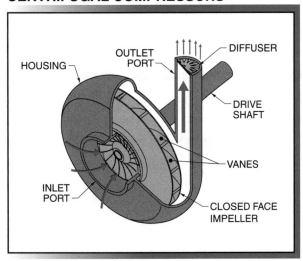

Figure 6-3. Centrifugal compressors are dynamic compressors that add kinetic energy to accelerate air and convert the velocity energy to pressure energy with a diffuser.

A *centrifugal compressor* is a compressor that uses centrifugal force to move air. In a centrifugal compressor, a closed-face impeller turns inside a housing. The air inlet port is located in the front of the housing at the center of the impeller. The air outlet port is located on the outer perimeter of the housing. The impeller turns at a high speed and draws air into the center of the impeller. The air is given kinetic energy as it flows through the vanes on the impeller, then thrown off the tips of the impeller by centrifugal force. The speed of the air increases as it is thrown to the perimeter of the impeller. As the air leaves the outer rim of the impeller, the speed is converted to pressure because the air is forced into a smaller opening (diffuser). Regardless of the compressor size or manufacturer, the compressed air must be clean, dry, oil-free, and at the correct pressure and volume to ensure trouble-free pneumatic control system operation.

Compressed Air Supply Particulate Removal

An air compressor station must provide clean air to the HVAC controls in a building. Clean air is air that is free of airborne particulate matter of a specific size. The particulate matter is usually measured in microns. A *micron* is a unit of measure equal to .000039".

Dirt in the control air supply accumulates in the air lines and eventually fouls the restrictors and ports at the controls. Fouled controls operate improperly or fail. Excessive dirt accumulates at the air compressor and may score the piston rings and cylinder walls. Scored piston rings and cylinder walls reduce the efficiency and performance of the air compressor. Due to the loss in efficiency, the air compressor motor must run longer to compress the same amount of air. As the compressor run time increases, heat builds, affecting the motor and the compressor. A hot compressor breaks down the compressor oil and pumps it past the piston rings, saturating the controllers with oil. An outside air intake and an intake air filter on the compressor are used to reduce particulate matter.

> *Dirt and particulate matter rarely travel to the HVAC controls. The primary effect of dirt and particulate matter is air compressor wear. Compressor run time is used to determine compressor service interval.*

Outside Air Intake. An air compressor is normally located in the equipment or mechanical room of a commercial building. The equipment or mechanical room normally contains the boilers, chillers, pumps, and air handling systems of the building. Equipment and mechanical rooms are normally the most dirty rooms in a commercial building. The effects of the airborne dirt may be reduced by obtaining the compressor air supply from the outside air instead of the mechanical room air. A supply pipe can be extended from the air compressor inlet to the outside air (OA). See Figure 6-4. For example, clean outside air can be obtained by tapping into the outside air inlet of an air handler downstream from the filter bank. The outside air contains less dirt than the air inside the mechanical room. The outside air intake should not be used if it is close to a source of dirt, for example, a location directly above a pile of fine industrial grit.

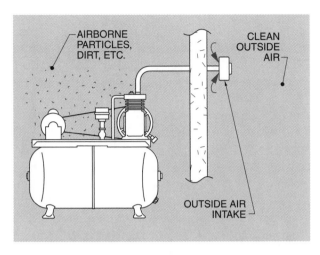

Figure 6-4. An outside air intake can reduce the particulate matter introduced to the air compressor.

Intake Air Filter. An intake air filter is commonly located in a metal housing at the intake of the air compressor. Construction of the filter varies widely depending on the size and capacity of the air compressor. Small units with a .5 HP or 1 HP motor may have a thin felt-type element that can be periodically removed, washed, dried, and replaced. Large air compressors may have one or more automotive-type air filters. See Figure 6-5.

Manufacturer recommendations must be followed for filter type and replacement schedule. Oil-saturated filtration elements should not be used for HVAC compressor intake air filters. Oil-saturated filtration elements are commonly used in industrial compressed air systems. Oil-saturated filtration elements are very efficient at removing dirt; however, they introduce high levels of oil into the air supply.

The appropriate filtration element must be used with an outside air intake. An outside air intake may have the filtration element located at the outside air inlet piping. Care must be taken to prevent an excessive pressure drop due to the length of piping. Normally, the pressure drop is reduced by increasing the pipe diameter as the pipe length increases. A safe guide is to increase the diameter of the inlet pipe 1″ for every 10′ in length.

Intake air filters are inspected and cleaned or replaced based on maintenance requirements.

Compressed Air Supply Moisture Removal

An air compressor station must provide moisture-free air to the HVAC controls in a building. One reason that air is compressed is to help squeeze the moisture out of it. Moisture may destroy the controls or cause them to operate improperly. In extreme cases, water leaks out of the controller bleed ports. Moisture also affects the operation of the air compressor. The moisture builds up inside the receiver, reducing the volume available for compressed air. This leads to an excessive number of compressor starts per hour, which increases the mechanical and electrical wear on the compressor. Moisture also causes the steel receiver to rust from the inside, which may cause a receiver to fail. An inspection port is provided in a receiver to check the condition of its interior. See Figure 6-6.

Figure 6-5. Intake air filters are used to trap particulate matter before it enters the air compressor.

RECEIVER INSPECTION PORTS

Figure 6-6. An inspection port on an air compressor receiver is used to check for corrosion inside the receiver.

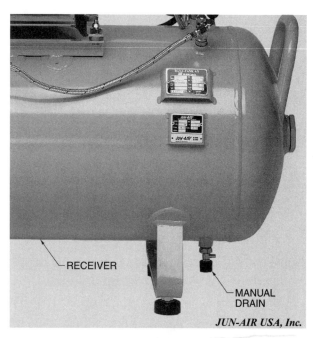

Figure 6-7. Most manufacturers recommend that the manual drain valve be opened to drain the moisture in a receiver a minimum of once every 30 days.

Air drawn into the air compressor inlet contains moisture. Heat is added to the air when the air is compressed. This high-pressure, high-temperature air is introduced to the receiver where it gives up its heat to the ambient air through the walls of the receiver. The cool air cannot hold the same amount of moisture that hot air can hold. For this reason, the moisture condenses to the bottom of the receiver. Multistage compressors have finned intercoolers between stages to help reduce moisture levels. Drains, driers, and filters are used to prevent and remove the moisture in a receiver.

Manual Drains. A *manual drain* is a device that is opened and closed manually to drain moisture from the receiver. A manual drain is attached to the lowest point of the receiver. All receivers have a manual drain installed as standard equipment. A manual drain is the minimum drain necessary for an air receiver application. See Figure 6-7. The disadvantage of a manual drain is that a person must be present to open the drain. A system that requires a person to open a drain also requires scheduled maintenance. This approach may not be practical in facilities in which staffing has been reduced. In these applications, automatic drains are required.

Automatic Drains. An *automatic drain* is a device that opens and closes automatically at a predetermined interval to drain moisture from the receiver. See Figure 6-8. An automatic drain contains an automatic drain valve, a float, and a manual valve. An *automatic drain valve* is a device that is normally piped from the lowest part of the receiver and opens based on differential pressure or moisture level buildup. The high-pressure receiver air forces the moisture out through the automatic drain valve to a floor drain.

The disadvantage of an automatic drain valve is that they may become fouled and plugged by rust, oil residue, and other contaminants in the moisture. Automatic drain valves may fail and allow the water level to rise in the receiver without being detected. During the air compressor preventive maintenance procedure, the operator can depress a manual valve button to discharge moisture into the floor drain. Depressing the manual valve button, dislodges any contaminants that may cause the automatic drain to fail. This procedure also checks the operation of the automatic drain valve.

An outside air intake may also be used to reduce the amount of dirt and moisture that enters a system. Mechanical room air normally contains high levels of moisture (humidity). Boilers, chillers, domestic water systems, and their pumps all contribute to the high level of moisture. For many buildings, the use of an outside air intake can dramatically reduce the amount of moisture in the air. A benefit derived from the use of outside air in northern climates is that the outside air is colder than the equipment room air during a large part of the year. Cold air is denser than warm air, enabling the air compressor to draw a greater amount of air per stroke of the piston, thus increasing the efficiency of the compressor.

AUTOMATIC DRAINS

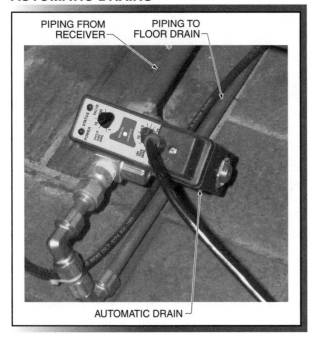

PIPING FROM RECEIVER

PIPING TO FLOOR DRAIN

AUTOMATIC DRAIN

Figure 6-8. An automatic drain may include a timer-actuated valve to automatically remove moisture from a receiver at preset intervals.

Refrigerated Air Driers. A *refrigerated air drier* is a device that uses refrigeration to lower the temperature of compressed air. Refrigerated air driers are located in the air discharge line from the receiver. Refrigerated air driers include small refrigeration systems that consist of compressors, condensers, evaporators, and electric controls. See Figure 6-9. A shell is provided in which the compressed air contacts cold refrigerant inside evaporator tubes. This process cools the compressed air and allows the moisture in the air to condense. The moisture is then removed by an automatic drain piped to a floor drain. Refrigerated air driers are very efficient. New models of refrigerated air driers have built-in activated charcoal filtration elements. Refrigerated air driers must be maintained properly. If the condenser or other parts of the drier become coated with dirt, the compressor can burn out.

REFRIGERATED AIR DRIERS

Figure 6-9. Refrigerated air driers lower the temperature of compressed air to remove its moisture.

Desiccant Driers. A *desiccant drier* is a device that removes moisture by adsorption. *Adsorption* is the adhesion of a gas or liquid to the surface of a porous material. Desiccant driers consist of a housing that is filled with silica gel or alumina that adsorbs moisture from the compressed air. See Figure 6-10.

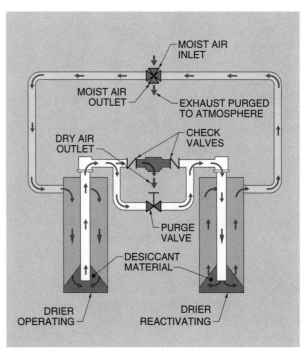

MOIST AIR INLET

MOIST AIR OUTLET

EXHAUST PURGED TO ATMOSPHERE

DRY AIR OUTLET

CHECK VALVES

PURGE VALVE

DESICCANT MATERIAL

DRIER OPERATING

DRIER REACTIVATING

Figure 6-10. A desiccant drier removes moisture from the compressed air by adsorption.

Desiccant driers contain fittings that allow them to be installed at any location in a pneumatic system. Some small packaged air compressor stations use desiccant driers as their primary method of moisture removal. Desiccant driers can be used with valves in multiple arrangements. These arrangements permit removal and changing of one desiccant material while enabling the other to remain in service with the air compressor station.

Air Line Filters. An *air line filter* is a device that consists of a housing containing a centrifugal deflector plate and a small filtration element. Air line filters are provided as a final filter against contaminants and moisture in a system. Particles are forced out of the air because of the swirling action inside the housing. Particles are also trapped by the filtration element. See Figure 6-11.

Air line filters are often used with modern pneumatic systems. The size of new controllers is continuously being reduced. These controllers are susceptible to damage from water and oil in the air. Air line filters are often used to protect these controllers because they are inexpensive and easily installed in the system.

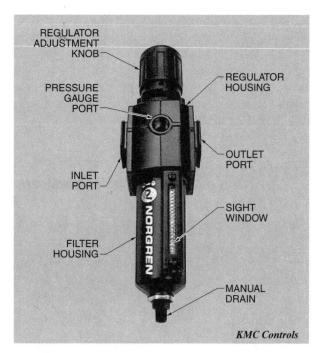

Figure 6-11. Air line filters and pressure regulators can be combined into a single unit.

Compressed Air Supply Oil Removal

An air compressor station must provide oil free air to the HVAC controls in a building. If the air is saturated with oil, pneumatic controllers may operate improperly or be destroyed because they contain small passageways, restrictors, and valves that can be plugged by oil.

Oil in a pneumatic control system originates at the air compressor, which is normally oil-lubricated. In some applications, usually in industrial environments, the intake air may be contaminated with oil. In these applications, an outside air intake is recommended.

The presence of excessive oil in a system indicates poor mechanical condition. Oil is minimized or eliminated by keeping the air compressor in top mechanical condition and by using oil removal filters (separators). Oil-less and oil free air compressors are also available. However, these units are normally high-speed, low-capacity units that have limited application.

Oil Removal Filters. An *oil removal filter (separator)* is a device that removes oil droplets from a pneumatic system by forcing compressed air to change direction quickly. Oil removal filters consist of a transparent housing bowl containing small filtration elements attached to the filter bodies. Oil removal filters are located in the air supply after the compressed air station. The filtration elements are made of glass fibers and rayon cloth. See Figure 6-12.

JUN-AIR USA, Inc.
Driers are often designed for continuous operation to ensure a constant supply of dry air.

OIL REMOVAL FILTERS

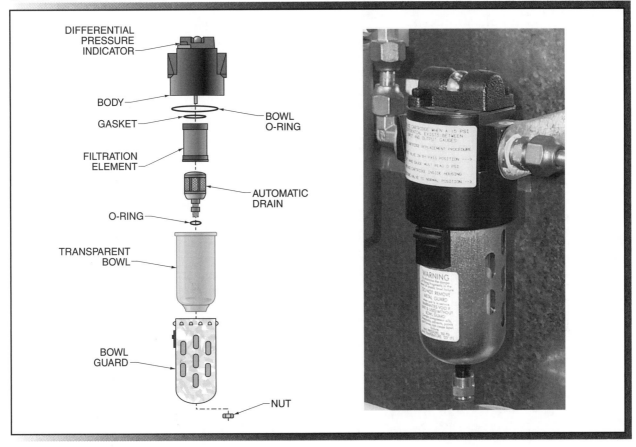

Figure 6-12. Regardless of the filter being used, its condition should be checked regularly, and the filtration element should be replaced when required.

New models of air driers are available with integral activated charcoal elements that remove oil and moisture particles. Regardless of the filter being used, its condition should be checked regularly, and the filtration element should be replaced when required. Some air compressor stations use multiple filters in series to provide a greater degree of oil removal.

The condition of a filtration element is determined by measuring the pressure drop across the element. The pressure drop across a filtration element is measured by installing pressure gauges before and after the filters. Some filter housings have a built-in green/red (good/bad) indicator that shows when the element requires replacement. Some filtration elements are factory designed to turn from white to brown as they become saturated with oil.

An HVAC technician must become familiar with the different filtration elements used in an air compressor station and follow the manufacturer recommendations regarding replacement. The transparent bowl is often constructed of a thick, durable plastic. The plastic material may crack and split under air pressure; therefore, care must be exercised when tightening fittings. These housings must be surrounded by a perforated metal shield (bowl guard). The shield protects technicians from harm due to the possible shattering of the housing. The shield also allows visual inspection of the filtration element.

Oil Measuring Stations. Oil contamination of the air supply must be prevented. Any oil in the air supply must be removed and measured. An oil measuring indicator can be installed in the air supply. The oil measuring indicator absorbs oil. The amount of oil is compared with an oil entrainment graph that provides a parts per million (PPM) reading of the oil in the air supply. See Figure 6-13.

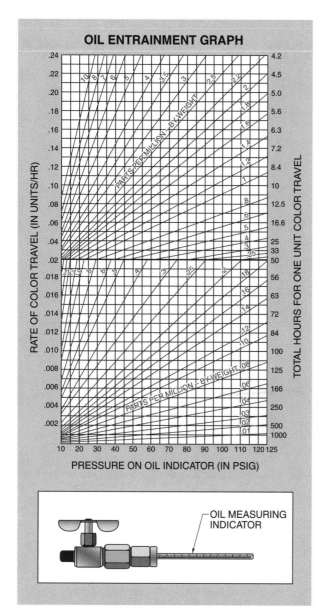

Figure 6-13. An oil measuring indicator shows the amount of oil in a system and is compared to an oil entrainment graph to determine the actual amount of oil present.

Compressed Air Supply Pressure and Volume

The correct pressure and volume of compressed air is required for proper operation of the control system. Controls can be damaged or destroyed if air at the incorrect pressure is present at the controllers. The controllers starve (lack air) and the air compressor operates excessively, causing erratic system performance and excessive air compressor wear. Devices used in the air compressor system to correct the pressure and volume of air include pressure switches, pressure regulators, and safety-relief valves.

Pressure Switches. A *pressure switch* is an electric switch operated by pressure that acts on a diaphragm or bellows element. Pressure switches start and stop an air compressor motor based on the pressure in the receiver. The pressure switch enables the compressed air station to deliver the correct system air pressure. The pressure switch is attached to the air compressor receiver. See Figure 6-14. Normal receiver pressure is approximately 80 psig. The cut-in and cut-out pressures are approximately 70 psig and 90 psig, respectively. Some compressors may operate at pressures above or below this range. The correctly sized receiver, operating at the appropriate pressure, enables the air compressor to efficiently provide compressed air for the pneumatic HVAC control system.

PRESSURE SWITCHES

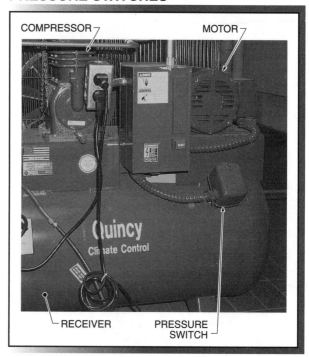

Figure 6-14. A pressure switch is mounted on the receiver and turns the compressor motor ON and OFF based on the pressure in the receiver.

Pressure Regulators. A *pressure regulator* is a valve that restricts and/or blocks downstream air flow. Pressure regulators allow the final adjustments of the air pressure to the controllers. See Figure 6-15. Pressure regulators deliver air to the controllers at a pressure between 15 psig and 25 psig. Air delivered to the controllers is referred to as main air or supply air.

PRESSURE REGULATORS

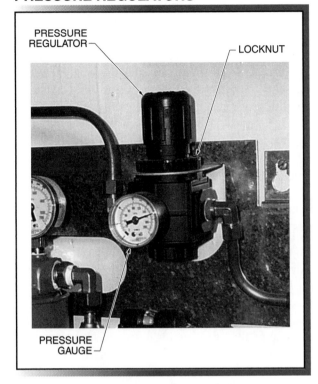

Figure 6-15. A pressure regulator allows a final adjustment of the air pressure to the controllers.

Single-pressure systems deliver one level of air pressure to the system at all times. Dual-pressure systems deliver more than one level of air pressure, depending on the air pressure selected at any given time. Dual-pressure systems are used to change the setpoints of room thermostats. When two pressure regulators are used, they are located in parallel and are connected to an air valve. The air valve is used to connect the proper pressure regulator to the system supply based on the required air pressure. In either single- or dual-pressure systems, the pressure regulators have an adjustable pressure stem and a locknut which prevents tampering and keeps the pressure regulator stem from moving out of adjustment. Pressure regulators are sized to handle a certain volume of air flow measured in standard cubic feet per minute (SCFM) of air.

Safety Relief Valves. A *safety relief valve* is a device that prevents excessive pressure from building up by venting air to the atmosphere. An air compressor must contain a safety relief valve to protect individuals and the system components against overpressurization. See Figure 6-16. Safety relief valves are normally not adjustable and have a ring that is pulled to manually release the air

pressure. A pneumatic HVAC control system should have a minimum of two safety relief valves. One valve is attached to the receiver and has a value of approximately 125 psig. The second valve is located downstream of the pressure regulator and has a value of 15 psig to 25 psig. The safety relief valve on the receiver protects the receiver against over-pressurization. The safety relief valve downstream from the pressure regulator protects the system controls against overpressurization.

SAFETY RELIEF VALVES

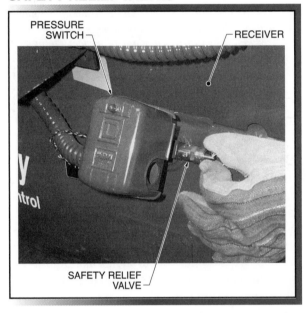

Figure 6-16. A safety relief valve protects the air compressor receiver and parts of the air distribution system against overpressurization.

Determining Required Compressed Air Volume. The correct air volume is normally calculated by the controls engineer when the controllers and air compressors are selected. Air compressor volume is measured in standard cubic feet per minute of compressed air. The volume of a pneumatic system is based on the receiver capacity, compressor, and drive motor.

To determine the correct capacity, the controls engineer adds the volume of air used by the components in the system, then multiplies that number by a factor of three. Multiplying by a factor of three provides added capacity to the air compressor station. The added capacity enables the compressor to shut off for periods of time to allow it to cool, thus extending compressor life. Adding a large number of controllers to the air compressor system affects the operation of the air compressor, possibly causing a loss of air pressure.

AIR LINES

The compressed air delivered to an HVAC control system must be clean, dry, and at the correct pressure and volume. If compressed air that contains moisture is exposed to low outside temperatures, air lines may freeze and split. The supply (main) air is delivered to the controllers in the building through air lines. The two most common air line materials are copper and poly (plastic) tubing.

> *Many old HVAC pneumatic control system installations use copper tubing. These installations may contain dozens of air lines running in the same direction. When turns are required, copper tubing is curved uniformly. The HVAC controls technician must be familiar with the installation procedures and fittings for poly and copper tubing because both are found in pneumatic system installations.*

Copper Tubing

Copper air lines were commonly used in the past in many areas of a building. Copper air lines are rugged and difficult to damage. Copper air lines also require more time for installation than poly tubing and thus, are more expensive. Today, it is common to see copper lines used in mechanical rooms and other areas where they are exposed to heat, humidity, impact, or other harsh conditions.

Poly Tubing

Poly (plastic) tubing is commonly used today to deliver air to the controllers and controlled devices of an HVAC control system. Poly tubing is inexpensive and simple to use in many applications. Poly tubing also provides greater flexibility than copper when relocating pneumatic devices and reconfiguring existing control panels. Poly tubing should not be used in locations where it may be damaged or exposed to adverse conditions. Poly tubing is often used in electrical conduits to protect the tubing from harsh conditions. Control panels may use surgical tubing instead of poly tubing. Surgical tubing has thicker walls than poly tubing and is almost impossible to kink. This makes it ideal for use in tight areas or when making small radius bends. Poly tubing, however, is prohibited by some fire codes.

Common Sizes

All air lines must be sized properly to provide adequate air flow for the controller to which they are connected. Air capacity for pneumatic control devices is measured in standard cubic inches of air per minute (SCIM). Copper and poly tubing are available in a variety of sizes. Available sizes include ½″, ⅜″, ¼″, and ⅚₄″. Main air runs normally use ½″ tubing to supply adequate air for the system. The ¼″ size is normally used for thermostat and other controller connections. The air lines should be routed away from known hazards such as steam lines or chemical feed piping. The air lines should be protected by installing them inside walls or floors when possible. Air lines can be run, and the concrete for floors can be poured on the air line runs, if necessary. The air lines can be run inside electrical conduit for added protection. In areas such as mechanical rooms, where lines are exposed, copper tubing is preferred due to its durability.

Air Line Installation

In some installations, pressure drops prevent controllers at the end of the lines from operating properly. This can be prevented by using an air loop configuration. In an air loop configuration, the compressed air enters both ends of the loop, reducing potential pressure drops over long runs. See Figure 6-17.

Care should be taken to avoid installations that can expose air line piping to damage by individuals changing light bulbs, or working on steam or hot water systems. High-pressure air (80 psig) may be piped to each floor of a building if excessive pressure drops are a problem. Pressure regulators are used to reduce the pressure to about 20 psig at the controllers. Always use the correct fittings for the type of tubing used. Joints should be tested under pressure to ensure proper operation after startup.

Copper air lines are more rugged than other air line materials and are used in areas exposed to high temperature and vibration.

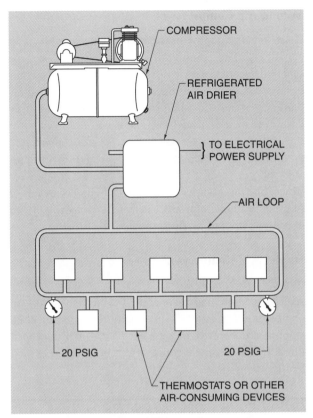

Figure 6-17. An air loop is used to minimize pressure drops in a compressed air distribution system.

AIR COMPRESSOR PERFORMANCE TESTS

Performance tests are used to check air compressor performance. Performance tests verify the proper selection and operation of the air compressor in the system. Air compressor performance tests include run time tests, pressure tests, and starts per hour tests.

Run Time Test

A *run time test* is an air compressor performance test that measures the percentage of time that a compressor runs to enable it to maintain a supply of compressed air to the control system. A run time test determines whether an air compressor is sized properly for the job. The run time test is performed under a normal compressed air load on the system. The length of time (in minutes and seconds) that the air compressor is on is measured with the length of time that the air compressor is off. The run time percentage is found by dividing the compressor on time by the compressor on time plus the compressor off time. This number is multiplied by 100 to convert to a percentage. Compressor run time percentage is found by applying the formula:

$$c_{RT} = \frac{c_{ON}}{c_{ON} + c_{OFF}} \times 100$$

where

c_{RT} = compressor run time (in %)

c_{ON} = compressor on time (in min)

c_{OFF} = compressor off time (in min)

100 = constant

For example, what is the compressor run time if an air compressor is on for 6 min and off for 12 min?

$$c_{RT} = \frac{c_{ON}}{c_{ON} + c_{OFF}} \times 100$$

$$c_{RT} = \frac{6}{6+12} \times 100$$

$$c_{RT} = \frac{6}{18} \times 100$$

$$c_{RT} = .33 \times 100$$

$$c_{RT} = \textbf{33\%}$$

Run time is only useful as a percentage of the total cycle time. For example, if an air compressor is on for 6 min and off for 18 min, the compressor run time is 25% {[6 ÷ (6 + 18)] × 100 = 25%}. If an air compressor is on for 6 min and off for 6 min, the compressor run time is 50% {[6 ÷ (6 + 6)] × 100 = 50%}. In both cases, the air compressor operates for 6 min, but the mechanical wear is much greater on the compressor having the 50% run time.

Most manufacturers recommend that a properly sized and installed air compressor have a high limit of 50% run time. Anything greater causes abnormal wear and premature air compressor failure. In addition to abnormal wear and premature air compressor failure, the oil carryover rate is greatly increased when an air compressor has a run time over 50%. See Figure 6-18.

The run time test is useful when analyzing air compressor performance on a month-by-month basis. The best use of a run time test is to perform it when the air compressor is installed and first brought on-line. In this way, a benchmark of performance is established that serves as a comparison for future air compressor operations. Run time test values are recorded as part of preventive maintenance and testing procedures, these records can indicate potential problems and future failures of the air compressor system. This enables the scheduling of a rebuild or replacement during a shutdown or weekend period rather than leading to an emergency situation where the comfort of the building occupants is adversely affected.

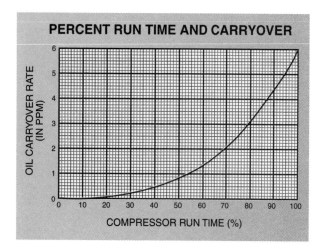

Figure 6-18. The amount of oil carryover increases greatly when a compressor has a run time over 50%.

Pressure Test

A *pressure test* is a test that determines the time required for an air compressor to reach the pressure switch shut-off pressure. A pressure test should be performed and recorded annually. A pressure test is performed with the compressor isolated from the system (under no load). This eliminates any effect from a fluctuating air load or air leaks in the building. In a pressure test, the air compressor receiver is drained, normally through the manual drain on the receiver. The manual drain is closed and the air compressor is started. The length of time required for the compressor to reach its maximum system pressure is recorded and compared to previous values to check compressor efficiency.

Starts Per Hour Test

A *starts per hour test* is a test that records the number of times an air compressor starts per hour under a standard load. The number of starts per hour should not exceed the manufacturer recommended limit for the compressor. Depending on the motor horsepower and capacity of the compressor, the number of starts per hour varies from approximately 4 starts per hour to a maximum of 10 starts per hour.

BACKUP COMPRESSED AIR SUPPLIES

A compressed air system must operate correctly for proper operation of the controllers and comfort of building occupants. The control of the entire building depends on the backup air supply. The backup air supply may consist of a second air compressor or duplex air compressor.

Second Air Compressor

A second air compressor is commonly used as a backup air supply. The compressors are of identical size so that either compressor can handle the load without excessive run time. In some applications, the second air compressor may be much smaller than the primary air compressor and sized for emergency backup only. In the dual-air compressor installations, the air compressors are isolated from each other by check valves which prevent the failure of a component at one air compressor from affecting both air compressors. The disadvantage of using two air compressors is that they may require an excessive amount of floor space.

Duplex Air Compressor

A *duplex air compressor* is an air compressor that consists of two air compressors and two electric motors on one common receiver. Duplex compressors reduce the amount of floor space needed, but a problem occurring at the common receiver could affect both compressors. See Figure 6-19.

DUPLEX AIR COMPRESSORS

Saylor-Beal Manufacturing Company

Figure 6-19. Duplex air compressors are used to reduce the wear on a single compressor while keeping floor space to a minimum.

Equalizing Compressor Run Time

Regardless of the backup air supply, (not all buildings have a backup air supply) the run time of the compressors should be equalized. Equalizing compressor run time is a common HVAC practice that applies to pumps, air compressors, and boilers. When equalizing compressor run time, the individual compressors should reach the same number of operational hours at approximately the same time. The operational life of two compressors can be extended by equalizing the run time of both compressors. If the run time is not equalized, the idle compressor may not start when the primary compressor fails. Methods used to equalize compressor run time include changing the compressor pressure switch settings, the use of a lead/lag switch, and the use of an alternator.

Changing Pressure Switch Settings. To equalize compressor run time, the setting on each receiver pressure switch can be manually adjusted so that the primary (lead) compressor becomes the backup (lag) compressor, and the backup (lag) compressor becomes the primary (lead) compressor. Manually adjusting receiver pressure switches should be done quarterly or semiannually. The primary compressor should have a normal cut-in pressure of 70 psig and a cut-out pressure of 90 psig. The backup compressor may have a normal cut-in pressure of 50 psig and a cut-out pressure of 70 psig. Since some pressure switches are not designed to regularly have their settings adjusted, regularly adjusting pressure switch settings can cause premature failure of the pressure switch.

Lead/Lag Switches. A *lead/lag switch* is a pressure switch that determines which compressor is the primary (lead) compressor and which compressor is the backup (lag) compressor. The lead compressor operates to maintain system pressure, and the lag compressor is used as the backup that turns on if the lead compressor fails. Lead/lag switches are commonly used with water pumps to designate which pump is the lead pump and which pump is the lag pump. The same logic can be used with air compressors. At an established point during the year, a technician can change the manual lead/lag switch and equalize the compressor run time. The disadvantages of a lead/lag switch are that they cost money to purchase and install, and they require a technician to change the switch and the compressor operation.

> ⓘ *Some alternator packages allow the servicing of one compressor while the other compressor continues to supply air to the building controls. This minimizes the effects of routine compressor maintenance on the building temperature controls.*

Alternators. A *compressor alternator* is a device that operates one compressor during one pumping cycle and the other compressor during the next pumping cycle. See Figure 6-20. Alternators also turn the backup air compressors on if the primary air compressors fail. Alternators are electrically complex and relatively expensive, but their operation is completely automatic and they almost perfectly equalize the run times of two compressors over a long time period. Alternators can be used with multiple air compressor installations and with duplex air compressors. Many duplex systems are shipped from the manufacturer with alternator packages.

ALTERNATOR

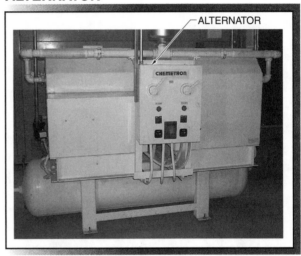

Figure 6-20. An alternator operates one compressor during one pumping cycle and the other compressor during the next pumping cycle.

JUN-AIR USA, Inc.

Adding compressors to a compressed air station greatly reduces downtime due to compressor malfunction or maintenance.

AIR COMPRESSOR PREVENTIVE MAINTENANCE

A preventive maintenance program is required to keep a compressed air station operating properly. The air compressor preventive maintenance program should be coordinated with preventive maintenance programs for other plant equipment. See Figure 6-21. Air compressor preventive maintenance is performed by applying the procedure:

1. Turn power OFF to the compressor. Follow lockout/tagout procedures.

2. Remove belt guard and check for cracking and glazing. Check belt tension using a tension tester.

3. Drain the tank manually. Check volume of oil in the water.

4. Manually operate the safety relief valves. This ensures that they are operational; it also prevents foreign matter from building up underneath the valves and preventing their proper operation when needed.

5. Check the oil level in the crankcase. This may be done using a dipstick or sight glass. Check the oil level after allowing the oil circulating in the compressor to return to the crankcase.

AIR COMPRESSOR PREVENTIVE MAINTENANCE

① Turn power OFF to compressor. Follow lockout/tagout procedures.

② Remove belt guard and check for cracking, glazing, and tension.

③ Drain tank. Check volume of oil in water.

④ Manually operate safety relief valves.

⑤ Check oil level in crankcase.

⑥ Properly replace belt and turn power ON to compressor.

Figure 6-21. A preventive maintenance program is required to keep a compressed air station operating properly.

If the oil is checked too soon after compressor operation, a false low may be indicated and unneeded oil may be added to the compressor. The excess oil may leave the air compressor and contaminate the system.

6. Properly replace belt and turn power ON to the air compressor.

Any belts that require replacement must be replaced as a matched set. In addition, check the pulleys for nicks or burrs that may damage the belts. Pulley alignment should be checked if continual belt problems occur. Follow manufacturer recommendations regarding pulley alignment. Check pulley alignment when installing a new air compressor and when performing any major disassembly or rebuild of an air compressor. Pulley wear can also be checked using a V belt sheave pitch gauge.

Always follow manufacturer recommendations regarding any air compressor maintenance, including oil change. Compressor oil should be changed at the intervals recommended by the manufacturer. Always use the oil type recommended by the manufacturer. The wrong oil may damage internal compressor parts. In addition, the proper amount of oil should be added. Underfilling may damage the compressor due to insufficient lubrication while excessive oil will migrate into the system and possibly damage the controller.

Changing Oil

Changing compressor oil is not part of the preventive maintenance program but is performed at the manufacturer recommended number of run hours. Air compressors use 60 W non-detergent oil. Detergent oil must not be used because the detergents contaminate the compressed air and damage O-rings and other controller parts. Old oil should be visually checked for contaminants.

Air Filter Replacement

The compressor intake air filter should be replaced at the interval recommended by the manufacturer. The intake air filter can be checked as part of the preventive maintenance program or changed at a specific interval. In a facility with a large number of air compressors, cost and time are saved by performing all oil changes and intake air filter changes at the same time. The proper oil, belts, and intake air filters should be pre-stocked in adequate quantities. In addition, the preventive maintenance performed and the results should be recorded. Many facilities use computer software that automatically calculates the proper maintenance interval for given equipment.

A good preventive maintenance program can use anything from wire tags to sophisticated personal computer (PC) software. The compressor may be scheduled for a teardown and rebuild of the major compressor parts if the testing and preventive maintenance of the air compressor indicates that it is mechanically unsound. The replacement parts may include piston rings, crankshafts, O-rings, valves, and gaskets. The cylinders and all parts should be checked for tolerances and machined if required. Rebuild kits, exploded views, and recommended tolerances are available from the manufacturer. See Figure 6-22.

AIR COMPRESSOR REPLACEMENT PARTS

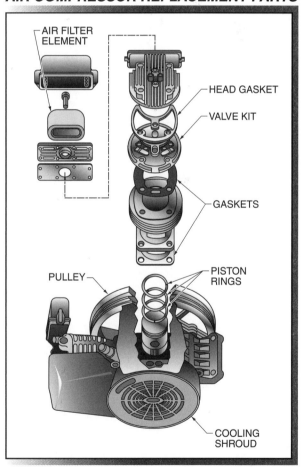

Figure 6-22. Air compressor replacement parts are available when compressor wear becomes excessive.

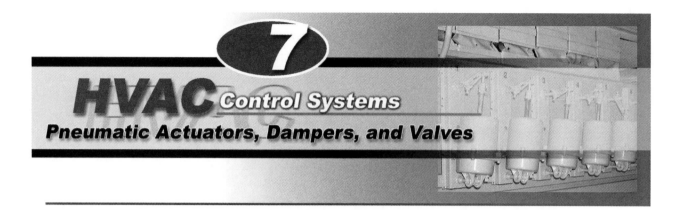

HVAC Control Systems
Pneumatic Actuators, Dampers, and Valves

HVAC control system technicians require knowledge of the operation of controllers and controlled devices. Controlled devices include actuators, dampers, and valves. Controlled devices regulate the flow of air, water, or steam to heat or cool a building. Knowledge of the controllers and controlled devices enables a technician to properly install and troubleshoot a commercial HVAC control system.

ACTUATORS

An *actuator* is a device that accepts a signal from a controller and causes a proportional mechanical motion to occur. In pneumatic control systems, the actuator accepts an air pressure signal from the controller and causes an actuator shaft or a valve stem to move. The movement of the actuator shaft or valve stem moves a device which regulates the flow of air across a damper or the flow of water or steam through a valve. Knowledge of the construction and operation of actuators is required because the actuators directly affect the temperature control of a building space. See Figure 7-1.

Actuators may be damper or valve actuators. The basic components of damper and valve actuators are the same, but their physical configurations are different. Damper actuator components are normally enclosed in the actuator body and have a longer stroke than valve actuators, often 6″ to 8″. A valve actuator shaft has a shorter stroke than a damper actuator, as little as 1″. Most valve actuator components are open to view (not enclosed by the actuator body). See Figure 7-2. Damper actuators may be connected to butterfly valves. Small terminal unit damper and valve actuators may consist of one piece of molded plastic. These actuators cannot be disassembled; they must be replaced as one unit.

ACTUATORS

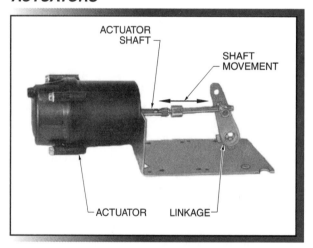

Figure 7-1. An actuator and linkage control damper position to regulate the temperature of a building space.

Some valve actuators are totally enclosed, with no access to the spring from the outside. These actuators are installed in libraries and schools to protect the actuator and prevent tampering and unauthorized operation. Enclosed valve actuators can be removed or repositioned by loosening a set screw.

DAMPER AND VALVE ACTUATORS

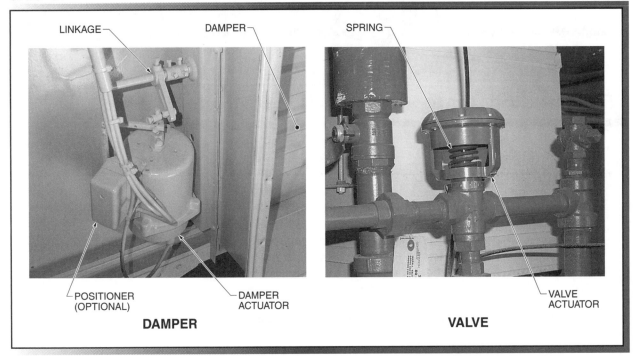

LINKAGE ── DAMPER ──

SPRING ──

POSITIONER (OPTIONAL) ── DAMPER ACTUATOR ──

DAMPER

VALVE ACTUATOR ──

VALVE

Figure 7-2. Actuators may be damper or valve actuators.

Actuator Components

The components of a pneumatic actuator include the end cap with air fitting, diaphragm, piston cup, and spring and shaft assembly. The end cap with air fitting provides a connection for the air line from the controller to the actuator. The end cap can be removed from the actuator body to provide access to the interior of the actuator. See Figure 7-3. The end cap bolts must be properly attached to ensure a tight, leak-free fit. Also, the air fitting should be tightened and sealed properly to prevent air leaks. Any time the actuator is disassembled or repaired, the end cap bolts and air fitting should be checked for leaks. The end cap with air fitting is normally made of aluminum or high-impact plastic. The actuator body provides a housing and support for the other parts of the actuator. The actuator body protects the other components from damage due to impact.

> *Many building automation systems continue to use pneumatic actuators because they are easy to troubleshoot and repair in the field, are relatively inexpensive, and create a great amount of force using compressed air.*

ALUMINUM END CAP

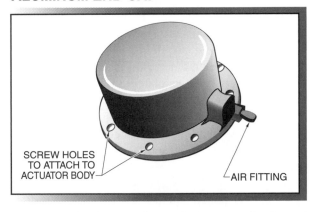

SCREW HOLES TO ATTACH TO ACTUATOR BODY ──

AIR FITTING

Figure 7-3. Caution must be used when removing the end cap from the actuator body because spring pressure on the end cap may cause it to separate suddenly.

A *diaphragm* is a flexible device that transmits the force of the incoming air pressure to the piston cup and then to the spring and shaft assembly. The diaphragm moves as the air pressure causes the actuator to move. The diaphragm has a lip on it that normally acts as a seal between the end cap and actuator body. See Figure 7-4. Care must be exercised when reassembling or servicing the diaphragm to prevent air leaks.

Over time, the diaphragm can develop cracks or small holes. This is a common problem that occurs in a pneumatic control system. The incoming air leaks through the crack, preventing the air pressure from building up against the piston cup, and the actuator cannot extend properly. The diaphragm can be replaced by removing the end cap. The diaphragm size also determines the force that the actuator can develop. A small diameter actuator can produce a small force. A large diameter actuator can produce a large force. Diaphragm force is found by applying the formula:

$$F = P \times A$$

where

F = force (in lb)

P = available air pressure from controller (in psig)

A = area of diaphragm (in sq in.)

For example, what is the force developed by a 4″ diameter diaphragm supplied with an air pressure of 20 psig?

Note: The area of the diaphragm is equal to $\pi \times r^2$. Area equals 12.57 sq in. ($3.1416 \times 2^2 = 12.57$).

$$F = P \times A$$
$$F = 20 \times 12.57$$
$$F = \textbf{251.4 lb}$$

NEOPRENE DIAPHRAGM

Figure 7-4. The diaphragm of an actuator transmits force and provides an air seal.

A *piston cup* is a device that transfers the force generated by the air pressure against the diaphragm to the spring and shaft assembly. See Figure 7-5. Normally, the piston cup requires no maintenance or replacement. The piston cup fits snugly inside the diaphragm.

ALUMINUM PISTON CUP

Figure 7-5. The piston cup transmits the force from the diaphragm to the spring and shaft assembly.

The spring and shaft assembly converts the air pressure change at the diaphragm into mechanical movement. In a damper actuator, the shaft is connected through a linkage to a damper. In a valve actuator, the shaft is the stem of the valve. The shaft can be damaged by overextension, improper connection to the damper or valve, or by careless maintenance procedures. See Figure 7-6. The spring is compressed when the air pressure acts against the diaphragm, piston cup, and shaft. The spring must overcome this force. The spring determines the amount of mechanical movement and the change in the controlled medium (air, water, or steam) through the valve or damper.

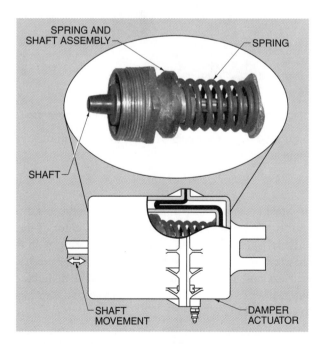

Figure 7-6. The spring and shaft assembly creates mechanical movement by responding to changes in air pressure.

Actuator Spring Range

An actuator controls the flow of the controlled medium. The force required to move the actuator is determined by the strength (tension) of the spring. A strong spring is hard to overcome with air pressure and a weak spring is easy to overcome. *Spring range* is the difference in pressure at which an actuator shaft moves and stops. The flow of controlled medium starts, stops, or is throttled through the valve or damper based on the spring range.

Spring Range Testing. An HVAC controls technician must field test the actuator spring range. A technician performs the spring range test, preferably without disconnecting or removing the actuator. A controller may be properly calibrated if the actuator spring range is known. A squeeze bulb or air supply with regulator may be used when testing actuator spring range. See Figure 7-7. The squeeze bulb is attached to the actuator and compressed, adding air to the actuator until its stroke is complete. The spring range test measures the spring range under actual field conditions including fluid pressures, mechanical wear, and age.

ACTUATOR SPRING RANGE TESTING

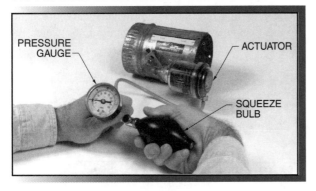

Figure 7-7. A squeeze bulb is used to determine the spring range of an actuator.

Most actuators have a nameplate that indicates the model number and spring range of the actuator. If the original spring has been replaced by a spring with a different spring range, or the nameplate is obscured, a squeeze bulb may be used. In addition, some manufacturers have color-coded springs that indicate the spring range. See Figure 7-8.

A pneumatic control diagram may be checked when an actuator is not accessible or cannot be checked otherwise. However, the pneumatic control diagram may be unreliable if the unit is old or has been modified. In addition, pneumatic control diagrams may not be legible or available for many units.

ACTUATOR COLOR CODES	
Color	**Spring Range***
Green	2 to 6
Black	4 to 8
Orange	8 to 13

*in psig

Figure 7-8. Manufacturers provide charts indicating actuator spring range color codes.

The actual temperature change may be used when no other means are available. In this method, the delivered air temperature is compared with the actuator air pressure. For example, an application has a delivered air temperature of 55°F at the hot water valve closed position corresponding to an actuator air pressure of 13 psig. The pressure is then reduced to 3 psig and the delivered air temperature rises to a maximum of 105°F. In this application, the actuator spring range is 3 psig to 13 psig. This method should be used primarily by technicians who are experienced with temperature/flow relationships.

Common Spring Ranges. A variety of spring ranges are used by pneumatic control equipment manufacturers. Common spring ranges include:

- 2 psig to 6 psig
- 3 psig to 9 psig
- 3 psig to 13 psig
- 3 psig to 15 psig
- 7 psig to 8 psig
- 9 psig to 13 psig

Spring Range Shift. Spring range is listed by the manufacturer of an actuator. The manufacturer spring range is referred to as the nominal (listed) spring range. After an actuator is installed in a system, it is exposed to influences, such as rust, corrosion, and mechanical friction, which alter the nominal spring range. For example, in most applications dampers must close against air flow. Therefore, the actuator must provide additional force using increased air pressure. After these influences have taken place, nominal spring range is referred to as the actual (operational) spring range. *Spring range shift* is the process by which the nominal spring range changes to the actual spring range. See Figure 7-9.

SPRING RANGE SHIFT

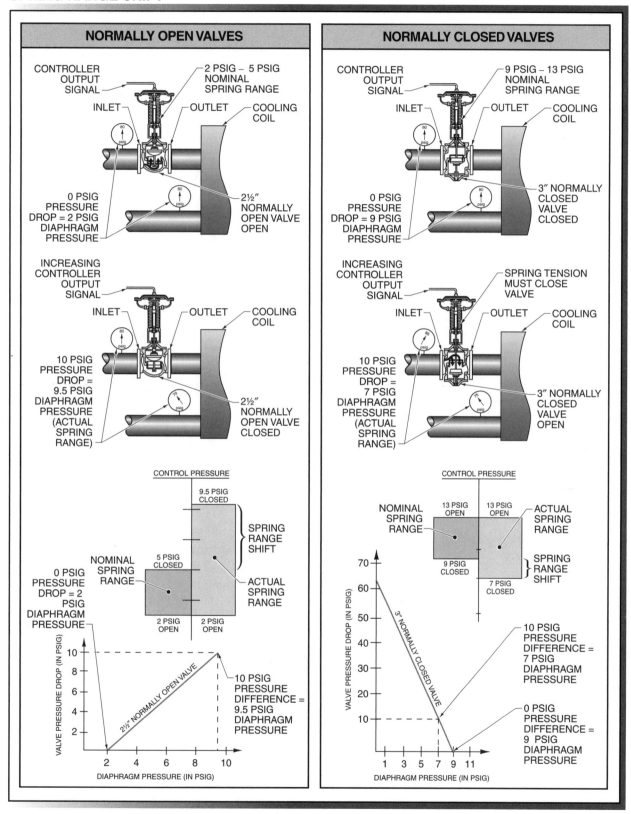

Figure 7-9. Spring range shift can be caused by mechanical wear and by closing against flow.

In the majority of HVAC control applications, spring range shift is so minor that it can be disregarded. However, in applications with high flow rates, poor maintenance, or detrimental environmental factors, spring range shift can cause the controller calibration to be inadequate for proper system performance. If significant spring range shift is suspected, the technician should use a squeeze bulb to check and compare the nominal spring range with the actual spring range. The controller calibration can be adjusted to allow for spring range shift, maintenance can be performed to remove rust and corrosion, or the actuator can be repaired. Spring range shift does not cause the majority of problems in pneumatic HVAC control systems.

Spring Range Overlap. Spring range shift can cause spring range overlap. *Spring range overlap* is a condition in which actuators with different spring ranges interfere with each other. Spring range overlap occurs in systems that have heating and cooling systems that have closely coordinated spring ranges in order to obtain close sequencing. For example, a heating valve may have a nominal spring range of 3 psig to 8 psig, and a cooling valve may have a nominal spring range of 8 psig to 13 psig. Both valves should not open at the same time; however, if the heating valve has an actual spring range of 3 psig to 10 psig, both valves could open simultaneously, wasting energy without providing additional comfort.

DAMPERS

A *damper* is an adjustable metal blade or set of blades used to control the flow of air. Dampers are available in different designs and sizes. Dampers are classified as parallel, opposed, and round blade dampers. See Figure 7-10.

Parallel Blade Dampers

A *parallel blade damper* is a damper in which adjacent blades are parallel and move in the same direction with one another. Parallel blade dampers are the most common damper used in HVAC systems. Parallel blade dampers are less expensive and provide better air mixture when installed to cause outside air and return air to collide. Parallel blade dampers are often used in systems that require a damper to be either fully open or fully closed and require less maintenance than opposed blade dampers. A disadvantage of parallel blade dampers is that they do not provide the same precise air control that opposed blade dampers provide. Many basic air handling units and packaged exhaust fans contain parallel blade dampers.

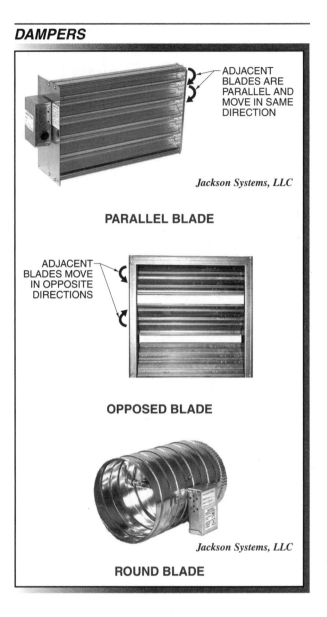

DAMPERS

ADJACENT BLADES ARE PARALLEL AND MOVE IN SAME DIRECTION

Jackson Systems, LLC

PARALLEL BLADE

ADJACENT BLADES MOVE IN OPPOSITE DIRECTIONS

OPPOSED BLADE

Jackson Systems, LLC

ROUND BLADE

Figure 7-10. Dampers are used in HVAC systems to control the flow of air and include parallel, opposed, and round blade dampers.

Opposed Blade Dampers

An *opposed blade damper* is a damper in which adjacent blades move in opposite directions from one another. Opposed blade dampers are more expensive than parallel blade dampers, but opposed blade dampers provide better flow characteristics than parallel blade dampers and precise air control. Opposed blade damper linkage is more complicated than parallel blade damper linkage because duplicate drive arms are required to drive adjacent blades in opposite directions, requiring increased maintenance.

Opposed blade dampers are used in applications that require two air streams to mix in order to prevent cold spots and possible freeze up.

Round Blade Dampers

A *round blade damper* is a damper that has a round blade. Round blade dampers are used in systems consisting of round ductwork. Round blade dampers are used primarily with small, variable air volume (VAV) terminal boxes. Round blade dampers use small diameter diaphragm actuators and are simple in operation. Round blade dampers have a nonlinear flow characteristic but are widely used in small terminal units such as VAV terminal boxes.

A variation from the round blade damper is the elliptical blade damper. Elliptical blade dampers have a slightly better flow characteristic than round blade dampers and also provide a tighter shutoff. Care must be taken with round or elliptical blade dampers to avoid strapping the duct too tightly at the VAV terminal box; doing so may bend the duct out of round and bind the damper. This binding may prevent the damper from working properly.

Damper Area

All dampers list the square feet of the damper blade area, pressure drop across the damper, and cubic feet per minute (cfm) of air flow at the given pressure drop. As the area of the damper blade is increased, a larger actuator diaphragm is required to produce the necessary force to open the damper. Therefore, a correlation exists between damper size and actuator size. In large applications, high air flow requirements can be met by joining small damper sections together into large sections and connecting the actuators to a common air line, thus driving all dampers at the same time.

Damper Construction and Linkage

Dampers are normally constructed from welded steel that is treated to resist corrosion and rust. Dampers also contain blade seals made of neoprene that seal when the blades contact each other. See Figure 7-11. Damper blade seals can become brittle and crack when exposed to temperature variations and humidity extremes. Cracked blade seals lead to excessive air leakage and poor temperature control. If needed, pilot positioners can be used to overcome external forces on a damper.

The damper blade linkage transmits the linear motion of the actuator into a rotary motion that turns the actuator

crank arm and rotates the drive blade to open or close the damper. The drive blade transmits the force to the other blades through a set of linkages. The linkage from the actuator must be connected to the drive blade only.

DAMPER SEALS

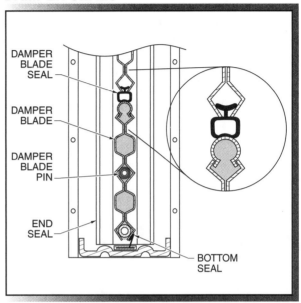

Figure 7-11. Damper seals provide a snug fit to minimize air leakage.

Some dampers are set up so that one damper blade drives another. This setup is normally used in an application where the actuator drives an outside air damper that is linked to a return air damper and/or exhaust air damper. This arrangement is used in small air handling units. In this application, the actuator must be large enough to handle the increased load of the additional dampers.

Greenheck Fan Corp.

Dampers are available in a variety of shapes and sizes for different pressure, velocity, and air flow requirements.

Damper Maintenance

Proper maintenance should be a primary focus of any plant preventive maintenance program. A major aspect of damper maintenance is the lubrication of the damper lubrication points. A dry, graphite type lubricant is often used to lubricate dampers at their lubrication points. See Figure 7-12. Avoid thick, viscous oils that may freeze in cold weather and attract dirt or foreign material. Damper edge seals also require replacement when they become worn. The frequency of damper maintenance depends on the unit run hours and damper service. All damper installations should follow the manufacturer's recommended procedures. Failure to do so will cause improper operation and premature damper failure.

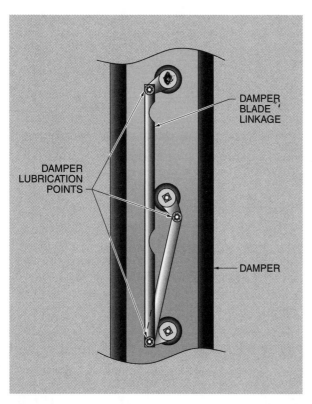

Figure 7-12. Manufacturer-recommended lubricant is used at damper lubrication points.

VALVES

A *valve* is a device that controls the flow of fluids in an HVAC system. The fluids may be water or steam. Hot water is used in heating systems and chilled water is used in cooling systems. Warm water is used in cooling towers or water source heat pumps. Steam is used in coils to provide heat in a building space, in heat exchangers to provide hot water, or discharged directly into air handlers to provide humidification. The type of fluid, temperature, and pressure are known as the service of a valve. The service of a valve cannot be changed without changing the capacity or function of the valve.

Valve Components

The components of an HVAC control valve include the valve body, stem, disc, and packing. See Figure 7-13. The valve body consists of the outer housing through which steam or water passes. The valve body also contains the means for attaching the valve to the piping. Valves are attached to piping using threads, flared fittings, or flanged fittings. The valve body is normally constructed of cast iron, bronze, steel, or stainless steel.

A *valve stem* is a valve component that consists of a metal shaft, normally made of stainless steel, that transmits the force of the actuator to the valve plug. The valve stem is attached to the piston cup and moves up and down, positioning the valve plug and disc. The valve plug allows a variable amount of fluid to flow through the valve. The shape of the plug determines the valve flow characteristics. The valve plug has a flat disc that contacts the valve seat. The flow through the valve is shut OFF when the disc contacts the valve seat.

The packing and packing nut prevent leakage of the fluid along the sides of the valve stem. Different types of packing are used for hot water, chilled water, and steam valves. Valve rebuild kits are available that contain packing and lubricant. Valves may be two-way or three-way valves.

Two-Way Valves

A *two-way valve* is a valve that has two pipe connections. See Figure 7-14. Two-way valves are available in a variety of pipe diameter sizes from ½″ to 27″, or larger. Two-way valves are normally open or normally closed. A *normally open (NO) valve* is a valve that allows fluid to flow when the valve is in its normal position. A *normally closed (NC) valve* is a valve that does not allow fluid to flow when the valve is in its normal position. Normally closed valves have a plug on the bottom of the valve body that enables the disassembly and removal of the disc and other internal parts.

A normally closed pneumatic valve has no flow when the valve is de-energized. A normally closed electrical valve has flow when the valve is de-energized.

HVAC CONTROL VALVE COMPONENTS

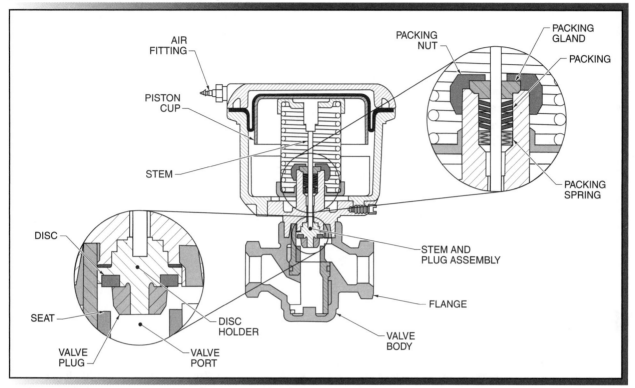

Figure 7-13. HVAC control valve components include the valve body, stem, disc, and packing.

TWO-WAY VALVES

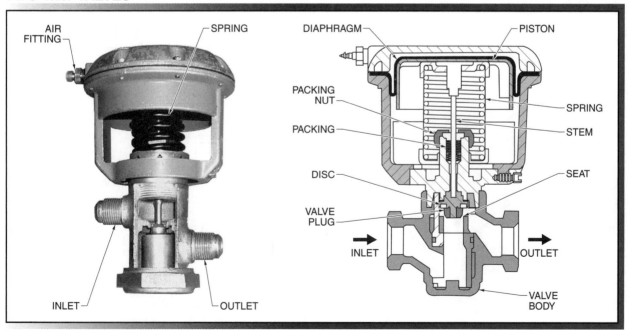

Figure 7-14. Two-way valves have one inlet and one outlet.

Two-way valves are used in steam, hot water, and chilled water applications. Steam valves are always two-way valves because steam does not require a return to the steam header or boiler. The disadvantage of using two-way valves in hot or chilled water applications is that there is no water flow through the valve when closed. This may cause pump wear or damage in large systems because the pump is forced to operate at an unstable load if a number of valves are closed at one time. Pump or system relief valves are installed to maintain a relatively constant supply and return pressure, which reduces pump wear. These systems can become complex and costly due to the required piping and control systems. In addition, the sources of the hot or cold water (boiler, converter, or chiller) require a constant water flow to operate at their maximum energy efficiency and to prevent damage. The use of two-way valves may also make water-balancing the system more difficult.

> *Water-balancing a system on startup is crucial to the proper operation of an HVAC system. What may appear as improper valve operation may be traced to lack of water flow at the valve inlet. Balancing reports are generated and turned in on equipment startup. Any anomalies should be identified at that time.*

Three-Way Valves

A *three-way valve* is a valve that has three pipe connections. Three-way valves are used in HVAC systems to control the flow of water because they provide a constant system pressure in the supply and return piping, and through the pumps, boilers, heat exchangers, and chillers. Three-way valves may be mixing valves or diverting (bypass) valves. See Figure 7-15.

A *mixing valve* is a three-way valve that has two inlets and one outlet. Mixing valve ports are referred to as common, normally open, and normally closed. Two of the ports are water inlets (normally open and normally closed). The common port is the outlet and the port designations cannot be switched. A mixing valve disc has an egg-shaped appearance.

When the pressure on the valve actuator is low, the normally open port is internally connected to the common port and the normally closed port is shut off to flow. When the pressure on the valve actuator is high, the normally closed port is connected to the common port and the normally open port is shut off to flow. At a part stroke position, the normally open and normally closed ports are partially open to the common port, mixing the hot and cold water. The actual pressure at which these ports are connected depends on the spring range of the valve actuator.

THREE-WAY VALVES

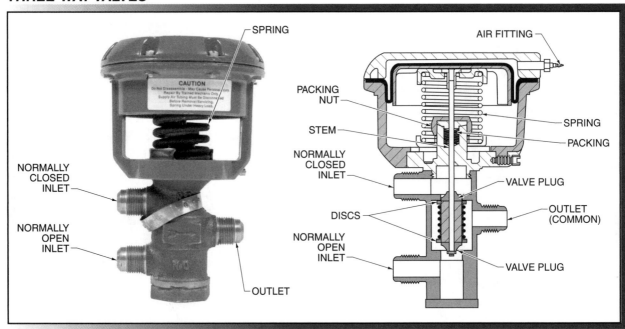

Figure 7-15. Mixing valves have two inlets and one outlet, diverting valves have one inlet and two outlets.

When used as a mixing valve, the normally open port is usually piped to the hot water supply and the normally closed port is piped to the chilled water supply. In this way, the valve fails to the heating position if a problem occurs. The piping may be reversed so the valve fails to the cooling position instead of the heating position.

A *diverting (bypass) valve* is a three-way valve that has one inlet and two outlets. Diverting valve ports are referred to as common, normally open, and normally closed. One outlet port is normally connected to a heat exchanger or coil (a cooling tower), and the other outlet port is connected to the system bypass.

Diverting valves are often used in applications that require two-position (ON/OFF) flow. Diverting valves cannot be used in mixing applications. Diverting and mixing valves may require balancing valves to be installed in the piping to balance the flow between supply and return. A diverting valve disc has an hourglass shape.

Butterfly Valves

A butterfly valve is a valve with a round plate that rotates to control flow. Butterfly valves are used in large piping systems to provide excellent flow characteristics in heat exchangers and other system devices. Butterfly valves appear similar to a round damper blade in that they consist of a round plate inside the valve that rotates when an actuator extends. See Figure 7-16. Butterfly valves are so similar to round dampers that they commonly use damper actuators for opening and closing instead of valve actuators. Butterfly valves are not as common as two- or three-way valves. Butterfly valves also have a nonlinear flow characteristic.

Valve Flow Characteristics

Valve flow characteristics is the relationship between the valve stroke and flow through the valve. Although the valve selection process is normally performed as part of the engineering process, the HVAC technician is required to know the valve flow characteristics in order to diagnose problems that may be due to a misapplied valve. HVAC control valves may have quick opening, linear, or equal percentage characteristics.

A *quick-opening valve* is a valve in which flow increases rapidly as soon as the valve is opened. Quick-opening valves are often used with steam valves and applications where a two-position, open/closed operation is desired. Quick-opening valves provide enough flow capacity to evenly heat a heat exchanger. This is required in steam systems that rapidly lose their ability to provide heat as the steam condenses, causing uneven heating and hot/cold spots of air passing over an air handler coil. In addition, quick-opening valves help protect valves from excessive wear and damage due to wire drawing. Wire drawing is valve erosion caused by steam molecules being accelerated by a valve operating in a cracked-open position for a long period of time.

BUTTERFLY VALVES

Leslie Controls, Inc.

Figure 7-16. Butterfly valves use damper-type actuators and mechanical movement to control water flow.

A *linear valve* is a valve in which the flow through the valve is equal to the amount of valve stroke. Linear valves have straight line flow characteristics. However, linear valve heat transfer characteristics are not linear. There is a mismatch between the valve flow characteristic and the heat exchanger output. A linear valve must be matched to a specific heat exchanger.

An *equal percentage valve* is a valve that provides incremental flow at light loads and large flow capabilities as the valve strokes (opens) farther. Equal percentage valves are widely used in many types of heat exchangers. Equal percentage valves are ideal for many applications because they closely match heat exchanger flow needs. In this way, the valve provides the proper flow at low and high load conditions. All of the three valve flow characteristics are accomplished by altering the shape of the valve plug. See Figure 7-17.

Checking an installed valve against the valve schedule can prevent system problems caused by the wrong valve being used in the system.

VALVE STEM LIFT AND FLOW

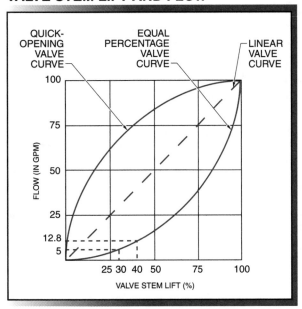

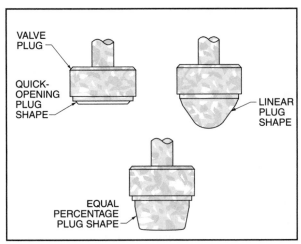

Figure 7-17. Valve flow characteristics vary based on the shape of the plug.

Valve Sizing and Selection

Valves used in HVAC applications must be properly sized and correctly applied. An improperly sized valve does not allow the proper amount of heating or cooling at maximum load. An incorrectly applied valve may slam open or closed, and may have packing that constantly leaks. The valve selection process begins with an accurate load determination. Once the load is determined, the heating/cooling equipment (boiler, chiller, heat pump) is selected. The mechanical equipment location is determined and a piping plan is laid out to deliver the proper

amount of water or steam to the equipment. The heat exchanger or coil is then selected to meet the heating or cooling requirements (in British thermal units per hour) of each building space.

Valves are sized after the heat exchanger or coil is selected. Valves are sized based on the medium (steam, hot water, cold water), the valve inlet temperature of the medium, the valve inlet and outlet pressures, the flow needed to meet the load (in gallons per minute), and the heat exchanger or coil pipe connections.

Every manufacturer provides a step-by-step guide for valve sizing and selection. The flow rate (in gallons per minute) of the medium is related to the C_v of the valve. C_v is the gallons per minute of flow through a valve at a 1 psig pressure drop across the valve. The desired valve C_v is compared to the available valves from the information supplied by the manufacturer. See Figure 7-18.

The required valve C_v rarely matches perfectly. For example, the sizing process may indicate a C_v of 6.3, but a manufacturer table only indicates a valve availability of 6 and 6.5. The smaller valve is commonly selected for use. The actuator is selected after the correct valve is selected. A common mistake is sizing the valve based on the inlet size of the heat exchanger or coil. The valve should be sized based on C_v and flow, not by pipe size; however, it is common for the valve size to be ± one pipe size from the coil inlet.

Valve Shut-Off Rating and Turn Down Ratio

Valve shut-off rating is the maximum fluid pressure against which a valve can completely close. Valve shut-off rating is the pressure at which the valve operates erratically or cannot shut off. Valve shut-off rating should be checked when troubleshooting an erratic system. *Valve turn down ratio* is the relationship between the maximum flow and the minimum controllable flow through the valve. Valve turn down ratio may affect valves that are operated for long periods of time at minimum flow. Valves that have unstable performance at minimum flow may indicate that the valve turn down ratio is being exceeded.

Valve Maintenance

Valve maintenance keeps valves in proper condition and can prolong the life of a valve. Common tasks performed during valve maintenance are valve repacking and valve rebuilding. Repacking or tightening of the packing nut is required when a valve body appears stained or water is visibly dripping from the packing nut. Valve rebuilding is required if the valve allows fluid to flow when closed. The valve may require disassembly and inspection if it vibrates or is excessively noisy.

Valves are sized by C_v – the amount of flow in gallons per minute at a 1 psig pressure drop across the valve. Valves are commonly sized one pipe size smaller than the coil inlet. For example, a 1½″ coil inlet may be fed by a 1″ or 1¼″ hot water valve.

Valve Repacking. *Packing* is a bulk deformable material or one or more mating deformable elements reshaped by manually adjustable compression. Packing prevents water or steam from leaking at the point where the valve stem penetrates the packing gland. Different types of packing material are used depending on the valve service. Valve repacking is performed when the valve begins to leak excessively at the valve stem and packing gland assembly. See Figure 7-19.

Most manufacturers provide valve repacking kits. The packing is normally lubricated with silicon-type grease included in the repacking kit. To repack a valve, the packing nut and gland are removed. The packing is then removed and replaced with the correct amount and type of packing. The packing gland and nut are then replaced. The packing nut should not be over tightened when the new packing is installed. Overtightening the packing nut binds the stem, especially in steam and hot water valves because the hot fluid expands the packing. The valve stem should not be nicked; a nicked stem damages the packing as it moves up and down. The stem should be checked to ensure it is clean and free to move properly.

C_v	3	5	7	9	12	15	20
0.2	0.3	0.4	0.5	0.6	0.7	0.8	0.9
0.4	0.7	0.9	1.1	1.2	1.4	1.5	1.8
0.6	1.05	1.35	1.6	1.8	2.05	2.3	2.7
0.8	1.3	1.7	2.1	2.4	2.7	3.0	3.5
1.0	1.7	2.2	2.6	3.0	3.5	3.9	4.5
1.2	2.0	2.6	3.1	3.6	4.1	4.6	5.25
1.4	2.4	3.1	3.7	4.2	4.8	5.4	6.1
1.6	2.8	3.6	4.2	4.8	5.5	6.2	7.1
1.8	3.1	4.0	4.7	5.4	6.2	6.9	8.1
2.0	3.4	4.4	5.2	6.0	6.9	7.7	9.0
2.2	3.8	4.9	5.8	6.6	7.6	8.5	9.75
2.4	4.1	5.3	6.2	7.1	8.2	9.25	10.8
2.6	4.5	5.8	6.8	7.8	9.0	10	11.9
2.8	4.8	6.3	7.4	8.4	10	11	12.5
3.0	5.2	6.7	7.9	9.0	10	12	13.5
3.2	5.5	7.1	8.4	9.6	11	12	14.2
3.4	5.9	7.6	9.0	10	12	13	15.2
3.6	6.2	8.0	9.5	11	12.5	14	16
3.8	6.5	8.4	10	11.5	13	15	17
4.0	6.9	9.0	11	12	14	15	18
4.5	7.7	10	12	14	16	17	20
5.0	8.6	11	13	15	17	19	22.5
5.5	9.5	12	15	17	19	21	24.2
6.0	10	13	16	18	21	23	27
6.5	11	14	17	19	23	25	29
7.0	12	16	19	21	24	27	31.9
7.5	13	17	20	23	26	29	34
8.0	14	18	21	24	28	31	36
8.5	14.5	19	22.5	25.9	29	33	38
9.0	15.5	20	24	27	31	35	40
9.5	16.5	21	25	28.5	33	36.5	42.5
10	17	22	26	30	35	39	45

HVAC VALVE WATER CAPACITY* — Differential Pressure†

* in gpm
† in psi

Figure 7-18. HVAC valves are selected by referencing the valve manufacturer water capacity chart.

Two-way valves are used for ON/OFF flow control for hot water reheat coil applications.

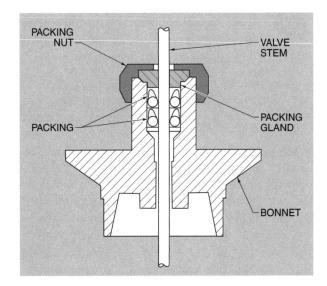

Figure 7-19. Valves are repacked when the valve begins to leak excessively at the valve stem and packing gland.

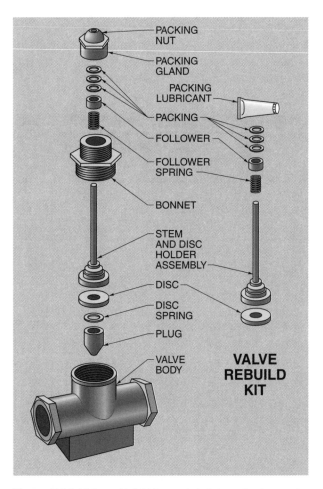

Figure 7-20. Valve rebuild kits contain internal valve parts, packing material, and packing lubricant.

Valve Rebuilding. Valve rebuilding is performed when a valve is mechanically damaged or the internal parts are worn by foreign material or wire drawing (the valve operating for long periods of time while cracked open). This partial opening acts like a nozzle and accelerates the steam across the valve. Over time, this high-velocity steam erodes the opposite side of the valve seat. Wire drawing may be present if a steam valve leaks when it is supposed to be closed. Wire drawing can be seen when the valve is disassembled as a small groove cut in the valve seat. Most valve manufacturers provide replacement parts and kits for valve rebuilding. See Figure 7-20. Valve seat erosion can be corrected by grinding the valve seat.

NORMALLY OPEN AND NORMALLY CLOSED DAMPERS AND VALVES

A normally open damper or valve allows fluid to flow when the damper or valve is in its normal position. Normally open dampers and valves fail to the open position (100% flow) if the air pressure is removed from the damper or valve actuator. A normally closed damper or valve does not allow fluid to flow when the damper or valve is in its normal position. Normally closed dampers and valves fail to the closed position (0% flow) if the air pressure is removed from the damper or valve actuator. The position of a damper or valve with no air pressure on the actuator is referred to as the fail-safe position of the actuator. See Figure 7-21.

In northern climates, heating system dampers and valves normally fail open to hot water, steam, or hot air. If the air compressor fails or a diaphragm breaks, heat is still supplied to the building. In northern climates, cooling system dampers and valves normally fail closed to flow because buildings are not normally damaged from the lack of cooling.

Cooling is more important in southern climates than in northern climates. For this reason, cooling system dampers and valves in southern climates normally fail open to flow to allow continued cooling if a problem occurs. In southern climates, heating system dampers and valves normally fail closed to flow because buildings are not normally damaged from the lack of heating. An exception is a computer room in which cooling system dampers and valves fail open, regardless of the climate, because computer rooms require extensive cooling.

NORMALLY OPEN AND NORMALLY CLOSED VALVES

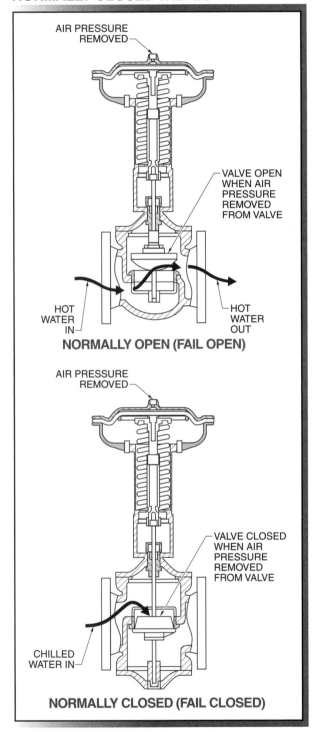

NORMALLY OPEN (FAIL OPEN)

NORMALLY CLOSED (FAIL CLOSED)

Figure 7-21. Normally open valves fail to the open position (100% flow) to allow the flow of steam or hot water in northern climate heating systems. Normally closed valves fail to the closed position (0% flow) to prevent the flow of chilled water in northern climate cooling systems.

Most air handling systems use normally closed outside air dampers because the outside air causes excessively cold conditions in the winter and excessively hot and humid conditions in the summer. Normally closed humidifier valves are used to prevent continual discharge of steam or water into the air stream and prevent dangerous levels of humidity in a building. See Figure 7-22.

DAMPER/VALVE APPLICATIONS

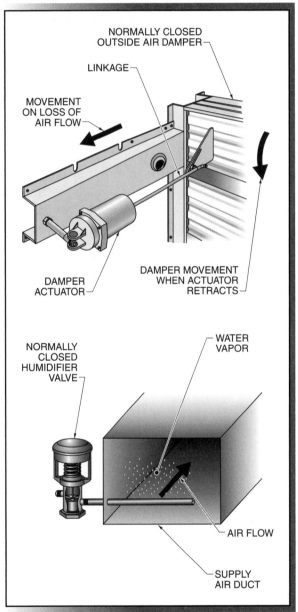

Figure 7-22. Normally closed dampers and valves prevent excessively hot, cold, or humid air from being introduced into a building.

> *Exercise caution when repacking hot water or steam valves. Do not overtighten the packing after replacement. Once the valve is back in service, the hot water or steam swells and expands the valve and packing. If a valve is overtightened when the packing is replaced, it may bind when subjected to hot water or steam. A solution is to leave the packing nut slightly loose, and tighten one-quarter turn at a time after the hot water or steam has been turned ON. Continue to tighten the valve until any leakage stops.*

Dampers are not designed to be normally open or normally closed by the manufacturer. Normally open and normally closed damper operation is determined by the attachment of the linkage to the damper blade.

Valves are designed by the manufacturer to be normally open or normally closed and cannot be altered. A valve is normally open if it is piped to have water flow through the coil if the air pressure to the actuator is removed. A valve is normally closed if it is piped to have no flow through the coil if the air pressure to the actuator is removed. This is separate of the port designations and is dependent on the system piping. See Figure 7-23.

NORMALLY OPEN AND NORMALLY CLOSED THREE-WAY VALVE APPLICATIONS

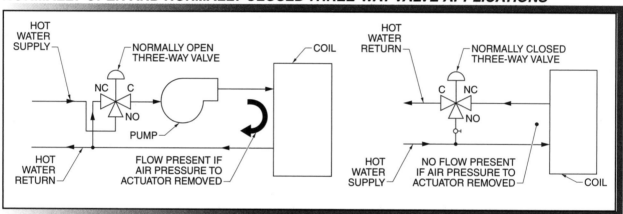

Figure 7-23. A normally open three-way valve allows flow if the air pressure to the actuator is removed. A normally closed three-way valve does not allow flow if the air pressure to the actuator is removed.

The two main types of pneumatic thermostats are bleed-type (one-pipe) and pilot bleed (two-pipe) thermostats. Additional thermostats available include single temperature direct-acting or reverse-acting, day/night, and winter/summer thermostats. Limit thermostats are used to prevent freeze-up or other undesirable conditions. Humidistats are used in building space humidity control systems. Pressure switches are used in duct or piping control systems.

PNEUMATIC THERMOSTATS

A *pneumatic thermostat* is a thermostat that uses changes in compressed air to control the temperature in individual rooms inside a commercial building. Pneumatic thermostats are available in many shapes and sizes, and with many options.

Pneumatic Thermostat Operation

All pneumatic thermostats operate on the same basic principles. In a pneumatic thermostat, a sensing device (bimetallic element) is mounted under a cover and exposed to the air in the room. A *bimetallic element* is a sensing device that consists of two different metals joined together. The different metals have different rates of expansion and contraction. One of the metals has a high rate of expansion and contraction while the other has a very low (almost zero) rate of expansion and contraction. When the room temperature increases, the bimetallic element bends toward the metal with the low rate of expansion and contraction. See Figure 8-1. The amount of bending is proportional to the change in temperature at the bimetallic element. The bimetallic element bends toward the metal with the high rate of expansion and contraction if the room temperature decreases.

BIMETALLIC ELEMENT SENSING DEVICE

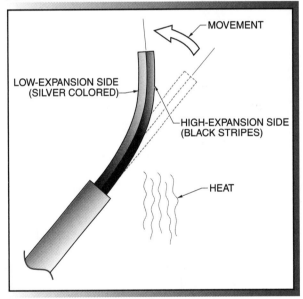

Figure 8-1. A bimetallic element is a sensing device that consists of two metals with different expansion and contraction rates joined together.

The bimetallic element is attached to the thermostat baseplate on one end, while the other end is free to bend. A *baseplate* is a flat piece of metal to which the thermostat components are mounted. The free end is used to cover or uncover a bleedport, depending on whether the room temperature increases or decreases. A *bleedport* is an orifice that allows a small volume of air to be expelled to the atmosphere. The bleedport is connected to the main (supply) air inlet. The bleedport has a small diameter and meters the air flow through it. If the room temperature is high, the bimetallic element bends toward the bleedport, preventing air from escaping. This increases the branch line pressure, causing the actuator to stroke out. *Branch line pressure* is the pressure in the air line that is piped from the thermostat to the controlled device. Normally, an increase in branch line pressure causes the temperature in the room to decrease. If the room temperature is too low, the bimetallic element bends away from the bleedport, allowing more air to escape and decreasing the branch line pressure. This fully retracts the damper, causing the actuator to return to its normal position. This causes the temperature in the building space to increase to its setpoint. *Setpoint* is the desired value to be maintained by a system.

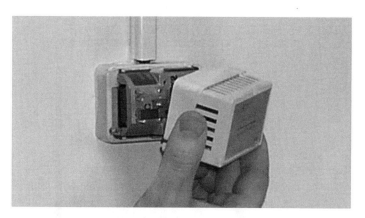

Pneumatic thermostats contain a sensing device mounted under a cover and exposed to the air in the room.

The room temperature change, the bimetallic element bending, the actuator stroke, and the amount of controlled medium added to the building space are proportional to one another. A bimetallic element may be direct-acting or reverse-acting. In a direct-acting bimetallic element, as the temperature increases, the branch line pressure increases, and vice versa. In a reverse-acting bimetallic element, as the temperature increases, the branch line pressure decreases, and vice versa.

Bleed-Type (One-Pipe) Thermostats

The simplest and most basic thermostat is a bleed-type thermostat. A *bleed-type thermostat* is a thermostat that

changes the air pressure to a valve or damper actuator by changing the amount of air that is expelled to the atmosphere. See Figure 8-2. A bleed-type thermostat works by the bending of a bimetallic element which covers or uncovers the bleedport. This varies the amount of air to an actuator, thus causing a change in position. A bleed-type thermostat is identified by its single air line connection at the back of the thermostat baseplate and the restrictor in the air line inlet to the thermostat. A *restrictor* is a fixed orifice that meters air flow through a port and allows fine output pressure adjustments and precise circuit control. The restrictor is most often in the form of a tee connection. Restrictors are usually made of plastic and are available in sizes of .007″ and .009″, or in standard cubic inches of air per minute (scim) capacity. Manually adjustable restrictors (needle valves) are used to calibrate controllers and affect system operation. Restrictors, because they can cause time delays, are often used with air reservoirs (accumulators or timing tanks) which can cause an air signal to change slowly over time. This response time of a system to change is referred to as hysteresis.

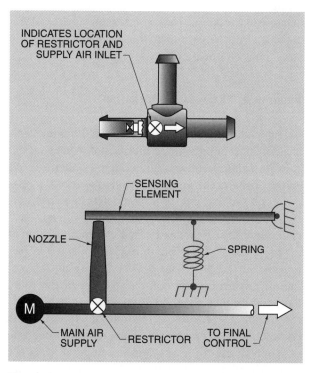

Figure 8-2. A bleed-type thermostat has a single air line connection at the back of the thermostat baseplate and a restrictor in the air line inlet to the thermostat.

While a restrictor is necessary to provide precise control in the system, the restrictor also reduces the amount of air available for actuator movement. The reduced amount of air may slow the control system response to unacceptable levels. This is similar to filling a 5 gal. bucket with a garden hose having a kink in the line. Bleed-type devices such as thermostats are also known as low-volume or non-relay thermostats. Bleed-type devices are not used in applications that require a fast control response or in applications that use large-volume actuators. Large-volume actuators require a large amount of air rapidly to provide proper control with a bleed-type thermostat. Although bleed-type thermostats may be smaller, cheaper, and simpler than pilot bleed (two-pipe) thermostats, they are not as popular. Bleed-type thermostats are available in direct-acting (DA) and reverse-acting (RA) models. A direct-acting thermostat increases the branch line pressure as the building space temperature increases and decreases the branch line pressure as the building space temperature decreases. A reverse-acting thermostat decreases the branch line pressure as the building space temperature increases and increases the branch line pressure as the building space temperature decreases. See Figure 8-3. Pneumatic thermostats also have a throttling range/sensitivity adjustment that can be changed to provide closer temperature control.

One use for a bleed-type thermostat is a light-troffer application. A *light-troffer application* is an application in which the thermostat is located in a return air slot that is integral to a light fixture. See Figure 8-4.

Bleed-Type Thermostat Calibration. A bleed-type thermostat is calibrated by applying the procedure:

1. Remove the thermostat cover.
2. Install a pressure gauge and thermometer.
3. Measure the building space temperature.
4. Adjust the setpoint dial to that temperature.
5. Adjust the output (branch line) pressure to the midpoint of the actuator spring range.
6. Test the thermostat response by adjusting the setpoint dial ± 2°F.
7. Adjust the setpoint dial back to the desired setpoint setting.
8. Remove the pressure gauge and reinstall the cover.

Pilot Bleed (Two-Pipe) Single-Temperature Thermostats

A *pilot bleed (two-pipe) thermostat* is a thermostat that uses air volume amplified by a relay to control the temperature in a building space or area. Pilot bleed thermostats are referred to as pilot bleed thermostats because of the relay that amplifies small changes in air pressure due to air being released to the atmosphere through the bleedport. Pilot bleed thermostats are also referred to as two-pipe thermostats because they have two air line connections at the rear of the thermostat subbase. See Figure 8-5. Pilot bleed thermostats are similar to bleed-type thermostats in that they both have the sensitive bi-

LIGHT-TROFFER APPLICATION

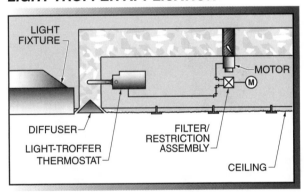

Figure 8-4. A light-troffer application has the thermostat located in a return air slot that is integral to a light fixture.

BLEED-TYPE THERMOSTATS		
Thermostat Action	**Output on Branch Line Thermostat When Temperature, Humidity, or Pressure Increases**	**Output on Branch Line Thermostat When Temperature, Humidity, or Pressure Decreases**
Direct-Acting	Increase	Decrease
Reverse-Acting	Decrease	Increase

Figure 8-3. A direct-acting thermostat increases the branch line pressure as the building space temperature increases. A reverse-acting thermostat decreases the branch line pressure as the building space temperature increases.

metallic element and port assembly. A pilot bleed thermostat has a relay that the bleed-type does not. The pilot bleed thermostat is often referred to as a relay thermostat while the bleed-type thermostat is often referred to as a non-relay thermostat.

PILOT BLEED THERMOSTATS

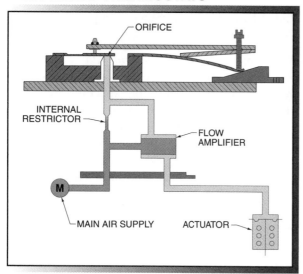

Figure 8-5. A pilot bleed thermostat has a flow amplifier relay that increases air capacity for larger actuator applications.

Pneumatic humidistats are used to sense and control the humidity in a building space or duct.

The relay in a pilot bleed thermostat acts like an amplifier. The relay increases the available volume at the thermostat branch line, thus reducing the volume problems of a one-pipe thermostat. Secondly, the relay provides restricted main air to the port assembly. The

restrictor is located inside the relay instead of in the air line. In addition, the relay provides adequate exhaust capability for the branch line when the pressure must be reduced quickly, retracting the actuator. Removing air from the branch line quickly is as important as adding air quickly. This air is vented to the atmosphere. The relay is attached to the thermostat and is non-field repairable and non-field replaceable.

Pilot bleed thermostats are superior to bleed-type thermostats because of their higher main air capacity. This means that they can be used in any thermostat application regardless of actuator air capacity requirements. Pilot bleed thermostats also respond quickly to any change in the building space temperature. Pilot bleed thermostats also retain the sensitive response of the bleed-type thermostat. Pilot bleed thermostats are available in many different types for different applications. The most common type is a single-temperature setpoint thermostat.

A *single-temperature-setpoint thermostat* is a pilot bleed thermostat that has one setpoint year-round for a building space or area. A single-temperature-setpoint thermostat may be direct-acting or reverse-acting as set by the factory. The bimetallic element is not reversible, but must be changed with an appropriate type to change the thermostat from direct-acting to reverse-acting operation. All thermostats should have their calibration checked a minimum of once a year.

Pilot Bleed Single-Temperature Thermostat Calibration. A single-temperature thermostat may need to be recalibrated because of mechanical wear, age, or tampering, as part of the troubleshooting process, or to ensure accuracy on a continuing basis. See Figure 8-6. A pilot bleed single-temperature thermostat is calibrated by applying the procedure:

1. Remove the thermostat cover.

2. Install the pressure gauge and thermometer. Wait 5 min.

3. Turn the setpoint dial to the ambient temperature.

4. Adjust the branch line pressure to the midpoint of the actuator spring range. *Note:* If the thermostat is operating properly, the pressure measured in step 2 can be used instead of the midpoint pressure.

5. Check the repeatability by turning the setpoint dial ± 2°F. The branch line pressure should return.

6. Turn the setpoint dial to the desired setpoint.

7. Remove the pressure gauge and thermometer. Reinstall cover.

PILOT BLEED SINGLE-TEMPERATURE THERMOSTAT CALIBRATION

 ① Remove thermostat cover.

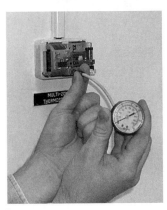 ② Install pressure gauge and thermometer. Wait 5 min. Measure air temperature near thermostat.

 ③ Turn setpoint dial to ambient temperature.*

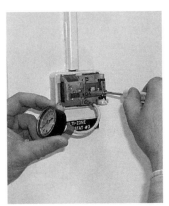 ④ Adjust branch line pressure to midpoint of actuator spring range.

 ⑤ Check repeatability by turning setpoint dial ±2°F.

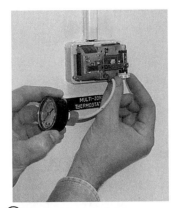 ⑥ Turn setpoint dial to desired setpoint.

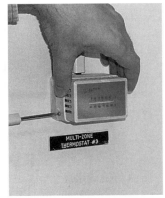 ⑦ Remove pressure gauge and thermometer. Reinstall cover.

Note: If the thermostat is operating properly, the pressure measured in step 2 can be used instead of the midpoint pressure.

Figure 8-6. Pilot bleed single-temperature thermostats require calibration to ensure their proper operation.

If possible, check the thermostat in actual operation. Different thermostats have different methods of removing the cover. Obtain the correct tools for performing the procedure. Different manufacturers use different types of gauges for measuring branch line pressure. Some popular types of gauges are the screw-in, football needle, and hypodermic types. See Figure 8-7. The pressure gauge used should be the exact type recommended by the manufacturer. Use a quality technician pocket thermometer to measure the exact building space temperature. Do not use the integral thermometer often supplied with the thermostat. Integral thermometers are inaccurate, especially on old thermostats. Also, ensure the pocket thermometer is checked monthly against a known standard to prevent drifting and improper thermostat calibration.

Different manufacturers have different methods for indicating the setpoint. Some manufacturers have whole-number indicators while others have dots or marks in between the whole numbers. Be sure that the indication is correct.

When adjusting the branch line pressure to the midpoint of the actuator spring range, note that the midpoint of the actuator spring range is the standard calibration point of all thermostats. The manufacturer's calibration procedures may instruct the technician to adjust the output (branch line pressure) to a specific number such as 8 psi. This is because the particular manufacturer prefers to use a specific spring range such as 3 psi – 13 psi. The midpoint is then 8 psi. However, not all spring ranges are consistent with this standard.

> ⓘ *All gauge components should be selected considering media and ambient operating conditions to prevent misapplication. Improper application can be harmful to the gauge, cause failure, and/or possibly cause personal injury or property damage.*

Checking the repeatability is not included in most manufacturer literature. The repeatability should be checked before leaving to avoid a callback. It is possible that the thermostat is calibrated properly but drifts or fails soon after calibration. To check repeatability, the thermostat is cycled up and down to check if it repeats its action. This normally catches any thermostats that are beginning to drift. Thermostats loose calibration as they begin to fail.

PRESSURE GAUGES

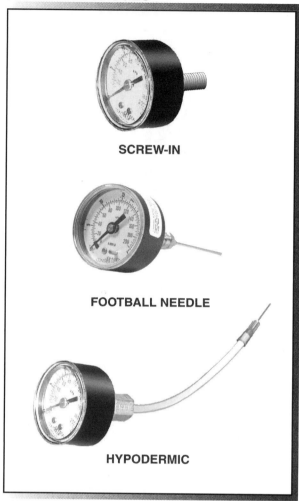

SCREW-IN

FOOTBALL NEEDLE

HYPODERMIC

Figure 8-7. Pressure gauges include screw-in, football needle, and hypodermic types.

The dial setpoint must be turned to the desired setpoint. If the building space temperature is too hot and the dial is not adjusted to the proper setpoint, the thermostat has not been calibrated.

Day/Night Thermostats

A *day/night (occupied/unoccupied) thermostat* is a thermostat that has two setpoints, one setpoint for day (occupied time period) and one setpoint for night (unoccupied time period). See Figure 8-8. The day setpoint is usually about 72°F. The night setpoint is usually about 60°F. Day/night thermostats are used in applications that require reduced building space temperatures at night to reduce energy consumption.

DAY/NIGHT THERMOSTATS

DAY/NIGHT

DIRECT-ACTING STRIP
SENSITIVITY SLIDER
NIGHT ADJUSTING SCREW
DIRECT-ACTING STRIP
DAY ADJUSTING SCREW

SUBBASE

DAY SETPOINT DIAL
DAY BIMETALLIC ELEMENT
DAY/AUTO LEVER
NIGHT SETPOINT DIAL
NIGHT BIMETALLIC ELEMENT

BACKPLATE

SWITCHOVER ADJUSTMENT SCREW
THERMOSTAT MULTISTAGE FILTER
DAY/AUTO LEVER
BACKPLATE SCREWS

Figure 8-8. Day/night thermostats have two setpoints, one for day (occupied time period) and one for night (unoccupied time period).

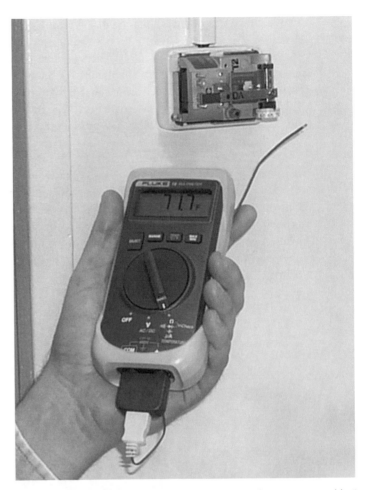

A multimeter with a thermocouple probe can be used to measure ambient temperature when calibrating a thermostat.

School buildings often use day/night thermostats. School building HVAC systems are usually heating-only with no mechanical cooling available. Day/night thermostats have two bimetallic elements, both normally direct-acting. Day/night thermostats also have two setpoint dials. It is common for the night setpoint dial to be smaller than the day setpoint dial. Day/night thermostats are actually two separate thermostats on one common body. Day/night thermostats also have a changeover relay that determines when the thermostat changes from the day temperature to the night temperature. A *changeover relay* is a relay that causes the operation of the thermostat to change between two or more modes such as day/night. The changeover relay is adjustable from the back of the thermostat with a screwdriver.

The changeover relay in a day/night thermostat is activated by changes in main air pressure from the air compressor station. See Figure 8-9. Normally, the night main air pressure is lower than the day pressure. The changeover relay can cause improper thermostat operation if it is not adjusted properly. The relay may hiss and do nothing if it is not changing the thermostat over properly or is hung up between the day and night modes. A day/night thermostat also has a day/auto lever. A *day/auto lever* is a lever that is used to override a thermostat to the day temperature during the night mode. This is done to accommodate night classes, etc. When the thermostat is in the night mode, the occupants move the switch to the day position, and the thermostat then controls the building space temperature to the day setpoint. The lever can be switched back to the auto position when the occupants are finished in the building space, or it switches back automatically when the main air pressure changes back in the morning.

Figure 8-9. Changes in main air pressure from an air compressor station are used to activate the changeover relay in a day/night thermostat.

Day/Night Thermostat Calibration. Day/night thermostats may need to be calibrated due to mechanical wear, tampering, or drifting from setpoint, or as part of the troubleshooting process. See Figure 8-10. A day/night thermostat is calibrated by applying the procedure:

1. Change the main air pressure to the night pressure.
2. Remove the cover and install a pressure gauge.
3. Adjust both setpoint dials to ambient temperature.

4. Adjust the night output to the midpoint of the actuator spring range. *Note:* If the thermostat is operating properly, the pressure measured in step 2 can be used instead of the midpoint pressure.
5. Adjust the night setpoint dial to the desired night temperature.
6. Change main air pressure to day pressure.
7. Adjust the day output to the midpoint of the actuator spring range. *Note:* If the thermostat is operating properly, the pressure measured in step 2 can be used instead of the midpoint pressure.
8. Adjust the day setpoint dial to the desired day temperature. Remove thermostat and pressure gauge. Reinstall cover.

The sensitivity and temperature range of a bimetallic element increase when the element is coiled.

The day/night thermostat calibration procedure calibrates two standard single-temperature two-pipe thermostats (one day and one night) in a row. It makes no difference which side of the thermostat (day or night) is calibrated first. At any given point, only one half of the thermostat should be active. If in the day pressure, the night side of the thermostat should be inactive, and vice versa. The side that is active can be easily tested by using the setpoint dials to find which one is active. In addition, the pressure reducing valves (PRVs) at the air compressor can be used to change the thermostat from night to day. The night pressure must be on the thermostat to accomplish this. This procedure can be accomplished while the thermostat is in normal operation. However, if any problems are detected, or if more in-depth adjustment is needed, it is best to remove the thermostat from the building space and install it in a bench-test application to avoid building space temperature-control problems. A spare or rebuilt thermostat can be installed in the building space during the bench test. Use the manufacturer's calibration form if available.

Winter/Summer Thermostats

A *winter/summer thermostat* is a pilot bleed thermostat that changes the setpoint and action of the thermostat from the winter (heating) to the summer (cooling) mode.

DAY/NIGHT THERMOSTAT CALIBRATION

1 Change main air pressure to night pressure.

2 Remove cover and install pressure gauge.

3 Adjust both setpoint dials to ambient temperature.

4 Adjust night output to mid-point of actuator spring range.*

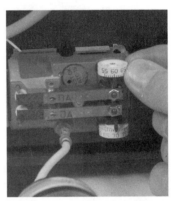

5 Adjust night setpoint dial to desired night temperature.

6 Change main air pressure to day pressure

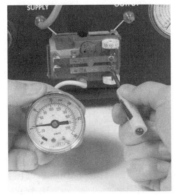

7 Adjust day output to mid-point of actuator spring range.*

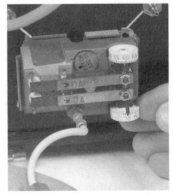

8 Adjust day setpoint dial to desired day temperature. Remove thermometer and pressure gauge. Reinstall cover.

*Note: If the thermostat is operating properly, the pressure measured in step 2 can be used instead of the midpoint pressure.

Figure 8-10. Day/night thermostats are calibrated to ensure proper function at day and night setpoints.

Winter/summer thermostats are also known as heating/cooling thermostats. Winter/summer thermostats have two bimetallic elements like day/night thermostats. However, unlike a day/night thermostat, one bimetallic element is normally direct-acting and the other is reverse-acting. This is because the mechanical systems are different. See Figure 8-11.

WINTER/SUMMER THERMOSTATS

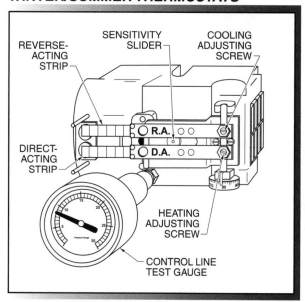

Figure 8-11. Winter/summer thermostats have two bimetallic elements. One is normally direct-acting and the other is reverse-acting.

Winter/summer thermostats are used in mechanical systems that switch from a heating medium in the winter to a cooling medium in the summer. An example of this is a fan coil unit that has hot water flowing through the coil in the winter and chilled water flowing through the same piping in the summer. See Figure 8-12. Because the valve does not change its operation, the action of the thermostat must change to accommodate the different medium. The use of one direct-acting bimetallic element and one reverse-acting bimetallic element is one way to identify a winter/summer thermostat. Winter/summer thermostats also have a changeover relay like day/night thermostats.

Some winter/summer thermostats have one setpoint dial to maintain the same temperature regardless of the season. Other manufacturers produce winter/summer thermostats that have two setpoint dials, one for winter (68°F) and one for summer (75°F). In this way, energy can be conserved regardless of the season.

WINTER/SUMMER THERMOSTAT APPLICATION

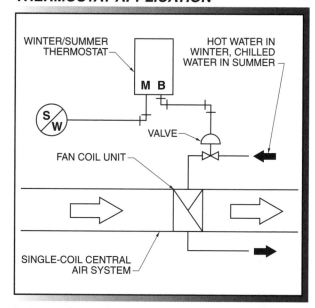

Figure 8-12. Winter/summer thermostats are used in mechanical systems that switch from a heating medium in the winter to a cooling medium in the summer.

Winter/summer thermostats also use the same type of dual-pressure air supply system as day/night thermostats. However, the method of building changeover is different. For example, the day/night system uses a time clock or a building automation system to cause the thermostats to change their setpoints. A winter/summer system uses an outside air thermostat, a manual switch, or a chilled-water thermostat strapped to the chilled-water-supply piping to cause the thermostats to change their setpoints. The mechanical system design has a great effect on the control system installation. Winter/summer thermostats do not have the day/auto lever that day/night thermostats have. Because the thermostat action changes from direct-acting to reverse-acting, the lever would cause a potentially harmful situation in the building. The unit may need heat and have hot water available, but the thermostat would actually try to shut the valve if it was reverse-acting. This could cause the unit to freeze up under cold outside-air conditions.

A loud hissing sound coming from a pneumatic thermostat may indicate that the main air connection is disconnected from the rear of the thermostat, or the actuator diaphragm has a large hole.

Winter/Summer Thermostat Calibration. Winter/summer thermostats may need to be calibrated due to mechanical wear, tampering, or drifting from setpoint, or as part of the troubleshooting process. A winter/summer thermostat is calibrated by applying the procedure:

1. Change the main air pressure to winter pressure.
2. Ensure the thermostat is in winter mode.
3. Calibrate the thermostat in winter mode.
4. Change the main air pressure to summer mode.
5. Ensure the thermostat is in summer mode.
6. Calibrate the thermostat in summer mode.

Winter/summer thermostats should be bench-calibrated because it is almost impossible to change the building from heating to cooling for calibration purposes. A rebuilt or spare thermostat can be used in the building space during the bench calibration. As with a day/night thermostat, it makes no difference whether the winter or summer mode is calibrated first. The thermostat mode can be checked by carefully using a pocket screwdriver to lift the bimetallic elements and determine which one is in operation. Use the manufacturer's calibration form if available.

Deadband Thermostats

The energy crisis of the mid to late 1970s caused all segments of U.S. society to search for ways to save energy. One of the results of this search was the development of deadband pneumatic thermostats. A *deadband pneumatic thermostat* is a thermostat that allows no heating or cooling to take place between two temperatures. See Figure 8-13. A *deadband* is the range between two temperatures in which no heating or cooling takes place. A deadband thermostat maintains a building space temperature between two values, such as 70°F and 74°F. Between these two values, the deadband thermostat does nothing.

Some deadband thermostats have two bimetallic elements, one for the lower value (usually for heating) and another for the higher value (usually for cooling). The bimetallic elements are individually calibrated so that the heating valve is fully closed at the lower value and the cooling valve is fully closed at the higher value. The cooling valve begins to open as the building space temperature increases above this value. Other deadband thermostats use an extra-large throttling range to achieve this result. Regardless of the type used, energy is conserved because no heating or cooling takes place in the system between the two temperatures. Deadband

thermostats are used most often in government buildings and installations. Use the manufacturer's recommended calibration procedures.

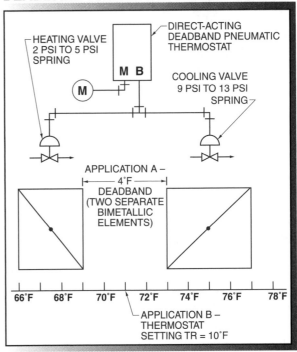

Figure 8-13. Deadband thermostats maintain the building space temperature within a range in which no heating or cooling takes place.

Master/Submaster Thermostats

A *master/submaster thermostat* is a pneumatic thermostat that has a third air connection piped to either a manual air regulator or an outside air transmitter. A master/submaster thermostat changes its setpoint based on the outside air temperature or allows an operator to change the setpoints from a remote location. Master/submaster thermostats are designed to conserve energy while maintaining relative comfort in the building space.

Master/submaster thermostats maintain a setpoint that maintains a specific temperature differential between the outside air temperature and the building space. If the outside air temperature is high, the occupants can tolerate a higher building space temperature as long as a differential is maintained. The opposite is true for a heating system. Follow the manufacturer's calibration instructions for master/submaster thermostats.

Limit Thermostats

A *limit thermostat* is a thermostat that maintains a temperature above or below an adjustable setpoint. Many HVAC systems such as air handling units, unit ventilators, etc., require the addition of a limit thermostat to provide safety in the control system. See Figure 8-14. If some type of limit protection is not provided, the system may suffer a freeze-up or some other problem. Limit thermostats are not used for primary building space temperature control but are used only as protection against failure.

Limit thermostats are usually one-pipe devices located in series with the primary controller such as a standard two-pipe thermostat. Limit thermostats are commonly located in the discharge of the heating coil and set at a minimum temperature that protects the coil or system. In a normally operating control system, limit thermostats do nothing. A limit thermostat causes the valves or dampers to go to a maximum condition to try to correct the situation if the temperature at the limit thermostat is too high or low, depending on the application. In the case of a discharge limit application, the normally open (NO) heating valve is forced wide open in an attempt to bring the discharge temperature up. The setpoint in this application is usually 35°F to 40°F. Limit thermostats normally use either a rod-type or a capillary-type sensing element. Limit thermostats have a setpoint dial, calibration screw, throttling range adjustment, and an integral air gauge port so that they can be easily checked by a technician.

Limit Thermostat Calibration. Limit thermostats may require calibration because of mechanical wear, or tampering, or as part of the troubleshooting process. See Figure 8-15. Limit thermostats are calibrated by applying the procedure:

1. Measure the ambient temperature at the limit thermostat with an accurate thermometer. Wait 5 min for temperature to stabilize.
2. Turn the setpoint dial to the ambient temperature.
3. Adjust the throttling range to the desired value. Adjust the output pressure to the midpoint of the spring range with the calibration screw. *Note:* If the thermostat is operating properly, the pressure measured in step 2 can be used instead of the midpoint pressure.
4. Turn the setpoint dial to the desired limit value.

Note: The desired setpoint and throttling range values should be determined from experience or a control diagram.

Pneumatic limit humidistats and pressure switches are used to prevent a damaging humidity or pressure level in a duct. Limit humidistats are normally located downstream of the humidifier and protect against high humidity levels in the duct. Limit pressure switches prevent overpressure in the duct, which can cause the

LIMIT THERMOSTATS

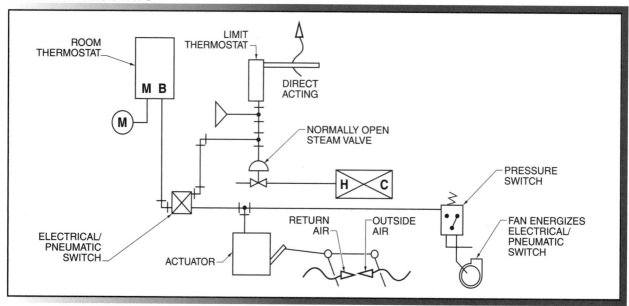

Figure 8-14. Limit thermostats are pneumatic thermostats that maintain a temperature above or below an adjustable setpoint.

LIMIT THERMOSTAT CALIBRATION

① Measure ambient temperature at limit thermostat. Wait 5 min for temperature to stabilize.

② Turn setpoint dial to ambient temperature.

③ Adjust throttling range to desired value. Adjust output pressure to midpoint of spring range.*

④ Turn setpoint dial to desired limit value.

Note: If the thermostat is operating properly, the pressure measured in step 2 can be used instead of the midpoint pressure.

Figure 8-15. Limit thermostats are calibrated in order to ensure limit operation at the proper temperature.

duct to burst. Pneumatic humidistats and pressure switches are calibrated the same as limit thermostats with the exception that humidity or pressure values are substituted for temperature.

Mounting, Piping, and Options

Pneumatic thermostats, humidistats, and pressure switches must be mounted according to manufacturer recommendations. Thermostats can be mounted in many ways. The most common mounting is a wall mounting. See Figure 8-16. Thermostats may also be mounted using an aspirator installation in which the thermostat is mounted inside the wall using a standard electrical handy (utility) box. This mounting is used in high-traffic public areas such as reception, elevator, and cafeteria areas. This mounting protects the thermostat from abuse or adjustment by the public. Room air is circulated through the box by means of an aspirator, or a small amount of restricted main air which is ejected into the room. This induces a small flow of room air through a slot in the thermostat box cover. In addition, ensure any holes in the wall directly behind the thermostat are plugged because the unconditioned air inside the wall may influence the thermostat.

Mounts are also available to install the thermostat in a return air grille or light-troffer. In all cases, the thermostat subbase is attached to the wall or duct. The thermostat is then attached to the subbase by screws

THERMOSTAT MOUNTING

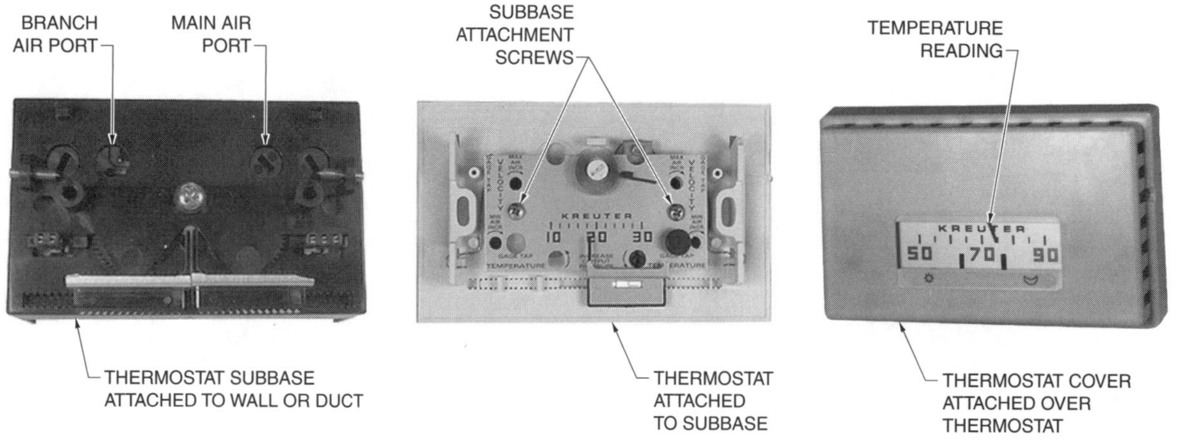

Figure 8-16. Thermostats are commonly mounted on a wall and include the subbase, thermostat, and cover.

or plastic clips. The cover is then attached over the thermostat by screws or clips. All manufacturers provide universal replacement kits that provide everything needed to adapt their thermostats to all of their competitors' installations. The pneumatic tubing is attached either to the subbase or the thermostat itself. In almost all cases, 5/32″ D poly tubing is used. Poly tubing is available in an integrated dual package to run the main and branch lines simultaneously.

Anti-kink springs should be inserted into approximately the first 8″ of the main and branch lines. Anti-kink springs prevent the poly tubing from kinking when the thermostat or subbase is attached to the wall. Clips are often attached to the air line at the point where it is attached to the thermostat or subbase. These clips prevent air lines from becoming loose and falling off, causing failure.

Many options are available for thermostats and humidistats. Many types of covers are provided, such as horizontal/vertical mount, with/without built-in mercury thermometer, and with/without setpoint view. Many colors and styles are available such as standard beige, white, or polished metal. Metal or plastic thermostat guards can be purchased to prevent unauthorized use. Another option available is an adjustable mechanical stop that prevents the setpoint from being adjusted more than 1°F or 2°F up or down.

Thermostat Troubleshooting

One of the most common skills required by a maintenance technician is basic HVAC control system troubleshooting. A maintenance technician must understand the most common problems that can occur with room temperature-control systems, the symptoms of those problems, and methods of correcting them.

To troubleshoot any type of system, the appropriate pressure gauge, a gauge adapter, and an accurate thermometer are required. The pressure gauge should always be installed in the thermostat when troubleshooting the system. An accurate thermometer, especially a new digital-readout model, removes any doubt about the actual temperature. The core of any pneumatic troubleshooting procedure is to install a pressure gauge on the thermostat output (branch line). The gauge reading gives an indication of the thermostat operation. Other tools required include a squeeze (aspirator) bulb and pocket screwdriver.

A common thermostat troubleshooting problem is that a building space is excessively cold or excessively hot. In this application, the mechanical system consists of a finned-tube steam heating system. See Figure 8-17. A pneumatic-actuated steam valve controls flow through the valve and finned-tube radiator. A thermostatic steam trap causes steam condensate to flow back to the return

FINNED-TUBE STEAM HEATING SYSTEM TROUBLESHOOTING CHART

Problem	Possible Causes	Suggestions	Corrective Actions
Building Space Too Cold	Thermostat out of calibration	Thermostat works at wrong temperature	Calibrate thermostat
	No steam	Check steam supply	Turn steam ON
	Valve stuck closed	Check gauge, check valve travel	Rebuild valve
	Trap failed closed	Valve open and lukewarm	Isolate and rebuild
	Thermostat has wrong setpoint	Check setpoint dial	Change setpoint dial
Building Space Too Hot	Thermostat out of calibration	Thermostat works at wrong temperature	Calibrate thermostat
	Hole in diaphragm	Use squeeze bulb, pressure is low	Disassemble and replace
	Valve wiredrawn	Valve is shut but leaks	Rebuild valve
	Thermostat has wrong setpoint	Check setpoint dial	Change setpoint dial

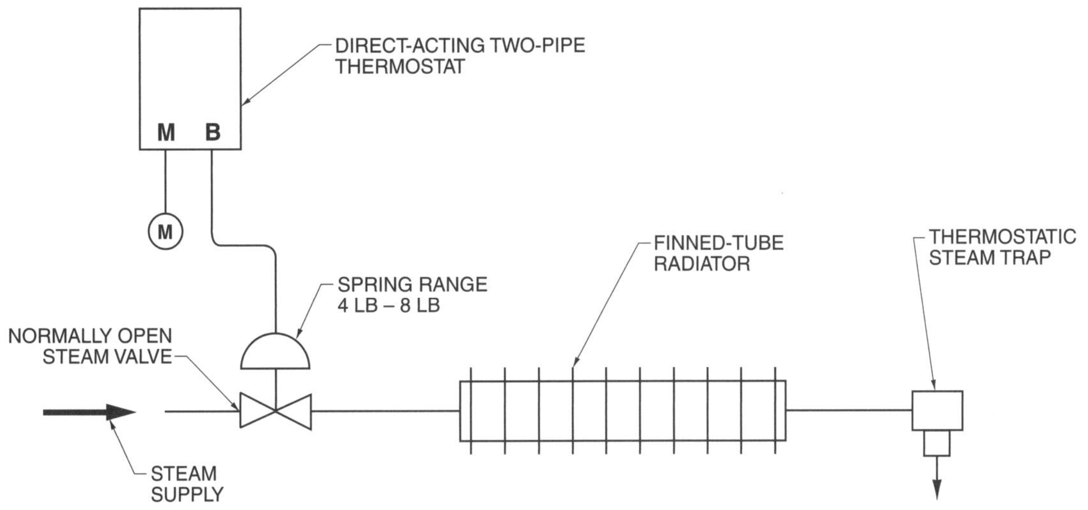

Figure 8-17. When troubleshooting a finned-tube steam heating system, the problem may be in the thermostat, steam valve, or steam trap.

system. The steam valve is controlled by a direct-acting two-pipe thermostat. Possible causes include thermostat out of calibration, no steam, valve stuck closed, etc. A troubleshooting chart is used to determine the possible causes, symptoms, and corrective action.

A building space that is excessively cold or hot may also occur in a system containing a dual-duct air handling unit. A dual-duct air handling unit supplies both hot and cold air to a sheet metal mixing box. A direct-acting two-pipe thermostat controls a damper actuator that opens and closes the hot and cold deck dampers inside the mixing box. This controls hot or cold air flow to the building space. See Figure 8-18. Tools required include a pressure gauge installed on the thermostat output (branch line), a squeeze (aspirator) bulb, pocket screwdriver, a ladder for reaching the mixing box above the ceiling, and tools to open and adjust the dampers inside the mixing box. Possible causes include thermostat out of calibration, a stuck damper, insufficient air flow, inappropriate air temperature, etc. A troubleshooting chart is used to determine the possible causes, symptoms, and corrective action.

MIXING BOX TROUBLESHOOTING CHART

Problem	Possible Causes	Suggestions	Corrective Actions
Building Space Too Cold	Thermostat out of calibration	Thermostat works at wrong temperature	Calibrate thermostat
	Damper stuck	Check movement	Free up or lube
	Air not hot enough	Check air handler unit temperatures	Fix air handler unit controls
	Not enough hot air	Volume damper/air balance	Adjust air balance
	Thermostat has wrong setpoint	Check setpoint dial	Change setpoint dial
Building Space Too Hot	Thermostat out of calibration	Thermostat works at wrong temperature	Calibrate thermostat
	Damper stuck	Check movement	Free up or lube
	Air not cold enough	Check air handler unit temperatures	Fix air handler unit controls
	Not enough cold air	Volume damper/air balance	Adjust air balance
	Thermostat has wrong setpoint	Check setpoint dial	Change setpoint dial
	Hole in diaphragm	Use squeeze bulb, pressure is low	Disassemble and replace

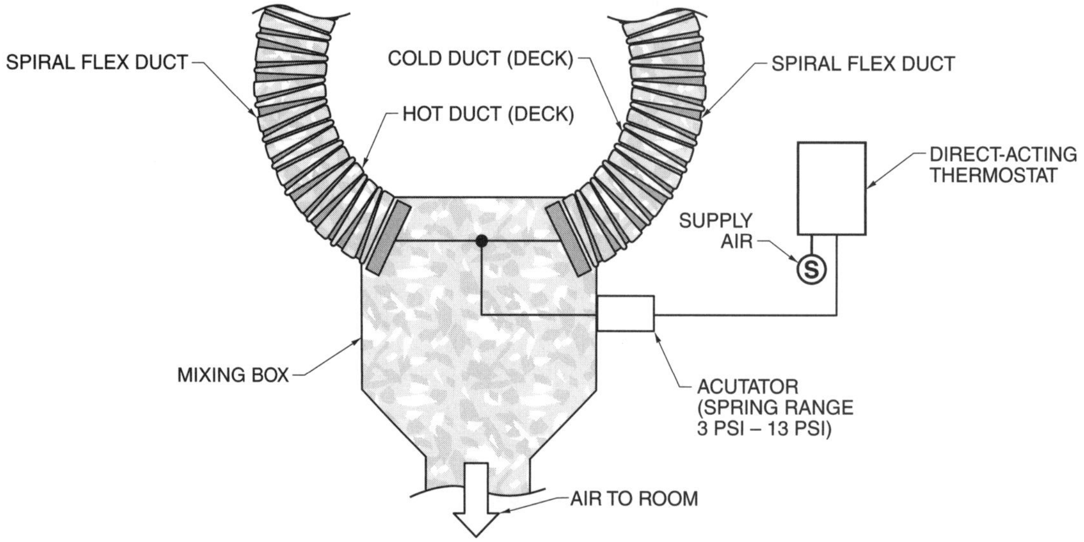

Figure 8-18. In a mixing box application, the thermostat drives a damper actuator that opens and closes a hot air or cold air inlet inside the box.

A building space that is excessively cold or hot may also occur in a finned-tube radiation steam heating system or a variable air volume (VAV) air cooling system. See Figure 8-19. In this application, a pneumatic-actuated steam valve controls flow through the valve and finned-tube radiator. A thermostatic steam trap causes steam condensate to flow back to the return system. The cooling is controlled by an air damper in a duct that changes position to provide various amounts of cool air flow into the building space for cooling. Possible causes of problems include thermostat being out of calibration or having a wrong setpoint, a damper or valve that is stuck, etc. A troubleshooting chart is used to determine the possible causes, symptoms, and corrective action.

The core of any pneumatic troubleshooting procedure is to install a pressure gauge on the thermostat output (branch line). The gauge reading gives an

DUAL-ACTUATOR SYSTEM TROUBLESHOOTING CHART

Problem	Possible Causes	Suggestions	Corrective Actions
Building Space Too Cold	Thermostat out of calibration	Thermostat works at wrong temperature	Calibrate thermostat
	Damper or valve stuck	Check movement	Free up or lube
	Thermostat has wrong setpoint	Check setpoint dial	Change setpoint dial
	No water flow	Cold coil	Check pumps
	Water not hot enough	Coil lukewarm	Check boiler
Building Space Too Hot	Thermostat out of calibration	Thermostat works at wrong temperature	Calibrate thermostat
	Thermostat has wrong setpoint	Check setpoint dial	Change setpoint dial
	Damper or valve stuck	Check movement	Free up or lube
	Hole in either diaphragm	Use squeeze bulb, pressure low	Disassemble and replace
	Air not cold enough	Check air handler unit temperatures	Fix air handler unit controls
	Not enough cold air	Check air flow rate	Air balance properly

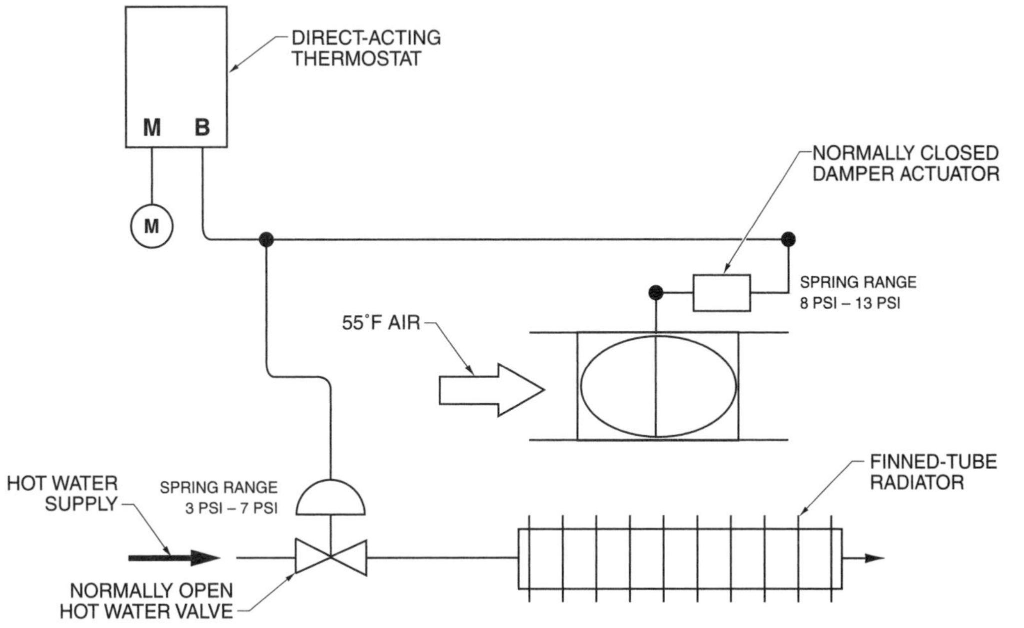

Figure 8-19. When troubleshooting an HVAC system with two actuators, both must be checked because a failure in either actuator may cause the system to operate improperly.

indication of the thermostat operation. Other tools required include a squeeze (aspirator) bulb, pocket screwdriver, a ladder to reach the VAV terminal damper, and tools to make mechanical adjustments to the damper assembly.

Do not overtighten a screw-in gauge adapter. Overtightening strips the threads of the adapter and/ or thermostat.

A thermostat can be out of calibration on either the high side or the low side. If an actuator has a hole in the diaphragm, the branch line pressure cannot build up properly and typically is at 3 lb or less.

If an actuator branch line is connected to more than one actuator, a hole in either diaphragm can cause the branch line pressure to be low. Both diaphragms must be checked. See Figure 8-20. Checking two diaphragms is accomplished by applying the procedure:

1. Remove the thermostat cover and the thermostat gauge plug.
2. Install the screw-in branch line pressure gauge in the thermostat. The reading should be 0 psig.
3. Use a squeeze bulb to pump up the normally closed VAV damper actuator.
4. Use a squeeze bulb to pump up the normally open hot water valve.

Note: Any mechanical device such as a damper or valve can stick in any position.

PNEUMATIC HUMIDISTATS

A *pneumatic humidistat* is a controller that uses compressed air to open or close a device which maintains a certain humidity level inside a duct or area. Pneumatic humidistats are used to sense and control the humidity in a building space or duct. Humidity in addition to temperature affects human comfort in a building. Discomfort may result if proper humidity levels are not maintained. Many products such as paper, textiles, and food production are also humidity-sensitive. Failure to maintain proper humidity levels in these environments may cause product damage, product loss, or production downtime.

Pneumatic Humidistat Operation

The construction of a pneumatic humidistat is similar to that of a thermostat, with the exception that the bimetallic element is replaced by a humidity-sensing (hygroscopic) element. See Figure 8-21. The hygroscopic element accepts moisture from, or rejects moisture to, the surrounding air. Pneumatic humidistats are available in both one-pipe and two-pipe versions as well as direct-acting or reverse-acting. Old humidistats used elements such as bi-wood or horsehair elements. New humidistats use manufactured elements such as cellulose acetate butyrate (CAB).

Pneumatic thermostats using manufactured elements are more accurate and consistent than the old models. Humidity sensing and control is not the same as temperature sensing and control. In general, all humidity

CHECKING TWO DIAPHRAGMS

① Remove thermostat cover and thermostat gauge plug.

② Install branch line pressure gauge.

③ Use squeeze bulb to pump up normally closed VAV damper actuator.

④ Use squeeze bulb to pump up normally open hot water valve.

Figure 8-20. Both diaphragms must be checked if an actuator branch line is connected to more than one actuator because a hole in either diaphragm can cause the branch line pressure to be low.

When checking for leaks, possible sources include the diaphragm, air line connections on the back of the thermostat baseplate, and the branch (output) line itself.

controllers and sensors have potential problems that temperature devices do not. These problems affect both pneumatic and electronic control systems and include dirt, corrosive gases, and drift. Dirt in duct systems and rooms coats the sensing element and reduces the ability of the material to accept moisture from, or reject moisture to, the ambient air. Some air handling units in applications such as hospitals and manufacturing plants circulate gases along with the air that can corrode and damage the sensing element. Virtually all humidity sensors change their operating characteristics over time (drift).

PNEUMATIC HUMIDISTATS

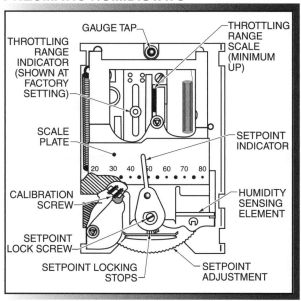

Figure 8-21. Pneumatic humidistats are used to sense and control the humidity in a building space or duct.

Humidistat Calibration and Troubleshooting

Humidistats are calibrated in exactly the same method as a standard single-temperature pilot bleed (two-pipe) thermostat. Humidistat calibration should be checked at least once a year. The difference between the calibration procedures is that the building space humidity is substituted in the procedure for temperature and an accurate humidity-sensing device must be used. With the advent of electronic sensing, a greater number of accurate humidity-sensing devices are available. Caution must be observed because any device calibrated to a false standard will not be accurate. The accuracy needed for human comfort is not as critical as that for industrial product applications. Industrial product ap-

plications may require a verification of the calibration-checking device by using another, certified device as a standard. Humidity control systems commonly use a second humidity controller as a limit in order to prevent the discharge of excessive humidity into a building space or duct.

Humidity that is excessively low or high may occur in a system that consists of a normally closed pneumatic-actuated steam valve that allows low-pressure steam to be metered into a duct. See Figure 8-22. The low-pressure steam increases the humidity of the air that flows into the room. The steam valve is controlled by a reverse-acting two-pipe humidistat. The core of a pneumatic troubleshooting procedure is to install a pressure gauge on the humidistat output (branch line). The gauge's reading gives an indication of the humidistat operation. Other tools required include a squeeze (aspirator) bulb and pocket screwdriver.

Almost all humidifier valves are normally closed in order to fail the humidifier closed. The location of the humidistat or limit controller affects system operation and could cause problems. Check with the manufacturer for recommendations. Calibration of humidity limit controllers is the same as for temperature limit controllers, except that the ambient humidity must be measured accurately instead of temperature.

PNEUMATIC PRESSURE SWITCHES

The third major variable that must be controlled in HVAC control systems is pressure inside a duct or building. A *pneumatic pressure switch* is a controller that maintains a constant air pressure in a duct or area. Pneumatic pressure switches are designed to sense and control the pressure inside a duct or building, which is measured in inches of water column (in. wc). See Figure 8-23. Many modern mechanical systems, such as variable air volume systems, rely on a specific air pressure inside a duct. The specific air pressure is used to supply terminal units, which open and close to maintain the temperature in the building space. Pressure switches may also be used to maintain duct air balance parameters and ensure that the internal building pressure is maintained at an acceptable level.

Pneumatic Pressure Switch Operation

Pressure switches are available in both bleed and pilot bleed (one- and two-pipe) as well as in direct-acting and reverse-acting models. The sensing element is normally a thin neoprene diaphragm that is connected to the duct or room by means of a copper or plastic tube.

HUMIDISTAT TROUBLESHOOTING CHART

Problem	Possible Causes	Suggestions	Corrective Actions
Humidity Too Low	Humidistat has wrong setpoint	Check setpoint dial	Change setpoint dial
	Humidifier out of calibration	Humidifier works at wrong humidity	Calibrate humidistat
	Valve stuck closed	Check movement	Free up or lube
	No steam or water for humidifier	Cold pipes	Check supply
	Hole in diaphragm	Use squeeze bulb, pressure is low	Disassemble and replace
Humidity Too High	Humidifier setpoint wrong	Check humidifier setpoint dial	Change dial
	Humidifier out of calibration	Humidifier works at wrong humidity	Calibrate humidifier
	Valve stuck open	Check movement	Free up or lube

HUMIDISTAT

M   M B

STEAM VALVE

STEAM SUPPLY

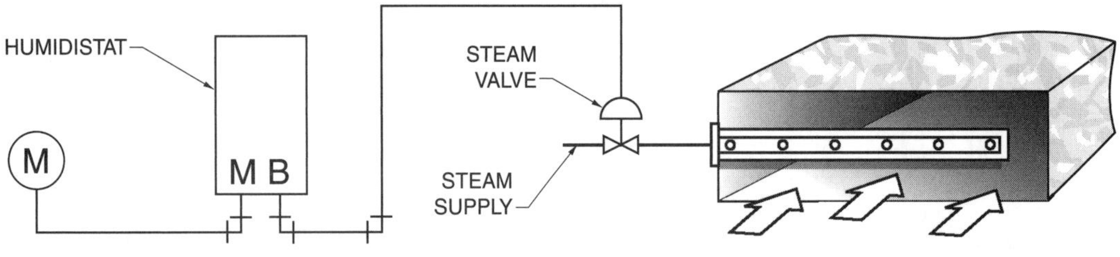

Figure 8-22. Humidity that is excessively low or high may occur in a system that consists of a normally closed pneumatic-actuated steam valve which allows low-pressure steam to be metered into a duct.

There are two connections on pressure switches that enable the pressure switch to measure a single pressure or the difference in pressure between ducts or areas. If the single-pressure system is used, the other connection is referenced to atmosphere. However, sophisticated pressure control is normally accomplished within receiver controllers and transmitters, not with simple pressure switches.

Pressure Switch Calibration and Troubleshooting

Pressure switches are calibrated using the same basic method as a thermostat or humidistat. A calibration pressure is piped into the pressure switch by a squeeze bulb or accurate pressure regulator. The output pressure adjustment is then set to the midpoint of the actuator spring range. Pressure switch calibration should be checked once a year.

Use accurate calibration equipment. Faulty, inaccurate equipment leads to inaccurate calibration. Pressure switches are often location-dependent. Check manufacturer literature for proper location. Check for leaky tubing and fittings. The pressure sensed will be incorrect if the pressure tubing is kinked or leaky.

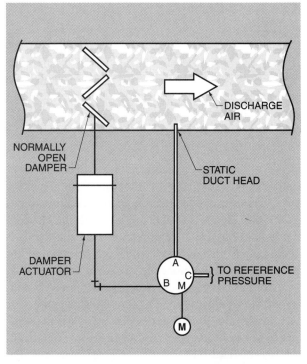

Figure 8-23. Pneumatic pressure switches are designed to sense and control the pressure inside a duct or building.

Pneumatic transmitters sense temperature, pressure, or humidity and output a proportional 3 psig to 15 psig pressure signal to a receiver controller. Temperature transmitters include room, rod-and-tube, inert-element, bulb-type, and averaging element transmitters. Humidity transmitters include room and duct transmitters. Pressure transmitters include duct and pipe transmitters.

PNEUMATIC TRANSMITTERS

A *pneumatic transmitter* is a device that senses temperature, pressure, or humidity and sends a proportional (3 psig to 15 psig) signal to a receiver controller. See Figure 9-1. Pneumatic transmitters provide the sensing function for HVAC air handling systems, boilers, chillers, and cooling towers. Pneumatic transmitters do not provide the system control. System control is provided by receiver controllers.

Pneumatic Transmitter Characteristics

Pneumatic transmitters are standardized from manufacturer to manufacturer. In most applications, a transmitter from one manufacturer may be replaced with a transmitter from another manufacturer.

Pneumatic transmitters may be one-pipe or two-pipe devices. A *one-pipe (low-volume) device* is a device that uses a small amount of the compressed air supply (restricted main air). The compressed air passes through an orifice (restrictor) that meters the air flow to the transmitter. Pneumatic transmitters used in building HVAC applications are all one-pipe (low-volume) devices. One-pipe devices have only one air connection to the transmitter. A *two-pipe (high-volume) device* is a device that uses the full volume of compressed air available. Two-pipe devices have separate main air and output air line connections to the transmitter.

Two-pipe (high-volume) transmitters are used in applications that require long piping runs and are rarely used in standard building HVAC applications.

Pneumatic transmitters are normally set up as direct-acting and only rarely as reverse-acting. All pneumatic transmitters have a linear output. The 3 psig to 15 psig pressure output of a pneumatic transmitter appears as a straight line between the endpoints of the pressure range. See Figure 9-2. This is the standard for all pneumatic transmitters and influences calibration and troubleshooting. This standard simplifies installation, calibration, and troubleshooting when compared to electronic sensor standards which can be 0 VDC to 10 VDC, 0 VDC to 20 VDC, or 4 mA to 20 mA.

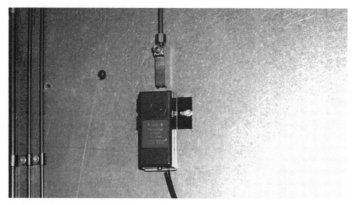

A pneumatic transmitter is often mounted directly to a duct, and its element is inserted into the air stream.

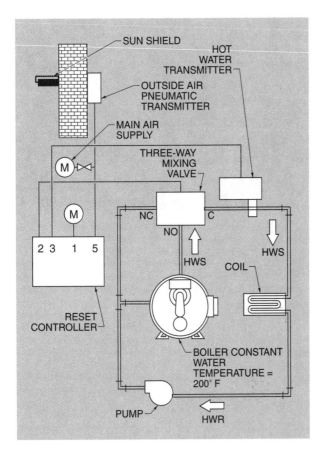

Figure 9-1. Pneumatic transmitters are used to sense various HVAC system conditions such as hot water temperature and outside air temperature.

When applied properly, pneumatic transmitters are relatively accurate. In addition, pneumatic transmitters are not designed to be calibrated in the field. In most cases, a transmitter is replaced if it is not working properly.

Pneumatic transmitters influence the calibration of receiver controllers because the output signals from the transmitters are normally piped to receiver controllers. For example, because nearly all pneumatic transmitters output a pressure between 3 psig and 15 psig, the receiver controllers can be calibrated using an air regulator that simulates these values.

Most pneumatic transmitters require a restricted main air supply obtained from inside the controller (internally restricted) or from a restrictor in the main air supply line (externally restricted). To determine internal or external restriction, the tube from the transmitter to the controller is removed. If air bleeds from the controller, it is internally restricted. If air bleeds from the tube, the controller is externally restricted.

Transmitter Range and Span. All pneumatic transmitters have a range and span. *Transmitter range* is the temperatures between which a transmitter is capable of sensing. All transmitters are designed with a sensing range between two temperatures, humidities, or pressures. Common transmitter temperature ranges include −40°F to 160°F, 0°F to 100°F, 0°F to 200°F, and 40°F to 240°F. Common transmitter humidity ranges include 0% rh to 100% rh, 25% rh to 95% rh, and 30% rh to 80% rh. Common transmitter pressure ranges include 0″ wc to 2″ wc, 0″ wc to 4″ wc, and .5″ wc to 1.5″ wc. These ranges are selected based on the desired application. The listed ranges should not be exceeded in normal use.

Transmitter span is the difference between the minimum and maximum sensing capability of a transmitter (number of units between the endpoints of the transmitter range). Transmitter span is found by subtracting the listed numbers for the range. For example, a transmitter with a range of 0°F to 200°F has a span of 200°F (200°F − 0°F = 200°F). Two transmitters with different ranges may have the same span. For example, a transmitter with a range of 40°F to 240°F also has a span of 200°F (240°F − 40°F = 200°F). These transmitters are not interchangeable unless the receiver controllers are recalibrated.

Transmitter Sensitivity. *Transmitter sensitivity* is the output pressure change that occurs per unit of measured variable change. Pneumatic transmitters respond predictably to a change in the variable they sense. When the variable changes, the output pressure changes. See Figure 9-3. Transmitter sensitivity is used in the calibration and setup formulae of receiver controllers. Transmitter sensitivity is found by dividing 12 psig by the transmitter span. The value of 12 psig is a constant derived from the transmitter pressure range of 3 psig to 15 psig. All 200°F span transmitters have a sensitivity of .06 psig/°F. All 100°F span transmitters have a sensitivity of .12 psig/°F. Leaks and piping problems must be avoided because of these small pressure changes. A 1 psig pressure drop can represent an error of approximately 19°F for a −40°F to 160°F transmitter.

Transmitter Receiver Gauges. In HVAC control systems, the current temperature, pressure, or humidity must be known. Transmitter receiver gauges are used to indicate current temperature, pressure, or humidity values. Transmitter receiver gauges are often mounted in the door of temperature control panels. See Figure 9-4. Transmitter receiver gauges are 3 psig to 15 psig gauges that are marked with the range of the appropriate transmitter. A standard gauge can be used for all transmitters because of the standard transmitter pressure range. For example, if an outside air temperature transmitter has a range of – 40°F to 160°F, then – 40°F corresponds to 3 psig and 160°F corresponds to 15 psig. Therefore, a transmitter receiver gauge can be piped to the transmitter line and used to observe system temperatures, pressures, and humidities. Transmitter receiver gauges are used for troubleshooting, system analysis, and calibration.

Temperature Transmitters

The majority of pneumatic control applications in buildings involve the sensing and control of system temperatures. Temperature transmitters measure the temperature inside rooms, ducts, and pipes. To meet this need, a wide variety of temperature-sensing elements are available. Temperature transmitter element types include room, rod-and-tube, inert-element, bulb-type, and averaging element transmitters.

Room Temperature Transmitters. A *room temperature transmitter* is a transmitter used in applications that require a receiver controller to measure and control the temperature in an area. Room temperature transmitters are not commonly used because a pneumatic thermostat can perform both sensing and control functions in a less expensive, less complex device. Room temperature transmitters are used in areas such as laboratories and product storage areas that require precise sensing and control. A room temperature transmitter is similar in appearance to a pneumatic thermostat. Room temperature transmitters and pneumatic thermostats both use a bimetallic strip for sensing. Room temperature transmitters, however, do not have a setpoint dial or adjustment. In addition, room temperature transmitters normally do not have an output pressure adjustment.

> *Transmitters are usually selected so that the variable they sense falls near the middle of their range. For example, mixed air applications have temperatures near 55°F, thus requiring a transmitter with a range of 0°F to 100°F.*

TEMPERATURE AND RELATIVE HUMIDITY vs TRANSMITTER OUTPUT PRESSURE

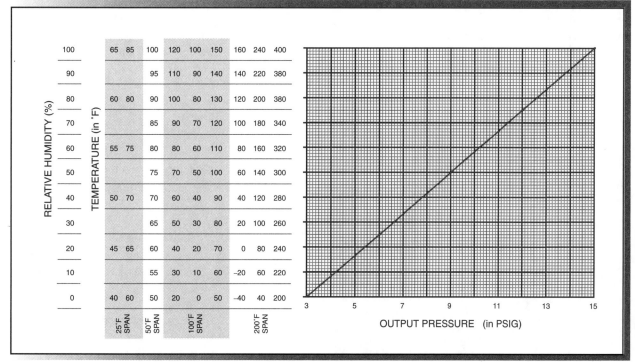

Figure 9-2. Pneumatic transmitters provide a linear 3 psig to 15 psig output.

TRANSMITTER CHARACTERISTICS

Temperature Range*	Span*	Sensitivity†
40 – 65 60 – 85	25	.48
50 – 100	50	.24
0 – 100 20 – 120 50 – 150	100	.12
40 – 240 –40 – 160 200 – 400	200	.06

* in °F
† in PSI/°F

Figure 9-3. Transmitter sensitivity is the output pressure change that occurs per unit of measured variable change.

Rod-and-Tube Temperature Transmitters. A *rod-and-tube temperature transmitter* is a pneumatic transmitter that uses a high-quality metal rod with precision expansion and contraction characteristics as the sensing element. See Figure 9-5. As the rod expands and contracts due to temperature changes, a strain gauge connection operates a flapper/nozzle assembly, which regulates the air pressure. A *strain gauge* is a spring or thin piece of metal that measures movement of a sensing element. The rod is enclosed in a brass tube which is inserted into the medium (air or water) being sensed. Rod-and-tube temperature transmitters may have short or long rods.

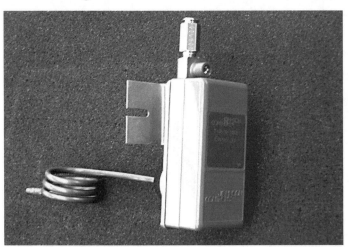

Bulb-type pneumatic transmitters use a capillary tube and bulb filled with a liquid or gas to sense temperature.

Figure 9-4. Multi-span transmitter receiver gauges are used for indication and calibration purposes.

Short-rod transmitters have sensing elements approximately 4″ long and are used to sense water temperatures in hot water, chilled water, or cooling tower systems. The sensing element is inserted into a well, which is installed in a pipe. See Figure 9-6. A thermal compound is used around the sensing element to ensure proper heat transfer. Proper location of short-rod transmitters is required for accurate temperature measurement. Short-rod transmitters are installed according to manufacturer recommendations.

Long-rod transmitters have sensing elements 6″ to 12″ long and are used to sense air temperature in air handling systems. Long-rod transmitters are located and installed based on manufacturer recommendations. Long-rod transmitters should not be installed in dead air spaces or locations that are affected by radiant or unwanted heat gains or losses.

> *Air leaks must be minimized because a transmitter with a sensitivity of .06 psig/°F indicates a temperature drop of 16°F with a 1 psig pressure drop.*

ROD-AND-TUBE TEMPERATURE TRANSMITTERS

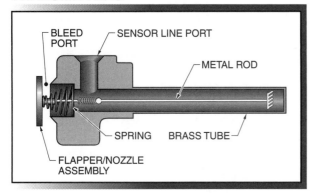

Figure 9-5. Rod-and-tube temperature transmitters use the expansion of a metal rod inside a tube to measure temperature changes.

Inert-Element Temperature Transmitters. An *inert-element temperature transmitter* is a pneumatic transmitter that is used to sense outside air temperatures in through-the-wall applications. See Figure 9-7. An inert-element transmitter has a long-rod transmitter section that is connected to a section of metal that has virtually a zero rate of expansion and contraction. The inert section acts as a spacer that enables the sensing rod to sense the outside air temperature without sensing the intermediate wall temperature. The inert section does not sense air temperature.

SHORT-ROD TRANSMITTERS

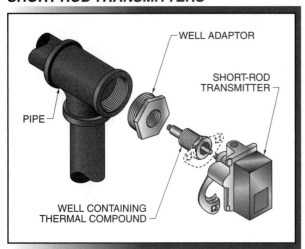

Figure 9-6. Short-rod transmitters are used to sense hot or chilled water temperature in HVAC piping systems.

INERT-ELEMENT TEMPERATURE TRANSMITTERS

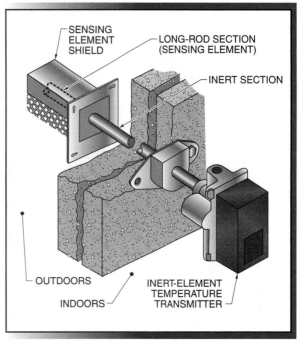

Figure 9-7. Inert-element transmitters are used to sense outside air temperatures.

Bulb-Type Temperature Transmitters. A *bulb-type temperature transmitter* is a pneumatic transmitter that uses a capillary tube and bulb filled with a liquid or gas to sense temperature. Bulb-type transmitters are used to sense water or air temperatures. The capillary tube is normally 1′ to 5′ long. The bulb senses the temperature and the capillary tube transmits the fluid force to a flapper/nozzle assembly. Manufacturer recommendations should be followed regarding location of the bulb in the desired medium.

Averaging Element Temperature Transmitters. An *averaging element temperature transmitter* is a pneumatic transmitter that uses a long tube filled with a liquid or gas to sense duct temperature. See Figure 9-8. The sensing element of averaging element transmitters consists of a capillary tube without a bulb. The sensing element may be 10′ to 25′ long. The sensing element is strung across the inside of a duct in a zigzag pattern. Care should be taken to prevent kinking or damage to the capillary tube. The temperature sensed is the average of the temperatures sensed across the entire capillary tube length. A spot check with a thermometer may indicate an improper reading because a thermometer checks the temperature at only one point in a duct, while an averaging element transmitter checks the temperature across a duct simultaneously.

AVERAGING ELEMENT TEMPERATURE TRANSMITTERS

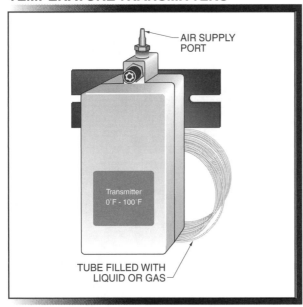

AIR SUPPLY PORT

Transmitter
0°F - 100°F

TUBE FILLED WITH LIQUID OR GAS

Figure 9-8. Averaging element transmitters use a long tube filled with a liquid or gas to sense duct temperature.

Averaging element transmitters are used in large duct (8′ D and greater) air handling applications. Long-rod transmitters lose their effectiveness because they can sense only a small part of the total duct airflow. An averaging element transmitter should be used if the duct diameter (long measurement for a rectangular duct) is more than four times the length of the rod element.

Pneumatic transmitters are normally piped to an input connection on a receiver controller.

> ⓘ *Exercise caution when installing an averaging element transmitter. A kink in the element may damage or destroy it. The element may be attached to a rigid frame such as electrical conduit for protection.*

Averaging element transmitters are used in mixed air, hot deck, cold deck, preheat, precool, and discharge air handling system applications. A variety of averaging element temperature transmitters are available based on sensing range, element length, and mounting configuration. See Figure 9-9.

AVERAGING ELEMENT TRANSMITTERS

Order No.	Sensing Range (Non-Adjustable)*	Element Length[†]	Mounting
MA825B 1003	–40 to 160	15	Duct
MA825B 1011	–40 to 160	27[‡]	Wall
MA825B 1029	40 to 240	15	Well
MA825B 1037	–40 to 160	15	Well
MA825B 1045	–40 to 160	7	Duct
MA825B 1052	40 to 240	7	Well
MA825B 1060	–40 to 160	7	Well
MA825B 1110	–20 to 80	15	Well
MA825B 1243	–20 to 80	15	Duct
MA825B 1250	–20 to 80	27[‡]	Wall
MA825B 1268	40 to 240	15	Duct
Order No.	Sensing Range (Non-Adjustable)*	Element Length[§]	Mounting
MA826B 1044	0 to 200	18½	Duct
MA826B 1051	0 to 200	8⅞	Duct
MA826B 1077	25 to 125	18½	Duct
MA826B 1085	25 to 125	8⅞	Duct

* in °F
[†] in in.
[‡] Active element 15″, inert section 12″
[§] in ft

Figure 9-9. Averaging element transmitters vary based on sensing range, element length, and mounting configuration.

Humidity Transmitters

A *pneumatic humidity transmitter* is a device that measures the amount of moisture in the air compared to the amount of moisture the air could hold if it were saturated. Humidity transmitters are used in different HVAC applications and contain the same manufactured sensing element as a humidistat. See Figure 9-10. The moisture-sensing (hygroscopic) element absorbs moisture from the air or releases moisture to the air. The moisture-sensing element is attached to a flapper/nozzle assembly that causes a 3 psig to 15 psig pressure signal in the air line. Humidity transmitters are affected by drift, corrosion, and dirt. Also, they are inaccurate at humidity extremes such as 0% to 10% rh and 90% to 100% rh. At these extremes, high-quality electronic sensors should be used. Humidity transmitters include room and duct humidity transmitters.

HUMIDITY TRANSMITTERS

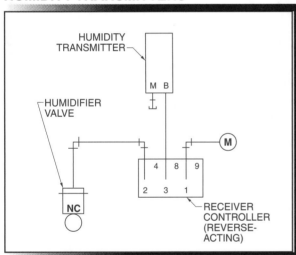

Figure 9-10. Humidity transmitters sense the moisture content of the air in a room or duct.

Room Humidity Transmitters. Room humidity transmitters are used to sense the relative humidity in a building space (room) and send a 3 psig to 15 psig signal to a receiver controller. The receiver controller provides the control setpoint and other functions. Room humidity transmitters are similar in appearance to humidistats. Room humidity transmitters do not have setpoint or other adjustments.

Duct Humidity Transmitters. Many air handling units require that humidity be sensed in a duct. Duct humidity transmitters are used to sense the relative humidity in a duct. Duct humidity transmitters are similar to room

humidity transmitters, with identical sensing elements and basic configuration. Many manufacturers allow the conversion of a room humidity transmitter to a duct humidity transmitter by adding a duct sampling tube kit. A duct sampling tube kit allows the air from the air stream to be forced across the humidity-sensing element.

> ⓘ *The accuracy of room humidity transmitters should be checked with an accurate test instrument. The test instrument may require checking against a consistent standard before use.*

Accurate humidity sensing depends on proper transmitter location and distance downstream from a heating coil, cooling coil, or humidifier. Also, good air mixing should be accomplished prior to the air passing over the transmitter. No humidity transmitters are able to sense average humidity levels. If averaging is desired, multiple humidity transmitters must be installed in the duct and their readings averaged.

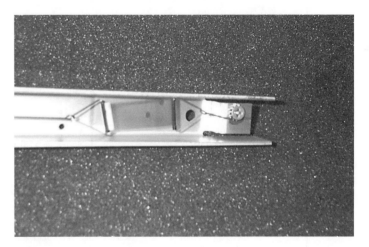

Humidity transmitters contain a moisture-sensing element that absorbs or releases moisture as the humidity level changes.

Pressure Transmitters

A *pressure transmitter* is a device used to sense the pressure due to air flow in a duct or water flow through a pipe. Pressure transmitters use sensing elements made of neoprene for air and a thin metal diaphragm or bellows for water. The sensing element is connected to a flapper/nozzle assembly through a strain gauge connection that causes the pressure in the air line to vary between 3 psig and 15 psig. Pressure transmitters include duct and pipe pressure transmitters.

Duct Pressure Transmitters. A *duct pressure transmitter* is a device mounted in a duct that senses the static pressure due to air movement. See Figure 9-11. A pitot tube is often used to provide proper air pressure sensing. A *pitot tube* is a device that senses static pressure and total pressure in a duct.

Many pressure-sensing and control problems are caused by improper transmitter location. A duct pressure transmitter should be installed at the proper location for good pressure control. For example, many variable air volume (VAV) control systems place the transmitter two-thirds of the distance of the longest run of ductwork after the fan. If mounted above the ceiling panels, a pressure transmitter can be used to sense the pressure inside a room as well.

Pipe Pressure Transmitters. Pipe pressure transmitters are used in HVAC piping systems to sense the pressure in water distribution systems. Pipe pressure transmitters use a metal diaphragm or bellows to sense the water pressure. Pipe pressure transmitters send a 3 psig to 15 psig pressure to a receiver controller which performs system control. As with duct pressure transmitters, proper location is required. All manufacturer recommendations must be followed regarding location and installation.

DUCT PRESSURE TRANSMITTERS

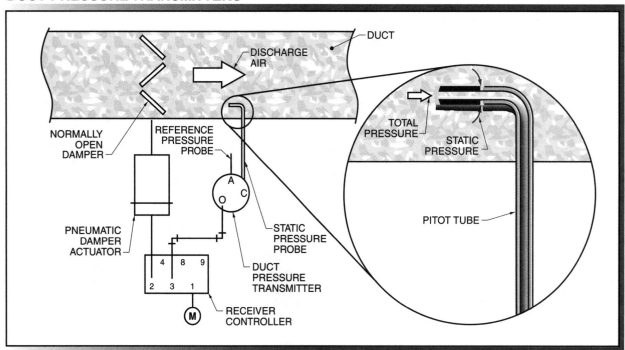

Figure 9-11. Duct pressure transmitters sense static pressure in a duct.

A receiver controller accepts one or more input signals from pneumatic transmitters and produces an output signal based on its calibration. A single-input receiver controller is designed to be connected to one transmitter and have one setpoint. A dual-input receiver controller changes its setpoint based on the change of another variable. Integration is a function that seeks to eliminate the remaining offset in a system by moving the controlled variable to setpoint.

RECEIVER CONTROLLERS

Thermostats, humidistats, and pressure switches (room controls) control the environment inside building spaces. Additional controls are required to achieve precise control of building central heating, ventilation, and air conditioning (HVAC) systems. HVAC systems include air handling units, boilers, heat exchangers, chillers, and cooling towers. Thermostats, humidistats, and pressure switches cannot provide the flexibility and accuracy needed to control these devices. A *receiver controller* is a device which accepts one or more input signals from pneumatic transmitters and produces an output signal based on its calibration. See Figure 10-1. Receiver controllers are used for HVAC system control. Receiver controllers are normally not used to control the temperature, humidity, and pressure in one room or zone of a commercial building. Receiver controllers replaced remote bulb controllers in the late 1960s.

Remote Bulb Controllers

Before the advent of receiver controllers, system control was performed by remote bulb controllers. A *remote bulb controller* is a device that consists of a controller mounted to a duct or pipe and that is connected by a capillary tube to a bulb that is inserted into the duct or pipe. See Figure 10-2. The outputs of remote bulb controllers were connected together to form sophisticated control sequences. Remote bulb controllers had many disadvantages. For example, the short length of the capillary tube meant that the controllers had to be mounted close to the duct or pipe. This prevented remote bulb controllers from being mounted in central, accessible control panels, making adjustment and servicing difficult.

Remote bulb controllers were also limited by the difficulty of performing ever more sophisticated control sequences. Remote bulb controllers could not perform the precise control demanded in energy-conscious modern commercial buildings. In addition, remote bulb controllers could not be used interchangeably with different temperature-, humidity-, and pressure-control systems.

> *Remote bulb controllers can be easily replaced by modern receiver controllers. The receiver controller replaces the remote bulb controller and a pneumatic transmitter replaces the remote bulb element. Supply air and output lines are attached to the new receiver controller and the new receiver controller is calibrated.*

RECEIVER CONTROLLERS

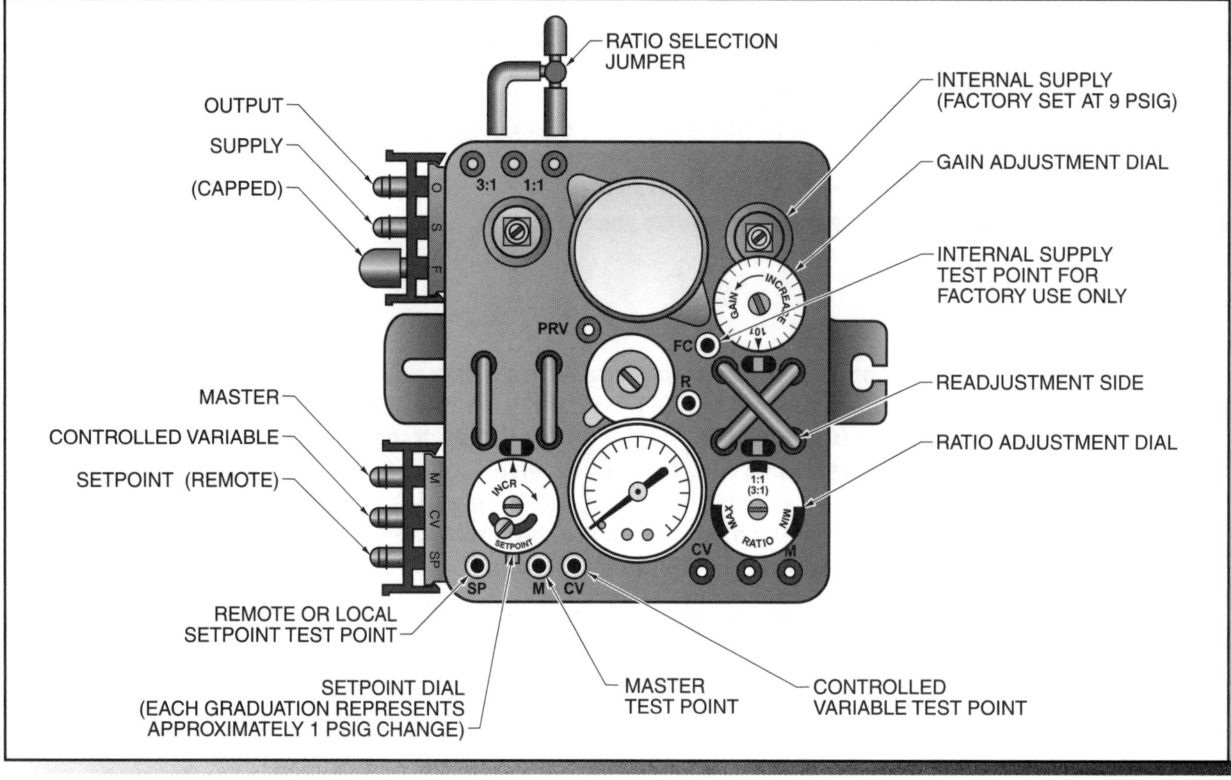

Figure 10-1. A receiver controller accepts one or more input signals from pneumatic transmitters and produces an output signal based on its calibration.

REMOTE BULB CONTROLLERS

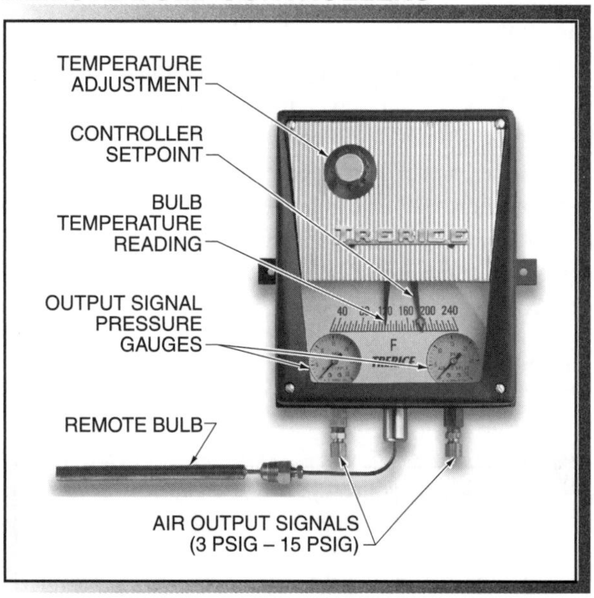

Figure 10-2. A remote bulb controller consists of a controller mounted to a duct or pipe that is connected by a capillary tube to a bulb that is inserted into the duct or pipe.

Receiver Controller Operation

Receiver controllers are available in a variety of different designs. Most receiver controllers use a force-balance design. A *force-balance design* is a design in which the controller output is determined by the relationship of mechanical pressures. The force-balance design is similar to a lever, with calibration and adjustments serving to change the fulcrum point in the middle of the lever. See Figure 10-3. Some receiver controllers use stacked diaphragms in place of the lever arrangement. The stacked diaphragm arrangement is considered to be the most reliable and durable design. The disadvantage of the stacked diaphragm design is that the moving parts may deteriorate and have increased friction over time, causing the controller to go out of calibration.

In the past, most receiver controllers were manufactured using metal. Today, most receiver controllers are manufactured using high-impact plastic.

FORCE-BALANCE RECEIVER CONTROLLERS

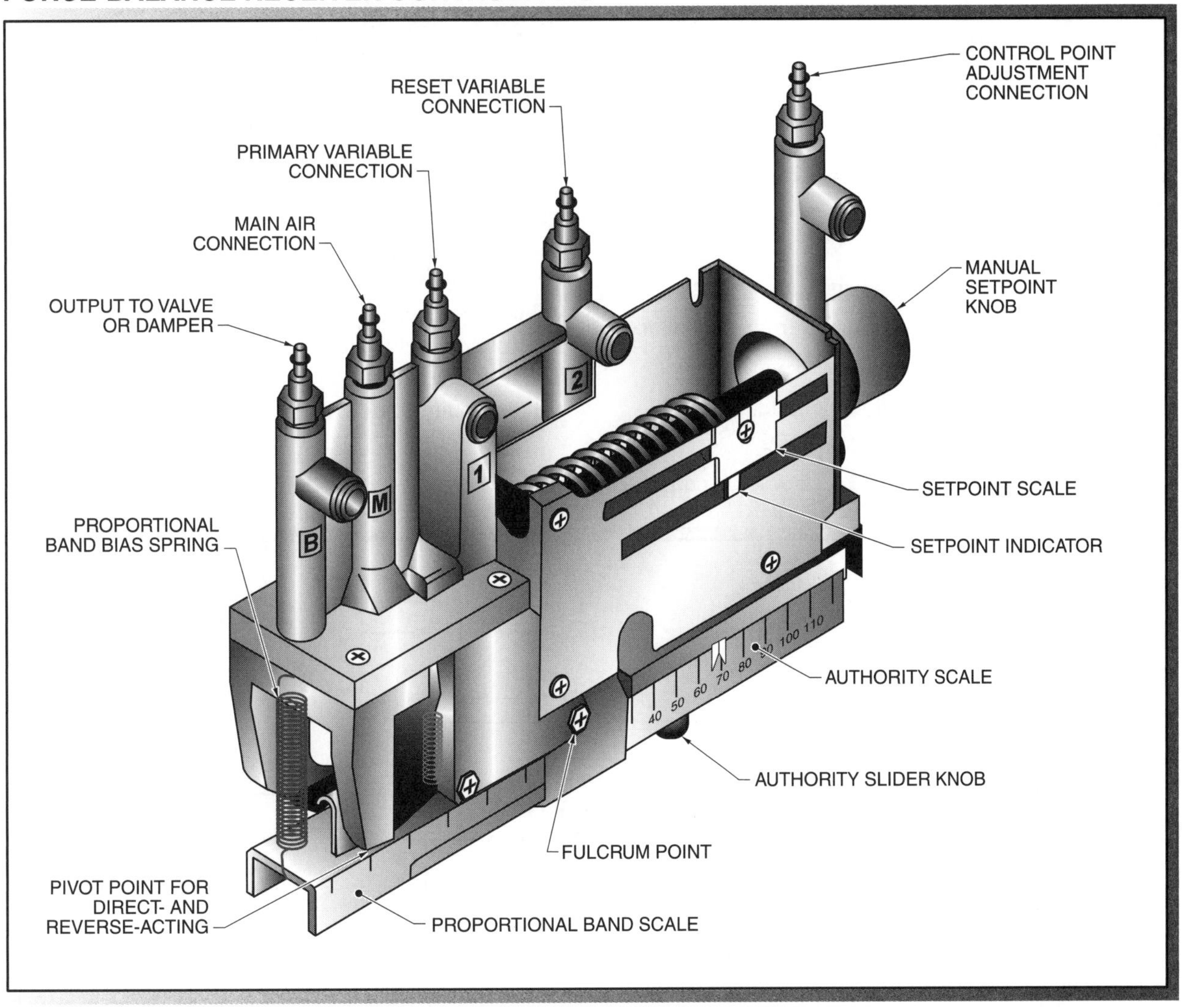

Figure 10-3. Force-balance receiver controllers have calibration and adjustments that change the fulcrum point of the lever.

Another receiver controller design used in the 1970s and 1980s was the fluidic controller. A *fluidic controller* is a controller that uses vector analysis to arrive at an output. *Vector analysis* is the calculation of outputs based on the direction and strength of forces. Fluidic controllers mathematically relate the different air pressures, setpoints, influences, etc., to arrive at an output. See Figure 10-4. An advantage of fluidic controllers is that they do not use moving parts. Disadvantages of fluidic controllers are that they are difficult to calibrate and are susceptible to water and oil in the air supply. In some cases, fluidic controllers go out of calibration when the main air is shut off overnight.

The disadvantages of fluidic controllers have made them unpopular and they were discontinued by the major manufacturers in the late 1980s. All receiver controllers sold today by the major manufacturers use the force-balance or stacked diaphragm design.

RECEIVER CONTROLLERS AND TRANSMITTERS

Receiver controllers use pneumatic transmitters to provide system control. A *transmitter* is a device that sends an air pressure signal to a receiver controller. The air pressure indicates a specific temperature, pressure, or humidity to the receiver controller. The receiver controller controls the central HVAC systems according to its calibration and adjustments.

Pneumatic transmitters sense a temperature, pressure, or humidity and send a proportional, 3 psig to 15 psig signal to a receiver controller. The general characteristics of pneumatic transmitters include the following:

• Most are direct-acting.

• All produce a 3 psig to 15 psig output signal.

• They are accurate when shipped from the factory.

• Most are one-pipe, low-volume devices.

• Most cannot be calibrated in the field.

Transmitter Restriction

Pneumatic transmitters are one-pipe (low-volume) devices. For this reason, they are similar to one-pipe thermostats, humidistats, and limitstats. All one-pipe pneumatic transmitters must have one restricted main air supply. A *restrictor* is a fixed orifice that is measured by size (inches diameter) or by the amount of air that can pass through it in one minute (standard cubic inches per minute). The transmitter bleeds off this restricted air to a pressure between 3 psig and 15 psig. This low volume of air makes the transmitter susceptible to small leaks at poorly attached air fittings, cracked tubing, etc. The restricted main air supply can be obtained by internal restriction or external restriction.

FLUIDIC CONTROLLERS

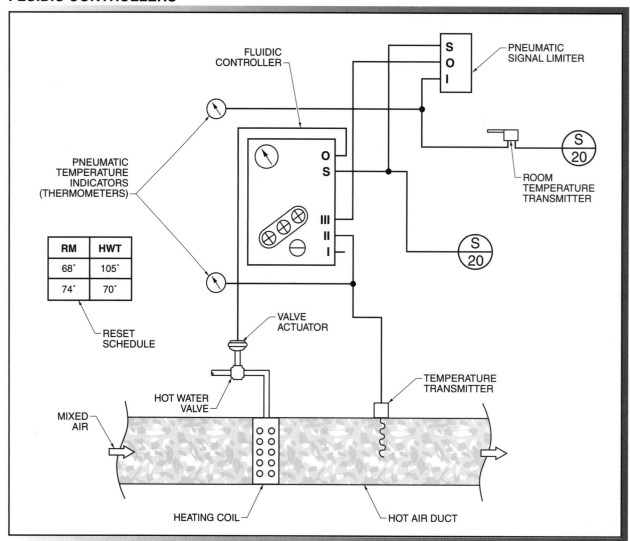

Figure 10-4. A fluidic controller uses vector analysis to arrive at an output.

Internal Restriction. An *internal restriction* is a restriction in which the receiver controller has an internally mounted restrictor. The restrictor can provide main air at the proper volume for the transmitter if needed. The restrictor taps into the main air supply that is piped into the controller. The majority of manufacturers provide this capability. The technician must decide to use the internal restrictor and turn it ON by a switch, dial indicator, or tubing jumper. Different manufacturers turn ON the internal restrictor in different manners. If this option is used at a transmitter port, the technician must be able to recognize this on a system print. An internal restriction on a receiver controller is represented on system prints by no symbol, a circle with a dot in the center, or a bow tie. See Figure 10-5. Each transmitter can have only one restricted main air supply. The transmitter is not able to bleed off the increased volume of air and the transmitter reads excessively high if a transmitter has more than one restricted main air supply.

This restrictor must be of the proper size and volume for the transmitter. Manufacturer literature should be checked for proper application. Many manufacturers use a color-coding system for restrictors. For example, red indicates a diameter of .007″.

The use of internal or external restriction is the choice of the design engineer or technician, but the transmitter can have only one main air supply. The internal restriction must be disconnected (shut OFF) if an external restriction main air supply is used. This is done by a switch or tubing jumper at the receiver controller. A method of recognizing this is required on a system print. Symbols used include a box around the transmitter port, a solid circle, or a bow tie with a line in the center. See Figure 10-6. All these symbols indicate that the internal restricted main air supply is disconnected.

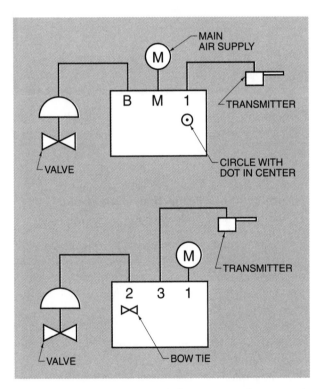

Figure 10-5. An internal restriction on a receiver controller is represented by no symbol, a circle with a dot in the center, or a bow tie.

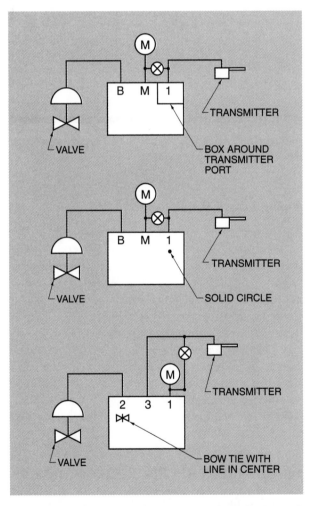

Figure 10-6. Symbols used to indicate that the internal restricted main air supply is shut OFF include a box around the transmitter port, a solid circle, or a bow tie with a line in the center.

External Restriction. An *external restriction* is a restriction in which a restrictor is placed outside the receiver controller. A separate restrictor is connected between the main air input and the transmitter air line.

Shared Transmitters

Pneumatic transmitters can be shared between multiple receiver controllers. Shared transmitters are common with outside air transmitters that are used in many control sequences. Only one receiver controller can have its internal restricted main air supply used if this arrangement is desired. All other internal restricted main air supplies must be disconnected. See Figure 10-7.

SHARED TRANSMITTERS

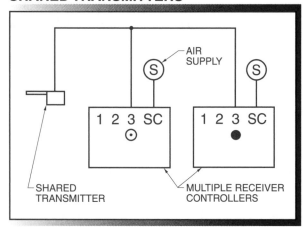

Figure 10-7. Pneumatic transmitters may be shared between multiple receiver controllers.

Piping Length Requirements

Under normal circumstances, the pressure drops due to long piping length to the transmitter do not affect transmitter indication. However, if extremely long piping runs are used (over 250′), an undue pressure drop may occur that affects the transmitter indication. In some cases, a repeater (signal booster relay) may be required to prevent the pressure drop. Check with the manufacturer concerning distance requirements and corrective measures. A signal booster relay can be installed if a problem is suspected. The signal booster relay does not adversely affect the system and may correct low transmitter indication due to long piping runs.

SINGLE-INPUT RECEIVER CONTROLLERS

A *single-input receiver controller* is a receiver controller that is designed to be connected to only one transmitter and to maintain only one temperature, pressure, or humidity setpoint. See Figure 10-8.

SINGLE-INPUT RECEIVER CONTROLLERS

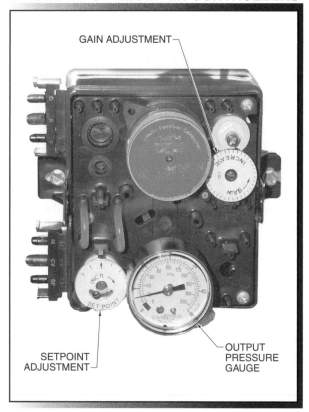

Figure 10-8. A single-input receiver controller is designed to be connected to only one transmitter and maintain only one temperature, pressure, or humidity setpoint.

Throttling Range/Proportional Band/Gain Adjustments and Calculations

In order to properly set up and calibrate a single-input receiver controller, the type of pneumatic transmitter, whether the controller is to be direct- or reverse-acting, and the quality of control that is desired must be known. Most often, this information is derived from a pneumatic print. In most cases, the best system control possible without hunting or cycling is desired. This means that the system should be very close to the setpoint without the damper or valve slamming open and closed. *Throttling range* is the number of units of controlled variable that causes the actuator to move through its entire spring range. Throttling range is the number of degrees that the temperature must change in order for the damper to go from full open to closed. See Figure 10-9. Precise control requires a small throttling range. A throttling range that is too small causes excessive actuator hunting or cycling.

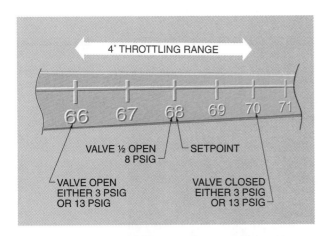

Figure 10-9. Throttling range is the number of units of controlled variable that causes the actuator to move through its entire spring range.

Many receiver controllers do not have a throttling range adjustment. Receiver controllers without a throttling range adjustment have a proportional band setting. *Proportional band* is the number of units of controlled variable that causes the actuator to move through its entire spring range. Proportional band is expressed as a percentage of the transmitter span.

Throttling range and proportional band mean the same thing but are expressed in different units of measure. For example, in an air handling unit mixed air control application, a mixed air temperature of 55°F, ± 2.5°F must be maintained. The total throttling range is 5°F. In this application, the transmitter range of 0°F to 100°F has a span of 100°F. A throttling range of 5°F represents 5% of the transmitter's range (5% proportional band). If the transmitter in the same application had a span of 200°F, the proportional band needed to obtain a throttling range of 5°F would be 2.5% (5°F ÷ 200°F = 2.5%). The relationship between throttling range, transmitter span, and proportional band can be given as follows: throttling range equals transmitter span times proportional band percentage, transmitter span equals throttling range divided by proportional band percentage, and proportional band equals throttling range divided by transmitter span. See Appendix.

In a dual-duct air handling unit application, the hot deck controller maintains a temperature of 140°F. A throttling range of 10°F (± 5°F) is desired. The transmitter span is 200°F from the listed range of 0°F to 200°F. The proportional band setting needed to obtain a 10°F throttling range is 5% (10°F ÷ 200°F = 5%).

Gain is the mathematical relationship between the controller output pressure change and the transmitter

pressure change that causes it. For example, a 5°F change on a 0°F to 100°F transmitter causes the controller output pressure to the actuator to change from 3 psig to 9 psig. In this example, the transmitter pressure changes .6 psig (5°F × .12 psig/°F = .6 psig). This .6 psig transmitter change causes a 6 psig pressure change to occur at the controller output. This relationship can be expressed as a 6 psig/.6 psig amplification (a gain of 10). Whole numbers such as 5%, 10%, and 20% proportional band are commonly used.

In some cases, a very small throttling range is desired. For example, a very small throttling range is desired (approximately 2°F to 4°F) if a receiver controller is used to switch a switching relay. This application demands a low proportional band. A low proportional band value normally causes the system to hunt. In this application, however, the receiver controller is used in a two-position (ON/OFF) control. This receiver controller is only used to switch other parts of the system operation. Whatever terminology is used, the desired throttling range, proportional band, or gain must be adjusted on the controller as part of the calibration procedure.

Single-Input Receiver Controller Calibration Procedure

Single-input receiver controllers should be calibrated a minimum of once per year. Single-input receiver controllers may be calibrated with or without a calibration simulator kit. A *calibration simulator kit* is a portable precision pressure-regulating kit with precise adjustment capabilities and gauges. Calibration simulator kits vary in size, shape, and cost. See Figure 10-10.

Single-input receiver controllers are adjusted by moving the proportional band slider to the appropriate vertical graduation that indicates the correct value.

CALIBRATION SIMULATOR KIT

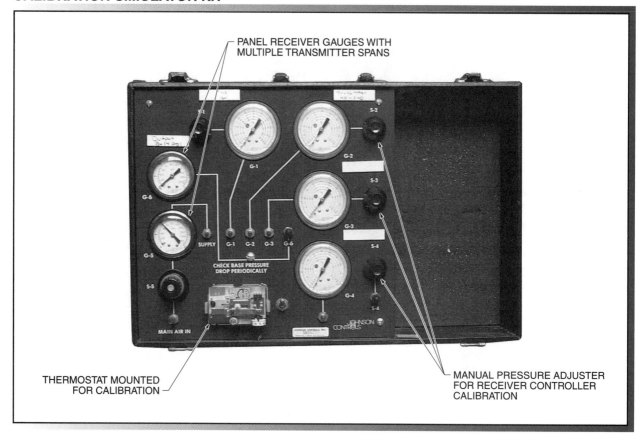

Figure 10-10. A calibration simulator kit is a portable precision pressure-regulating kit with precise adjustment capabilities and gauges.

The transmitter air line is removed temporarily and the simulator is connected to the transmitter port. The technician then adjusts the pressure-reducing valves. This simulates the presence of a transmitter and enables the controller to be calibrated or checked. See Figure 10-11. A single-input receiver controller is calibrated using a calibration simulator kit and applying the procedure:

1. Determine the transmitter range, desired setpoint, throttling range, gain (proportional band), and actuator spring range.
2. Disconnect the transmitter air line and connect the simulator to the transmitter port.
3. Enter the desired throttling range (proportional band) on the controller.
4. Adjust the simulator pressure-regulating valve to the desired setpoint on the receiver gauge.
5. Adjust the output pressure of the controller to the midpoint of the actuator spring range.

6. Check the throttling range and gain (proportional band) adjustment by adjusting the pressure-reducing valve to ± limits. Readjust throttling range and gain (proportional band) if necessary, then recheck.
7. Remove the simulator, reattach the air lines, and check the actual operation.

Parts of the HVAC system may have to be shut down temporarily to prevent overheating or overcooling of the building space due to the controller being out of service. Hand valves may be used to shut down specific affected heating or cooling coils. Exercise caution when removing the existing transmitter air line. Mark the transmitter air line appropriately so that it can be located for reconnection when the calibration is complete. Many temperature control cabinets have previously used air lines hanging loose in the cabinet. If the air lines are not marked, the wrong air line may be reattached, causing control system failure.

SINGLE-INPUT RECEIVER CONTROLLER CALIBRATION

(1) Determine transmitter range, setpoint, throttling range, gain, and actuator spring range.

(2) Disconnect transmitter air line and connect simulator to transmitter port.

(3) Enter desired throttling range and gain on controller.

(4) Adjust simulator pressure-regulating valve to desired setpoint.

(5) Adjust output pressure of controller to midpoint of actuator spring range.

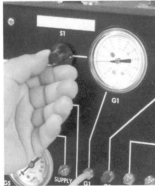

(6) Check throttling range and gain adjustment. Readjust if necessary.

(7) Remove simulator, reattach air lines, and check actual operation.

Figure 10-11. A calibration simulator kit allows the technician to adjust the pressure-reducing valves to simulate the presence of a transmitter and enable the controller to be calibrated or checked.

A single-input receiver controller is calibrated without a simulator kit by applying the procedure:

1. Determine the desired setpoint, actuator spring range, and desired throttling range.
2. Adjust the throttling range and gain (proportional band) to the desired level.
3. Accurately measure the actual temperature at the transmitter.
4. Loosen the setpoint dial and change the dial to the actual temperature.
5. Tighten the dial and move it to the desired temperature.

Calibration of a single-input receiver controller without a simulator kit is not as precise as when using a simulator. The simulator method allows the technician to check throttling range and controller operation more completely.

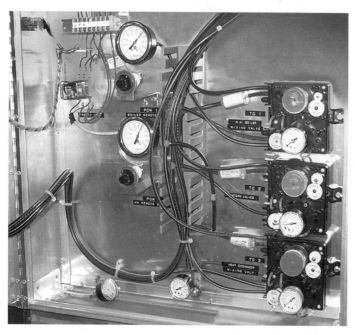

Pneumatic receiver controllers are usually mounted inside control panels along with receiver gauges and other control equipment. Accurate labeling of the control equipment is required for efficient calibration and troubleshooting.

Single-Input Receiver Controller Applications

Many air handling units and central plant systems use single-input receiver controllers to maintain specific desired setpoints in the building mechanical systems. Air handling units may use single-input receiver controllers for mixed air, hot deck, cold deck, or discharge air applications, depending on the mechanical system. Central plant hot water and chilled water systems typically use single-input receiver controllers for maintaining a specific constant setpoint.

Mixed Air Control. Most air handling units mix outside air and return air to obtain a specific mixed air temperature. This mixed air is used for cooling when the system is in the economizer mode. For example, a controller is set for direct acting, has a 55°F setpoint, and a 2% proportional band. See Figure 10-12. The duct mounted averaging element transmitter has a 0°F to 200°F range. The throttling range is 4°F (2% × 200°F span = 4°F). As the mixed air temperature drops below the 55°F setpoint, the transmitter output pressure drops as well (direct acting). The transmitter output pressure at setpoint is the midpoint of the actuator spring range, or 8 psig for a 3 psig to 13 psig setpoint. As the mixed air temperature rises, the output pressure rises, allowing the outside air damper to open, admitting more outside air. *Note:* This example does not represent a complete air handling unit control system.

> ⓘ
> *To check whether a receiver controller is internally or externally restricted, remove the air tubing attached to the transmitter port of the receiver controller. Air flowing out of the receiver controller indicates that the receiver controller is internally restricted. Air flowing out of the tubing indicates that the receiver controller is externally restricted.*

Hot Deck Control. Warm air is delivered to building spaces to provide heat for the building. The correct temperature of the warm air must be maintained when delivered to the building. In this application, a single-input receiver controller, a duct-mounted temperature transmitter, and a hot water valve are used to maintain the correct air temperature.

Dual-duct and multizone air handling units have a split duct configuration. One of the ducts contains hot air, the other cold air. For example, a single-input controller is set up to be direct acting, has a 120°F setpoint, and a normally open hot deck valve spring range of 3 psig to 6 psig. See Figure 10-13. If a throttling range of 10°F is desired, the desired gain must be found. A 10°F throttling range means a transmitter pressure change of .6 psig (10°F × .06 psig/°F transmitter sensitivity = .6 psig). This transmitter change causes an output change of 3 psig (6 psig − 3 psig). Three pounds per square inch gauge divided by .6 psig equals a controller gain of 5. *Note:* This example does not represent a complete air handling unit control system.

MIXED AIR CONTROL

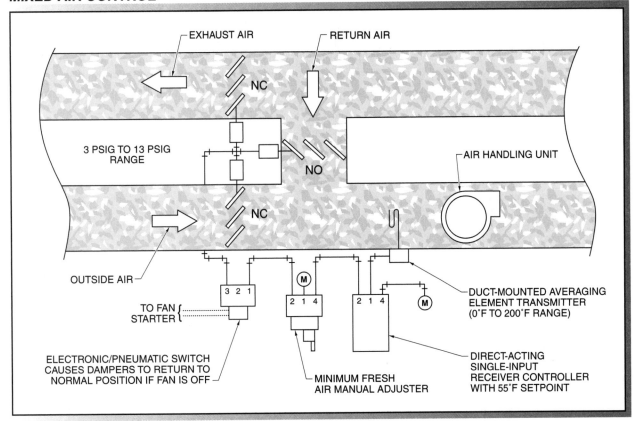

Figure 10-12. Most air handling units mix outside air and return air to obtain a specific mixed air temperature.

Hot Water Control. Many commercial buildings use central hot water heating systems to heat the building. For example, a receiver controller is set up to be direct acting, has a 140°F setpoint, and controls a normally open steam valve with a spring range of 3 psig to 13 psig. See Figure 10-14. The transmitter is mounted in a well in the hot water supply line. The transmitter range is 40°F to 240°F. The proportional band is set at 5%, making the throttling range 10°F (5% × 200°F transmitter span = 10°F). The steam valve opens and admits steam to the heat exchanger, which transfers heat to the water pumped within the tubes. As the water temperature drops, the controller causes the valve to open, admitting steam and increasing the temperature of the water. As the water temperature increases, the controller causes the valve to close, reducing the steam flow into the heat exchanger and causing the water temperature to drop. The symbol under the number 1 port indicates that internal restriction is used. *Note:* This example does not represent a complete hot water control system.

HOT DECK CONTROL

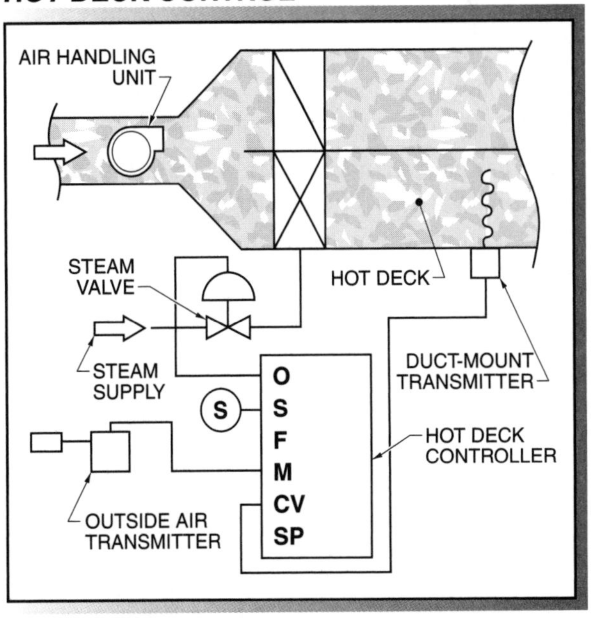

Figure 10-13. Dual-duct and multizone air handling units have a split duct configuration.

HOT WATER CONTROL

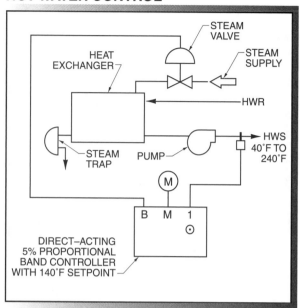

Figure 10-14. Many commercial buildings use central hot water heating systems to heat a building.

Troubleshooting Single-Input Receiver Controller Systems

Single-input receiver controllers are often used to control vital mechanical systems in a building. If one of these controllers operates improperly, it can lead to lack of comfort, reduced productivity, damaged products, or building damage. The tools needed for troubleshooting single-input receiver controllers include a simulator, squeeze bulb, pneumatic print, and accurate thermometer for temperature control applications.

Receiver Controller Troubleshooting – Heating Water Supply Too Hot. Many commercial buildings pump hot water through the building in order to heat the building spaces. In this application, the building temperature is too hot due to a problem in the hot water mechanical or control system. Troubleshooting a single-input receiver controller when the heating water supply is too hot is accomplished by applying the procedure:

1. Check the water temperature with the thermometer at the outlet of the heat exchanger. Open the strainer outlet to run water on the thermometer. The water temperature should be within the desired setpoint and throttling range.
2. Calibrate the controller if the water temperature is too high. The controller calibration should be checked whenever any problem occurs.

3. Check the valve operation with a squeeze bulb if the controller is calibrated properly. The valve may be wiredrawn. Wiredrawing of a steam valve occurs when the valve is open a slight amount for a long time. The small opening between the valve disk and seat acts like a nozzle, increasing steam velocity and eroding the valve parts. The valve seats on the downstream side of the valve will have a small, clean V-shaped area of material eaten away. The appearance is as if a wire saw was pulled (drawn) across the valve seat.
4. Check the transmitter indicated value with a thermometer if the water temperature is still too high. The transmitter may need to be replaced if the transmitter indicated temperature and the thermometer reading are different.
5. Check the controller to ensure that the action is correct. The controller forces the valve wide open if the controller is reverse acting instead of direct acting.

Receiver Controller Troubleshooting – Mixed Air Too Cold. Commercial air handling units bring in fresh outside air and mix it with return air. This mixture is then passed across the heating and cooling coils and into the building space. If the temperature of this mixture is too cold, it may cause the building occupants to be excessively cold. In some cases, the cold mixture can cause a coil downstream to freeze and break, leading to extensive building damage. Troubleshooting a single-input receiver controller when the mixed air is too cold is accomplished by applying the procedure:

1. Check the mixed air temperature with a thermometer in the duct. The air temperature should be within the desired setpoint and throttling range.
2. Calibrate the controller if the air temperature is too low. The controller calibration should be checked whenever any problem occurs.
3. Check the damper operation with a squeeze bulb if the controller is calibrated properly. The damper may be stuck open mechanically, admitting too much cold outside air.
4. Check the dampers for proper maintenance and damper blade edge seals. Cold outside air may be admitted if the damper blade edge seals are missing.
5. Check the transmitter indicated value with a thermometer if the air temperature is still too low. The transmitter may need to be replaced if the transmitter indicated temperature and the thermometer reading are different.

6. Check the controller to ensure that the action is correct. The controller forces the dampers open to outside air if the controller is reverse acting instead of direct acting.

Receiver Controller Troubleshooting – Hot Deck Air Temperature Too Hot. Air handling units such as dual duct and multizone units have a split duct design. One duct carries cold air to the building space, the other hot air. If the hot deck temperature is too low, the building may be too cold. If the hot deck temperature is too high, the occupants may complain of being too warm. A hot deck air temperature that is too high results in higher than required energy expenditures. Troubleshooting a single-input receiver controller when the hot deck air temperature is too hot is accomplished by applying the procedure:

1. Check the hot deck air temperature with a thermometer in the duct. The air temperature should be within the desired setpoint and throttling range.
2. Calibrate the controller if the hot deck air temperature is too high. The controller calibration should be checked whenever any problem occurs.
3. Check the valve operation with a squeeze bulb if the controller is calibrated properly. The valve may be stuck open mechanically or wiredrawn, admitting too much steam or hot water.
4. Check the transmitter indicated value with a thermometer if the air temperature is still too high. The transmitter may need to be replaced if the transmitter indicated temperature and the thermometer reading are different.
5. Check the controller to ensure that the action is correct. The controller forces the valve wide open if the controller is reverse acting instead of direct acting.

DUAL-INPUT RECEIVER CONTROLLERS

Single-input receiver controllers are designed to maintain one setpoint. This setpoint can be changed only by manually adjusting the setpoint dial or in some rare cases by using remote control point adjustment. In the majority of applications, the controller maintains only one setpoint year-round. This is acceptable for system mixed air and discharge air, but not desirable for hot deck and heating hot water supply. Comfort is enhanced and energy saved if the setpoint can be automatically changed to match different conditions.

In many systems, the setpoint of the receiver controller must be changed automatically as another measured variable changes. This is usually done for comfort and energy conservation purposes. For example, a hot water heating system must change its setpoint as the outside air temperature changes. If the outside air temperature decreases, the hot water setpoint must increase in order to compensate for the increased heat loss through the building structure. As the outside air temperature increases, the hot water setpoint is lowered, saving energy and money and increasing comfort.

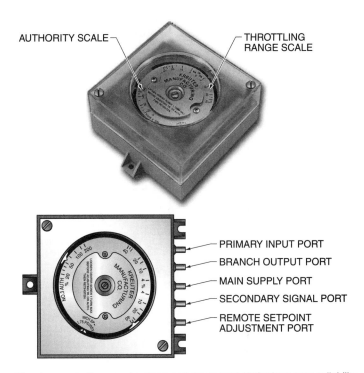

Receiver controllers require clean and dry control air for long-term reliability and performance.

A *dual-input (reset) receiver controller* is a receiver controller in which the change of one variable, commonly outside air temperature, causes the setpoint of the controller to automatically change (reset) to match the changing condition. One transmitter measures the variable that is being controlled, such as hot water. The other transmitter measures the variable that causes the controlled variable to change, such as outside air. The controller setpoint change occurs automatically as the outside air temperature changes. Additional applications that may use dual-input controllers include the highest zone temperature automatically changing the cold deck setpoint, the lowest zone temperature automatically changing the hot deck setpoint, and the chilled water return temperature automatically changing the chilled water supply setpoint.

Reset Schedules

A *reset schedule* is a chart that describes the setpoint changes in a pneumatic control system. A reset schedule is usually listed on the pneumatic system print. A reset schedule consists of the extremes of the setpoint change listed for each variable. A reset schedule can be shown in tabular or a graphical form. See Figure 10-15. These forms can be copied and taped inside the temperature control panel for reference by the technician. The graph enables the technician to easily reference the hot water setpoint at each corresponding outside air temperature, thus simplifying troubleshooting.

RESET SCHEDULES

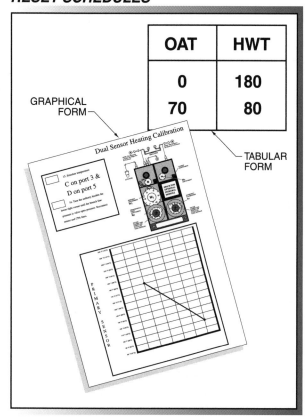

OAT	HWT
0	180
70	80

GRAPHICAL FORM

TABULAR FORM

Figure 10-15. A reset schedule describes the setpoint changes in a system and consists of the extremes of the setpoint change listed for each variable.

Authority and Ratio Adjustments and Calculations

If a dual-input receiver controller is to perform the reset schedule, there must be a method of taking the information from the reset schedule and setting up the controller to execute it. This is performed by determin-

ing the authority of the receiver controller. *Authority* is the relationship of the primary variable change to the secondary variable change, expressed as a percentage. The primary variable is the HVAC variable, such as hot water, hot deck, cold deck, etc. The secondary variable is the reset variable (variable that causes the primary variable to change). The most common secondary variable is outside air temperature. The technician must calculate the authority and enter it into the controller as part of the calibration procedure. Different manufacturers use different authority formulas. See Figure 10-16.

Due to the lack of throttling range, the non-throttling range formula produces a slightly lower authority than the throttling range authority for the same application. A particular formula must be used with a particular manufacturer. The throttling range formula is used with Honeywell controllers. The non-throttling range formula is used with Landis and Gyr Powers controllers. The controller does not function properly if the wrong authority formula is used.

Some manufacturers use a ratio for controller adjustment. Ratio is expressed as a decimal. The ratio formula is used with Johnson Controls controllers. The ratio can be derived from the non-throttling range authority by eliminating the multiplication factor of 100. Slide rules are available from manufacturers to help simplify this task.

> *Authority calculations may be used repeatedly in different parts of a building. For example, four separate hot water controllers for different zones may have the same reset schedules and therefore, the authority calculations are the same and do not need to be recalculated.*

Readjustment

Readjustment is the relationship between the primary and secondary (reset) variables from the schedule. *Reverse readjustment (winter reset)* is a schedule in which the primary variable increases as the secondary variable decreases, and decreases as the secondary variable increases. *Direct readjustment (summer reset)* is a schedule in which the primary variable increases as the secondary variable increases, and decreases as the secondary variable decreases. Either readjustment can be easily derived by reviewing the schedule. For example, a schedule indicates that the hot water temperature increases as the outside air temperature decreases. This schedule indicates reverse readjustment.

AUTHORITY/RATIO CALCULATION

Throttling Range Formula	Non-Throttling Range Formula	Ratio Formula
$A = [(\Delta T_1 \times S_1 + TR \times S_1) \div (\Delta T_2 \times S_2)] \times 100$ where: A = authority (in %) ΔT_1 = change in temperature at primary variable from schedule (in °F) S_1 = sensor one variable change (in psig/°F) TR = throttling range of controller ΔT_2 = change in temperature at secondary (reset) variable from schedule (in °F) S_2 = sensor two variable change (in psig/°F) 100 = constant	$A = [(\Delta T_1 \times S_1) \div (\Delta T_2 \times S_2)] \times 100$ where: A = authority (in %) ΔT_1 = change in temperature at primary variable from schedule (in °F) S_1 = sensor one variable change (in psig/°F) ΔT_2 = change in temperature at secondary (reset) variable from schedule (in °F) S_2 = sensor two variable change (in psig/°F) 100 = constant	$R = (\Delta T_1 \times S_1) \div (\Delta T_2 \times S_2)$ where: R = ratio ΔT_1 = change in temperature at primary variable from schedule (in °F) S_1 = sensor one variable change (in psig/°F) ΔT_2 = change in temperature at secondary (reset) variable from schedule (in °F) S_2 = sensor two variable change (in psig/°F)

Authority/Ratio Example:

Outside air transmitter range −40°F – 160°F
Hot water transmitter range 40°F – 240°F

Schedule:

Outside air temperature 0°F – 60°F
Hot water temperature 180°F – 80°F
Proportional band 5%

$A = [(\Delta T_1 \times S_1 + TR \times S_1) \div (\Delta T_2 \times S_2)] \times 100$ $A = [(100°F \times .06\ psig/°F + 10°F \times .06\ psig/°F)$ $\quad \div (60°F \times .06\ psig/°F)] \times 100$ $A = [(6 + .6) \div 3.6] \times 100$ $A = (6.6 \div 3.6) \times 100$ $A = \textbf{183%}$	$A = [(\Delta T_1 \times S_1) \div (\Delta T_2 \times S_2)] \times 100$ $A = [(100°F \times .06\ psig/°F)$ $\quad \div (60°F \times .06\ psig/°F)] \times 100$ $A = (6 \div 3.6) \times 100$ $A = \textbf{166%}$	$R = (\Delta T_1 \times S_1) \div (\Delta T_2 \times S_2)$ $R = (100°F \times .06\ psig/°F) \div (60°F \times .06\ psig/°F)$ $R = 6 \div 3.6$ $R = \textbf{1.66}$

Authority/Ratio Example:

Outside air transmitter range −40°F – 160°F
Hot deck transmitter range 0°F – 200°F

Schedule:

Outside air temperature 0°F – 70°F
Hot deck temperature range 120°F – 70°F
Proportional band 5%

$A = [(\Delta T_1 \times S_1 + TR \times S_1) \div (\Delta T_2 \times S_2)] \times 100$ $A = [(50°F \times .06\ psig/°F + 10°F \times .06\ psig/°F)$ $\quad \div (70°F \times .06\ psig/°F)] \times 100$ $A = [(3 + .6) \div 4.2] \times 100$ $A = (3.6 \div 4.2) \times 100$ $A = \textbf{84%}$	$A = \Delta T_1 \times S_1 \div \Delta T_2 \times S_2 \times 100$ $A = [(50°F \times .06\ psig/°F)$ $\quad \div (70°F \times .06\ psig/°F)] \times 100$ $A = (3 \div 4.2) \times 100$ $A = \textbf{70 %}$	$R = (\Delta T_1 \times S_1) \div (\Delta T_2 \times S_2)$ $R = (50°F \times .06\ psig/°F) \div (70°F \times .06\ psig/°F)$ $R = 3 \div 4.2$ $R = \textbf{.70}$

Figure 10-16. Authority is the relationship of the primary variable change to the secondary variable change, expressed as a percentage.

Old receiver controllers were only reverse readjustment models. Modern receiver controllers can be switched between reverse and direct readjustment.

Direct readjustment and reverse readjustment are different than direct acting and reverse acting. Direct acting and reverse acting describe the relationship between the primary variable and the output pressure. Direct readjustment and reverse readjustment refer to the relationship between the secondary (reset) and primary variables on the reset schedule. Different combinations are available, with a controller being either direct acting/reverse readjustment, reverse acting/reverse readjustment, direct acting/direct readjustment, or reverse acting/direct readjustment. The combination used should be indicated on the controller, with direct acting/reverse readjustment being the most common.

Dual-Input Receiver Controller Calibration Procedure

All receiver controllers should be recalibrated a minimum of once per year. Normal controller wear and readjustment by operating personnel take their toll during the year. Comfort, energy usage, and utility costs are affected if calibration is ignored. Some receiver controllers that are not accessible to operating personnel have operated properly and kept their calibration for years. Always follow the manufacturer's specific calibration procedure as listed in their literature. Also, dual-input receiver controllers, unlike single-input receiver controllers, should not be calibrated without a calibration simulator kit. The controller complexity requires the use of a calibration simulator kit.

A dual-input receiver controller is calibrated by applying the procedure:

1. Determine the transmitter ranges (actuator spring range, desired proportional band, throttling range, or gain), schedule, and desired authority/ratio.

2. Remove the transmitter air lines and connect the simulator to the transmitter ports on the controller.

3. Adjust the controller to the proper proportional band, throttling range, or gain. Adjust the controller to the calculated gain or authority/ratio.

4. Adjust the simulator to the midpoints of the schedule for the primary and secondary (reset) inputs.

5. Adjust the output pressure to the midpoint of the actuator spring range.

6. Check the proportional band, gain, or throttling range settings by adjusting the simulator through the calculated throttling range and observing the controller operation. Adjust if needed.

7. Adjust the simulator to one set of endpoints of the schedule. The receiver controller output pressure should return to the midpoint of the actuator spring range if the calibration is proper.

8. Adjust the simulator to the opposite endpoints of the schedule. Check whether the controller output pressure returns to the midpoint of the actuator spring range. If it does, the calibration is proper. Adjust the authority slightly and return to step 7 if the controller does not return to the midpoint of the actuator spring range.

9. Remove the air lines from the simulator and reattach the transmitter air lines if the calibration is proper.

10. Wait for controller to stabilize, then check for proper operation.

The technician should ensure that the transmitter lines are marked properly when they are removed. If not marked properly, the transmitter lines may be misidentified when replaced and cause improper controller operation.

The midpoint of the schedule is halfway between the listed temperatures on the schedule. For example, if the schedule is listed as a 0°F to 60°F outside air temperature change causing a hot water setpoint change of 180°F to 60°F, the schedule midpoints are 30°F outside air and 120°F hot water, respectively.

The technician must be careful to adjust the primary and secondary inputs when checking the authority setting. The inputs operate as logical pairs. The technician must also be careful not to misidentify the primary and reset inputs when adjusting the simulator dials. If possible, wait for the control system to stabilize when checking for proper operation. Check the actual variable with an accurate thermometer or other instrument.

Dual-Input Receiver Controller Applications

Many air handling units and central plant systems use dual-input receiver controllers to automatically change setpoints in the building mechanical systems. Air handling units may use dual-input receiver controllers for cold deck, hot deck, or discharge air applications, depending on the mechanical system. Central plant hot water and chilled water systems typically use a dual-input receiver controller to save energy and increase comfort by changing setpoints when outside air temperature or load conditions change.

Cold Deck Reset from Warmest Room. The amount of cooling delivered by an air handling system should match the highest demand for cooling by a zone in the building. If a simple reset based on outside air temperature is performed, energy may be wasted due to overcooling, or comfort may be compromised by not having sufficient cooling available. One solution is to automatically change the amount of cooling delivered based on the actual amount demanded by the highest zone. See Figure 10-17.

Dual-duct and multizone air handling systems have a split duct configuration. One duct contains hot air and the other duct contains cold air. The duct-mounted averaging element transmitter has a 25°F to 125°F range. The throttling range is therefore 5°F (5% × 100°F = 5°F). The reset transmitter is the output of a high multiple signal selector. This selector is connected to the output signals of multiple room thermostats. The highest signal represents the zone with the greatest demand for cooling. This high demand zone resets the cold deck temperature in the duct. The reset schedule is as follows: If the high zone signal is 13 psig, the cold deck setpoint is 55°F. If the high zone signal is 5 psig, the

Zone thermostats are commonly used to reset heating and cooling systems. The zone thermostats must be calibrated for proper control of the heating or cooling system.

cold deck setpoint is 65°F. The output pressure at setpoint is the midpoint of the actuator spring range, or 8 psig for a 3 psig to 13 psig setpoint. As the mixed air temperature rises, the output pressure rises, allowing the outside air damper to open and admitting more outside air. *Note:* This example does not represent a complete air handling unit control system.

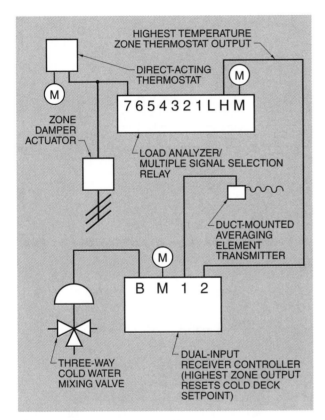

Figure 10-17. The amount of cooling delivered by an air handling system should match the highest demand for cooling by a zone in the building.

Hot Deck Reset Control. As the temperature outside a building increases, heat losses due to the temperature difference decrease. If a constant setpoint is used for the hot deck, the system is set for the worst-case scenario, which is maximum demand for heat. This can lead to energy waste and possible lack of comfort. One solution is to automatically change the amount of hot air delivered based on the outside air temperature. See Figure 10-18. Dual-duct and multizone air handling units have a split duct configuration. One duct contains hot air and the other duct contains cold air. In this application, the dual-input receiver controller is set for direct acting/reverse readjustment. The desired schedule is as

follows: as the outside air temperature varies from 10°F to 65°F, the hot deck temperature varies from 130°F to 80°F. The normally open hot deck valve has a spring range of 3 psig to 6 psig. If a throttling range of 10°F is desired, the desired gain must be calculated. A 10°F throttling range represents a transmitter pressure change of 10°F × .06 psig/°F transmitter sensitivity. This means that the transmitter pressure change required is .6 psig (10°F × .06 psig = .6 psig). This transmitter change causes an output change of 3 psig (6 psig – 3 psig = 3 psig). Three pounds per square inch gauge divided by .6 psig equals a controller gain of 5. Using the ratio formula, the required ratio is 1.0. *Note:* This example indicates only the hot deck part of an air handling unit control system.

HOT DECK RESET CONTROL

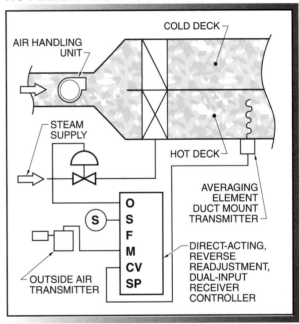

Figure 10-18. Hot deck reset control automatically changes the amount of cooling delivered based on the outside air temperature.

Outside Air Reset of Hot Water Control. Many commercial buildings use central hot water heating systems to heat the building. The central hot water heating system may be controlled by a receiver controller set up to be reverse acting/reverse readjustment controlling a normally closed steam valve with a spring range of 3 psig to 13 psig. See Figure 10-19. The controller schedule is as follows: As the outside air temperature varies from 0°F to 60°F, the hot water setpoint varies

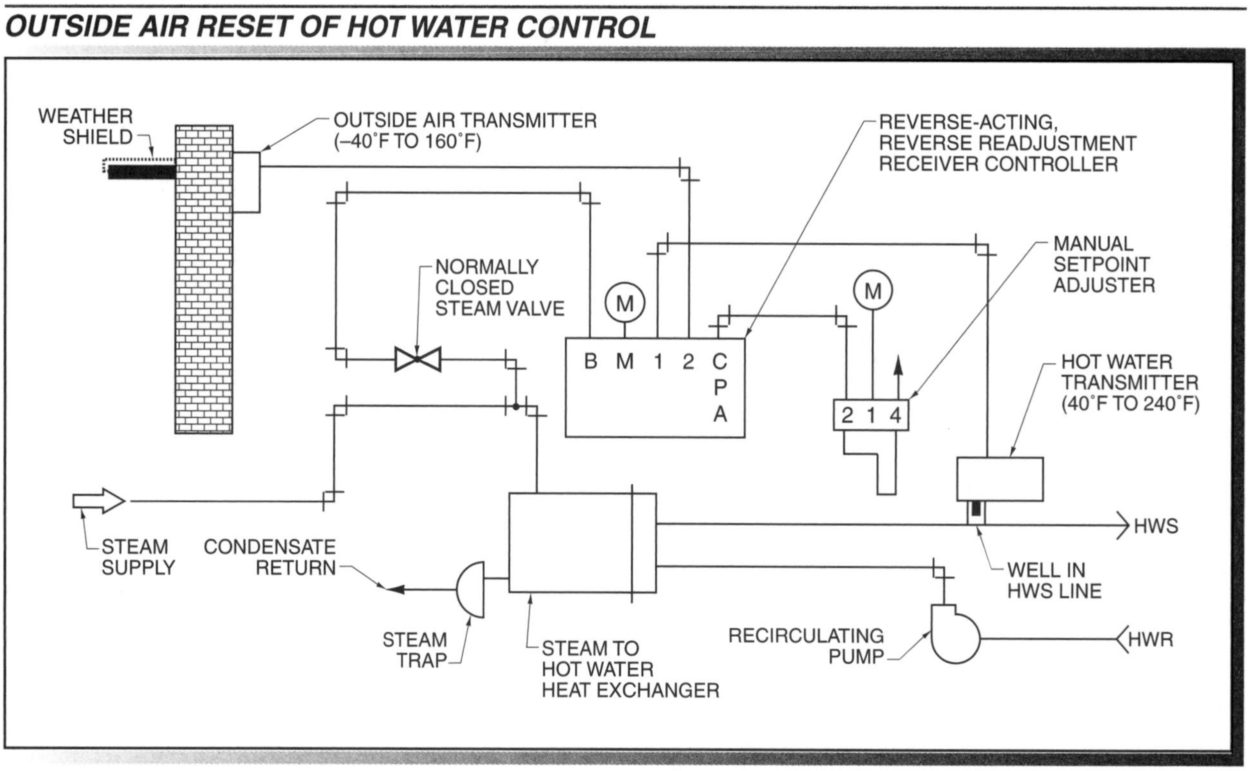

OUTSIDE AIR RESET OF HOT WATER CONTROL

Figure 10-19. Many commercial buildings use a central hot water heating system to heat the building.

from 180°F to 80°F. The hot water transmitter is mounted in a well in the hot water supply line. The hot water transmitter range is 40°F to 240°F. The outside air transmitter is an inert-element type with a range of –40°F to 160°F. The proportional band is set at 5%, which makes the throttling range 10°F. The throttling range is found by multiplying 5% by the transmitter span of 200°F. Using the throttling range formula, the authority is 183%.

The steam valve opens and admits steam to the heat exchanger, which transfers heat to the water pumped within the tubes. As the water temperature drops, the controller causes the valve to open, admitting steam and heating the water. As the water temperature increases, the controller causes the valve to close, reducing the steam flow into the heat exchanger and causing the water temperature to drop. *Note:* This example does not represent a complete hot water control system.

Troubleshooting Dual-Input Receiver Controllers

Dual-input receiver controllers are often used to control vital mechanical system applications in a building. If one of these controllers operates improperly, it can

lead to lack of comfort, reduced productivity, damaged products, or damage to the building. The tools needed for troubleshooting dual-input receiver controllers include a calibration simulator kit, squeeze bulb, pneumatic print, and a thermometer.

Receiver Controller Troubleshooting – Heating Water Supply Too Hot. Many commercial buildings pump hot water through the building to heat it. In this application, the building temperature is too hot due to a problem in the hot water mechanical or control system. Troubleshooting a dual-input receiver controller when the heating water supply is too hot is accomplished by applying the procedure:

1. Measure the outside air temperature and find the corresponding hot water temperature setpoint on the reset schedule. Check the water temperature with a thermometer at the outlet of the heat exchanger. Open the strainer outlet to run water on the thermometer.

2. Recalibrate the controller if the water temperature is not within the desired setpoint and throttling range. The most common dual-input controller problem is improper system operation due to an incorrect au-

thority adjustment on the controller. The controller calibration should be checked whenever any problem occurs.

3. Check the valve operation with a squeeze bulb if the controller is calibrated properly. The valve may be wiredrawn.

4. Check the outside air and hot water transmitter indicated value with a thermometer if the water temperature is still too high. The transmitter may need replacement if the transmitter indicated temperature and the thermometer reading are different.

5. Check the controller to ensure that the action is correct. The controller forces the valve wide open if the controller is direct acting instead of reverse acting.

Receiver Controller Troubleshooting – Cold Deck and Room Temperature Too Hot. In this application, the highest zone demand causes the setpoint of the receiver controller to change. This receiver controller controls a chilled water valve in order to maintain a setpoint in the cold deck. Troubleshooting a dual-input receiver controller when the cold deck and room temperature is too hot is accomplished by applying the procedure:

1. At the high signal selector, determine the zone input with the highest pressure. Use the schedule to determine the proper cold deck temperature at that zone pressure. Check the cold deck temperature with a thermometer in the duct. The air temperature should be within the desired setpoint and throttling range.

2. Calibrate the controller. The controller calibration should be checked whenever any problem occurs.

3. Check the valve operation with a squeeze bulb if the controller is calibrated properly. The valve may be stuck closed mechanically, not allowing chilled water to flow through the coil.

4. Check the calibration of the zone thermostats. The indicated high zone may be off if the zone thermostats are improperly calibrated.

5. Check the transmitter indicated value with a thermometer if the air temperature is still too high. The transmitter may require replacement if the transmitter indicated temperature and the thermometer reading are different.

6. Check the controller to ensure that the action is correct. The controller forces the dampers open to outside air if the controller is reverse acting instead of direct acting.

7. Check the readjustment of the controller to be certain it is correct. If the readjustment is not correct, the controller increases the setpoint of the cold deck as the zone temperature increases.

Receiver Controller Troubleshooting – Hot Deck Air Temperature Too Hot. Air handling units such as dual-duct and multizone units have a split duct design. One duct carries cold air to the building space, the other duct carries hot air. If the hot deck temperature is too cold, the building may be too cold. If the hot deck temperature is too high, the occupants may complain of being too warm. This results in excessive energy expenditures, uncomfortable occupants, and lack of productivity. In this application, a dual-input receiver controller changes the hot deck temperature as the outside air temperature changes. Troubleshooting a dual-input receiver controller when the hot deck air temperature is too hot is accomplished by applying the procedure:

1. Check the outside air temperature and determine the proper hot deck temperature from the reset schedule. Check the hot deck air temperature with a thermometer in the duct. The air temperature should be within the desired setpoint and throttling range.

2. Calibrate the controller if the air temperature is too high. The most common problem with dual-input receiver controllers is the misadjustment of the authority/ratio settings. The controller calibration should be checked whenever any problem occurs.

3. Check the valve operation with a squeeze bulb if the controller is calibrated properly. The valve may be stuck open mechanically or wiredrawn, admitting too much steam or hot water.

4. Check the indicated values of both transmitters with a thermometer if the air temperature is still too high. One or both transmitters may require replacement if the transmitter indicated temperatures and the thermometer readings are different.

5. Check the controller to ensure that the action is correct. The controller forces the valve wide open if the controller is reverse acting instead of direct acting.

6. Check the readjustment of the controller. The hot deck temperature increases as the outside air temperature increases if the controller is set for direct readjustment on a reverse readjustment schedule.

REMOTE CONTROL POINT ADJUSTMENT

Receiver controllers are expected to maintain their calibration setpoints under various conditions. Situations arise in commercial buildings that demand the capability of changing the setpoint to match certain events or temperatures. For example, a high solar load on a building in the winter may temporarily allow a lower than normal heating water temperature. Extremely high outside air temperatures may demand a lower than normal cold deck temperature. Two methods are used to change the setpoints to allow for these conditions. The first method uses operating personnel to manually adjust the setpoint at the control panel. The second method uses remote control point adjustment. *Remote control point adjustment* is the ability to adjust the controller setpoint from a remote location. Changing the setpoint of the receiver controller is accomplished by changing an air pressure. The air pressure change is typically accomplished by the use of a pressure-reducing valve station. See Figure 10-20. Changing the controller setpoints can be accomplished manually from a central building maintenance office, or automatically.

Remote Control Point Adjustment Calculations

Remote control point adjustment operation is defined by each manufacturer. The first number that is defined is the total pressure change of the remote control point adjustment. Different standards are used by different manufacturers, but most have settled on a pressure input range of 3 psig to 15 psig. The control point adjustment appears as another transmitter input to the controller. The next defined number is the total amount of remote control point adjustment change allowed. This is expressed as a percentage of the primary transmitter span, usually 10% or 20%. This means that a remote control point adjustment controller with a primary transmitter range of 40°F to 240°F has a total remote control point adjustment change available of either 20°F or 40°F, depending on the standard used. The remote control point adjustment causes this change as its pressure changes from minimum to maximum. For example, a Honeywell controller uses a 3 psig to 15 psig remote control point adjustment pressure input. This 12 psig pressure change causes a 20% total setpoint change for a transmitter that has a

REMOTE CONTROL POINT ADJUSTMENT

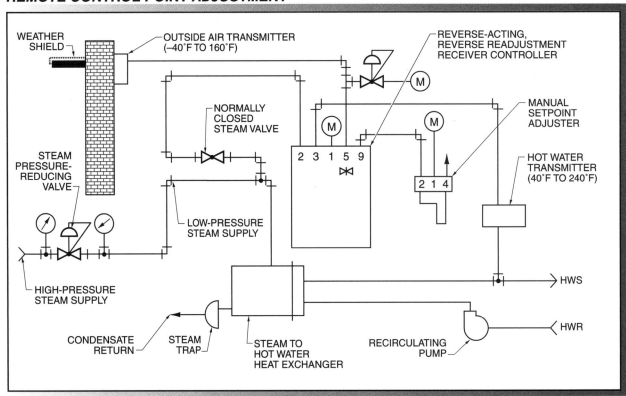

Figure 10-20. Remotely changing the setpoint of a receiver controller is accomplished by changing the air pressure at a pressure-reducing valve station.

span of 200°F. This means each time the remote control point adjustment pressure changes 1 psig, the setpoint changes 3.3°F (20% × 200°F / 12 psig = 3.3°F). The remote control point adjustment change may be positive or negative, meaning that the remote control point adjustment can raise or lower the setpoint. The relationship of the remote control point adjustment pressure and the setpoint is direct, meaning that the setpoint increases as the remote control point adjustment pressure increases.

> *Remote control point adjustment can become complex. In the past, control rooms were constructed that enabled building operators to adjust the setpoints remotely. These control rooms were elaborately constructed and included air handlers, pumps, chillers, and boilers represented on inlaid panels. Receiver gauges indicated temperatures and other building conditions.*

Remote Control Point Adjustment Calibration

In single- or dual-input receiver controllers, the standard calibration procedure should be followed, with one major change. The controller must be calibrated with the control point adjustment pressure midpoint connected to the control point adjustment port. This midpoint pressure is 9 psig for a 3 psig to 15 psig system or 8 psig for a 3 psig to 13 psig system. The rest of the calibration procedure is the same as the standard calibration procedure. Because remote control point adjustment is optional, a controller does not have to use it. A controller must be totally recalibrated in order to take remote control point adjustment into account if the controller does not use remote control point adjustment but it is desired at a later date.

Remote Control Point Adjustment Applications

Remote control point adjustment is used with a dual-input hot water controller with outside air reset. This controller is a dual-input controller with a primary transmitter in the hot water supply line and an inert element transmitter in the outside air. The remote control point adjustment reducing valve is located in the maintenance operator's office. The maintenance operator can increase or decrease the controller setpoint from this remote location. There may be multiple remote control point adjustment adjusters for different zones or controllers in the building.

Remote control point adjustment is also used with single-input cold deck controllers. This controller is a single-input controller with an averaging element transmitter in the cold deck of the dual-duct air handling system. The maintenance operator can adjust the controller setpoint from the remote control point adjustment pressure-reducing valve station in the maintenance office. See Figure 10-21.

REMOTE CONTROL POINT ADJUSTMENT COLD DECK CONTROLLER APPLICATION

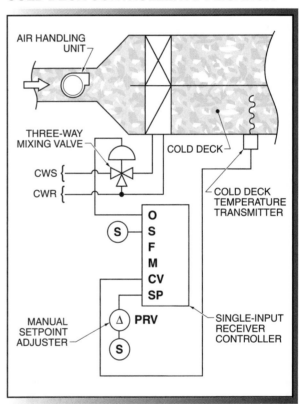

Figure 10-21. Remote control point adjustment is used with a single-input cold deck controller to allow adjustment of the controller setpoint at the pressure-reducing valve station in the maintenance office.

Troubleshooting Remote Control Point Adjustment

Troubleshooting remote control point adjustment systems is the same as troubleshooting standard single- or dual-input receiver controllers, with only one major difference. The controller must be calibrated with the remote control point adjustment attached. The complexity of the calibration and adjustments requires that the controller be precisely calibrated and checked. Any controller that has not been using the remote control point adjustment function must be recalibrated before remote control point adjustment can be used. Also, ensure that the remote control point adjustment is returned to its midpoint if it is changed.

PROPORTIONAL/INTEGRAL CONTROL

The demands placed on modern commercial control systems have grown enormously. In order to conserve energy and save money, greater control of system operation is required. Modern variable air volume HVAC systems are sometimes difficult to control without excessive hunting, cycling, or excessive variation from setpoint (offset). In order to address this need, receiver controllers were developed that use integration to achieve precise control. *Integration* is a function that calculates the amount of difference between the setpoint and control point (offset) over time. Integration calculates the area underneath a standard time-temperature curve.

Standard receiver controllers are proportional controllers in that the controller output is proportional to the amount of offset in the system. This works fine for many systems. In some systems, excessive offset is present in the system. Integration seeks to eliminate the remaining offset in the system by moving the controlled variable to setpoint. See Figure 10-22. The integration function is listed as a time function. The time function is the time that elapses before the controller drives the variable to the setpoint. Controllers that add integration are referred to as proportional/integral (PI) controllers. When calibrated properly, these controllers are more accurate than standard receiver controllers. However, proportional/integral controllers require greater knowledge by the technician.

Proportional/Integral Controller Calibration

Controllers using integration are calibrated the same as a standard receiver controller, except for the addition of the integral function. The controller should first be calibrated with the integration set at 0. The throttling range and gain (proportional band) should be doubled from their previous settings. The controller should be set up on an operational HVAC system under worst-case conditions if possible. If the HVAC system is a cooling system, it is best if it is calibrated at a maximum outside air temperature and internal load. If the controller works properly, then it should work properly under a lower load condition. Each HVAC system is different, so each controller is custom-calibrated.

PROPORTIONAL PLUS INTEGRAL CONTROLLED VARIABLE vs. TIME

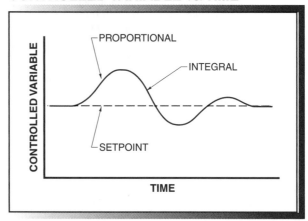

Figure 10-22. Integration seeks to eliminate the remaining offset in a system by moving the controlled variable to setpoint.

A longer integration time causes the system to respond more slowly, but its stability is increased. A shorter integration time causes a faster response but leads to a higher possibility of hunting. The best setting is that of the shortest time possible without inducing hunting or cycling. The setting can be reduced slowly until the system becomes unstable, and then increased slightly. This method can be used to tune the controller to achieve optimum calibration.

Auxiliary devices change the direction of air flow, change the amount of air flow, or interface with other types of control systems. Auxiliary devices are normally located between the controller and the controlled device, and change the control operation in some way. Auxiliary devices include relays, positioners, electric/pneumatic switches, pneumatic/electric switches, transducers, and low-limit thermostats.

AUXILIARY DEVICES

An *auxiliary device* is a device used in a control system that produces a desired function when actuated by the output signal from a controller. Auxiliary devices are used to change the direction or amount of air flow, or used as interfaces. An *interface* is a device that allows two different types of components, voltage levels, voltage types, or systems to be interconnected. Auxiliary devices are commonly located between the controller and controlled device.

Auxiliary devices are available in a variety of shapes and sizes for different applications. In addition, two devices from different manufacturers can have the same function yet different appearances. Finally, an auxiliary device may perform more than one function, depending on how it is piped. Confusion is minimized by studying the application closely, using a print or sketching the application if necessary, and having the proper manufacturer literature.

Auxiliary devices are categorized based on their function. Auxiliary device functions include changing flow direction, changing pressure, and interfacing between two devices. Auxiliary devices that change the direction of air flow in a system include switching relays, booster relays, and signal selection relays. Auxiliary devices that change the direction of air flow do not normally change the air pressure. Auxiliary devices that change air pressure include averaging relays, minimum position relays, reversing relays, positioners, and biasing relays. Auxiliary devices that change pressure do not change the direction of air flow. Interface auxiliary devices include electric/pneumatic switches, pneumatic/electric switches, electronic/pneumatic transducers, pneumatic/electronic transducers, and low-limit thermostats. Pneumatic control systems require interface auxiliary devices to enable the control of electric and electronic output devices.

Switching Relays

A *switching relay* is a device that switches air flow from one circuit to another. Switching relays are used to change systems from summer to winter operation and create pneumatic logic circuits to enable and disable control system functions.

Switching Relay Operation. Switching relay ports (air connections) include a pilot (P) port, common (C) port, normally open (NO) port, and normally closed (NC) port. See Figure 11-1. The pilot pressure is the input pressure from a controller that causes the relay output to switch from one port to another. The pilot port is a dead-ended port, meaning that the air at the pilot port does not go through the relay. If the pilot pressure is above the relay switch point, the normally closed and common ports are connected. If the pilot pressure is below the relay switch point, the normally open and common ports are connected. The relay switch point may be adjustable or fixed at the factory. Manufacturers use port designations such as P, NO, NC, and C, or combinations such as A, B, C, and D, or 5, 6, 7, and 8. Manufacturer literature is used to determine the port nomenclature of a particular switching relay.

SWITCHING RELAYS

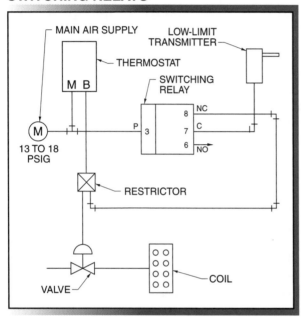

Figure 11-1. Switching relays are used to change air flow from one circuit to another.

Switching Relay Application. Switching relays are used to create pneumatic logic circuits. Switching relays are commonly used in applications that require switching between two conditions such as summer and winter. For example, a switching relay can be used to force an outside air damper closed. See Figure 11-2. In this application, an outside air transmitter is piped to a single-input receiver controller (changeover control). This controller is direct-acting and has a 2% proportional band and a 60°F setpoint. The small throttling range (2%) causes the controller to act as a two-position controller, with the output pressure being 0 psig when the outside air temperature is below 60°F and 20 psig when the outside air is above 60°F. The output of the controller is piped to the pilot port of the switching relay. In this application, the receiver controller must be calibrated using the switching relay switch point as the controller calibration point.

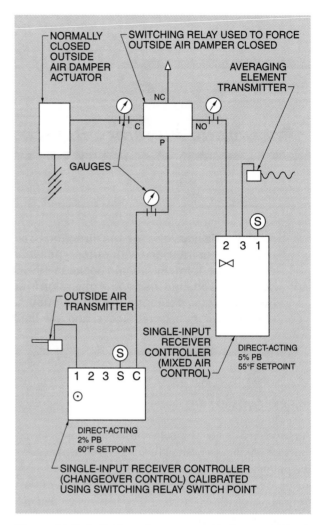

Figure 11-2. Switching relays are used to force outside air dampers closed.

The second single-input receiver controller (mixed air control) measures mixed air through an averaging element transmitter mounted in the mixed air plenum. This controller is direct-acting and has a proportional band of 5% and a 55°F economizer setpoint. The throttling range is 5°F. The output of this controller is piped to the normally open (NO) port of the switching relay. The common port of the switching relay is piped to the normally closed outside air damper actuator. The normally closed (NC) port of the switching relay is left open to vent to the atmosphere. If the outside air temperature is below 60°F, the output of the changeover control is 0 psig. The switching relay normally open port and common port are connected. This means that the mixed air control output is connected to the outside air dampers through the switching relay. The mixed air control modulates the dampers to maintain its setpoint. When the outside air temperature is above 60°F, the output of the changeover control is 20 psig. The switching relay normally closed port and common port are connected. The mixed air control does not control the dampers. The air at the dampers is vented to the atmosphere, causing the dampers to close.

Switching relays can also be used to enable or disable cooling systems. This application uses a multiple signal selection relay that is connected to the output pressures of six two-pipe, direct-acting pneumatic thermostats. See Figure 11-3. The multiple signal selection relay outputs the highest pressure (warmest zone) to the normally closed port of the switching relay. The common port of the switching relay is piped to two pneumatic/electric switches. The normally open port of the switching relay is vented to the atmosphere. The pilot port of the switching relay is piped to the output pressure of the changeover control. The changeover control measures outside air temperature and has a proportional band of approximately 2% and a 60°F setpoint. If the outside air temperature is above 60°F, the output pressure from the changeover control to the switching relay is 20 psig. The switching relay normally closed port and common port are connected. This means that the output signal from the multiple signal selection relay (warmest room) energizes cooling stages 1 and 2 through the pneumatic/electric switches. When the outside air temperature is below 60°F, the output pressure from the changeover control to the switching relay is 0 psig. The switching relay normally open port and common port are connected. The air at the pneumatic/electric switches is vented to the atmosphere. This stops the cooling process.

Booster Relays

A *booster (capacity) relay* is a device that increases the air volume available to a damper or valve while maintaining the air pressure at a 1:1 ratio. Booster relays are used in HVAC control systems that require a higher-than-normal air capacity to quickly move valves or dampers. Some systems also have multiple dampers and valves that are opened and closed simultaneously. Multiple dampers and valves can cause a time delay and slow operation due to the high demand for air. Booster relays increase air volume, not air pressure. The pressure to a damper or valve does not change so the setpoint is not affected. Booster relays have no adjustments. Booster relay ports include input, output, and main air ports.

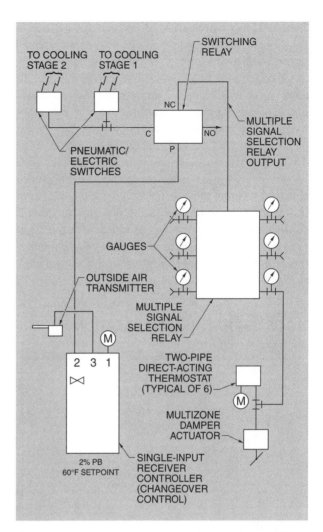

Figure 11-3. Switching relays can be used to enable or disable cooling systems.

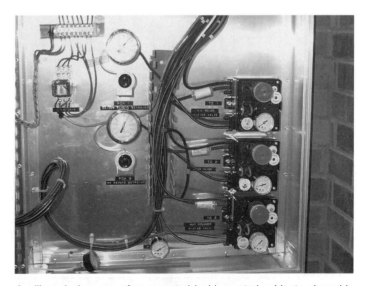

Auxiliary devices are often mounted inside control cabinets alongside receiver controllers.

Booster Relay Application. Booster relays are used to amplify air capacity. For example, a booster relay can be used with a low-volume limit thermostat and damper or valve actuator. See Figure 11-4. This is a mixed air application, with a single-input receiver controller having an averaging element transmitter in the mixed air plenum. The controller output is connected to the input of booster relays 1, 2, and 3. Each booster relay drives one actuator. The main air connection at each booster relay allows main air to quickly drive each actuator. Without the booster relays, the receiver controller must drive all three actuators simultaneously, causing a time delay due to insufficient air capacity at the controller. The outputs of the relays are piped to the damper actuators. Each booster relay output pressure is the same as the input pressure.

BOOSTER RELAY AMPLIFICATION APPLICATION

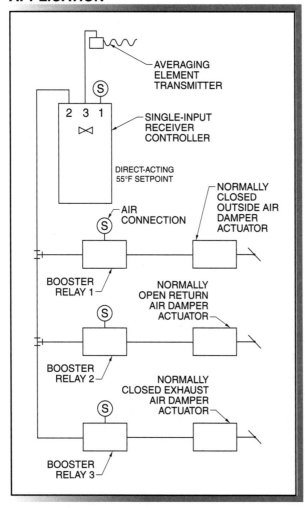

Figure 11-4. Booster relays are used to amplify air capacity.

Signal Selection Relays

A *signal selection relay* is a multiple-input device that selects the higher or lower of two pneumatic signal levels. The higher or lower signal level is piped to a controller or controlled device. Signal selection relays have two or more inputs and one output. See Figure 11-5. Signal selection relays are used to measure actual building heating and cooling demand. This accurate measurement is used to provide the minimum amount of heating and cooling necessary to satisfy the building spaces. Signal selection relays are installed per manufacturer recommendations and include high signal selection, low signal selection, and multiple signal selection relays.

SIGNAL SELECTION RELAYS

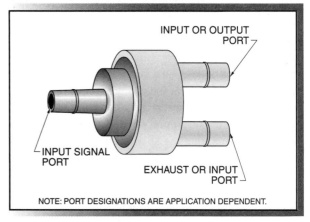

NOTE: PORT DESIGNATIONS ARE APPLICATION DEPENDENT.

Figure 11-5. Signal selection relays have multiple ports that must be piped properly for different applications.

High Signal Selection Relays. A *high signal selection relay* is a signal selection relay that selects the higher of two input pressures and outputs the higher pressure to a controlled device. A high signal selection relay has three ports. The two input ports are piped to the outputs of controllers located in different building zones. The controllers are normally direct-acting thermostats. The output port of a high signal selection relay is connected to the controlled device that supplies cooling for both zones. The higher pressure is the zone with the higher demand for cooling. The individual zones retain their own temperature control.

A high signal selection relay is often used to allow the warmest zone to control the cooling valve. See Figure 11-6. In this application, direct-acting pneumatic thermostats T-1 and T-2 are located in separate zones or rooms. Both thermostats are piped to the high signal selection relay. The branch line pressure of direct-acting thermostats increases as the temperature increases. The input line to the high signal selection relay that has the higher pressure is output to the normally closed cooling valve.

HIGH SIGNAL SELECTION RELAYS

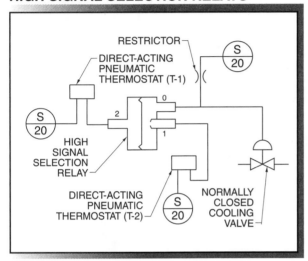

Figure 11-6. High signal selection relays allow the warmest zone to control the cooling valve.

> *Care must be taken in the piping and setup of auxiliary devices. Use of the wrong port can result in an operation exactly the opposite of the intended operation. Some auxiliary devices can be used in 10 or more different applications. Always consult the manufacturer's data sheet for the particular device. Do not assume that auxiliary devices are installed properly. The differences between connections can be so subtle that a system may be installed incorrectly and give poor service for years before the improper installation is discovered.*

Low Signal Selection Relays. A *low signal selection relay* is a signal selection relay that selects the lower of two input pressures (coolest zone) and outputs the lower pressure to a controlled device. In this application, only enough heat is provided to satisfy the coldest zone, saving energy. Low signal selection relays have two input ports and one output port. The individual zones retain their own temperature control. See Figure 11-7.

LOW SIGNAL SELECTION RELAYS

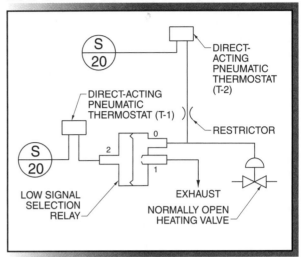

Figure 11-7. Low signal selection relays allow the coldest zone to control the heating valve.

Multiple Signal Selection Relays. High and low signal selection relays are limited to measuring the temperature in only two zones. In most modern commercial buildings, the temperature in more than two zones must be measured. A *multiple signal selection relay* is a relay that provides measurement of a large number of zones to ensure accurate zone signal measurement. See Figure 11-8.

MULTIPLE SIGNAL SELECTION RELAYS

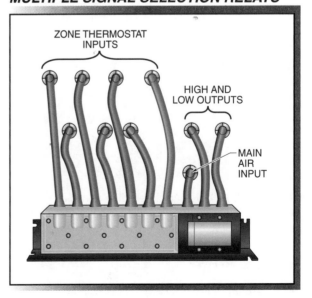

Figure 11-8. Multiple signal selection relays are used when many zones require temperature measurement.

Multiple signal selection relays are available with different numbers of zones. Most multiple signal selection relays have a six- or seven-zone capability. Some devices are modular, with the ability to add zones as required. A multiple signal selection relay measures the highest and lowest pressures simultaneously. The high and low outputs are piped to the heating and cooling controlled devices as needed. Any unused inputs are connected to a zone thermostat to prevent the lowest input from being 0 psig due to an unused input. Each zone input should have a labeled gauge attached to its line so the maintenance technician can determine the condition of each zone. A main air input is normally provided at the controller to provide power.

Multiple signal selection relays are used in applications in which the warmest zone is selected to operate the cooling valve. See Figure 11-9. In this application, the warmest of four zone inputs is connected to the normally closed cooling valve. The warmest zone controls the cooling available at the air handling unit. The coolest of the four zone inputs is piped to the normally open heating valve. The coolest zone controls the heating available at the air handling unit.

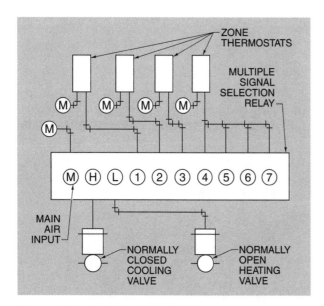

Figure 11-9. Multiple signal selection relays are used in applications in which the warmest zone is selected to operate the cooling valve.

Averaging Relays

An *averaging relay* is a relay that calculates the mathematical average of two input pressures and sends this pressure to a controlled device. An averaging relay is used in applications where two zones with different HVAC needs are served by the same damper or valve. The result is often a compromise, with neither zone being completely satisfied.

Averaging relays are used to determine the average signal from two room thermostats. See Figure 11-10. In this application, a variable air volume (VAV) terminal box serves a large area. Normally, there is one thermostat controlling the terminal box. This may lead to lack of comfort due to poor thermostat location. Individuals near the thermostat may be comfortable while individuals in a different part of the area may not. This problem is solved by installing a second thermostat at the problem area and using an averaging relay to control both thermostats.

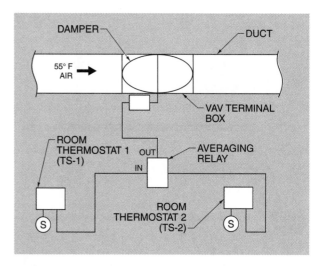

Figure 11-10. Averaging relays are used to determine the average signal from two room thermostats.

Minimum-Position Relays

A *minimum-position relay* is a relay that prevents outside air dampers from completely closing. Outside air dampers cannot be permitted to close completely because of ventilation air requirements. Minimum-position relays allow a minimum amount of outside air to flow through the air handling unit and into the building space.

A minimum-position relay commonly has three ports, an input port, output port, and main air input port. Minimum-position relays are low-pressure limit devices in that the device does not allow the output pressure to fall below a predefined, adjustable low limit. Above this pressure, the input and output are identical. Minimum-position relays have a removable adjustment dial that is usually mounted through the door of a temperature control panel. In this manner, the minimum-position limit can be easily adjusted. Minimum-position relays are available with a decal that indicates the percentage of outside air flow of the setpoint.

Minimum-Position Relay Setup. The proper minimum air setpoint must be determined when installing and setting up a minimum-position relay. To determine minimum air setpoint, the current outside air temperature (in °F) is multiplied by 10% and the current return air temperature (in °F) is multiplied by 90%. These values are added to determine the minimum air setpoint. The relay setpoint dial is adjusted until the proper mixed air temperature is reached. Minimum air setpoint is calculated by applying the following formula:

$$M_{AT} = 10\% \times O_{AT} + 90\% \times R_{AT}$$

where

M_{AT} = minimum air setpoint

10% = constant (outside air percentage)

O_{AT} = current outside air temperature (in °F)

90% = constant (return air percentage)

R_{AT} = current return air temperature (in °F)

For example, what is the minimum air setpoint if the outside air temperature is 85°F and the return air temperature is 72°F?

$$M_{AT} = 10\% \times O_{AT} + 90\% \times R_{AT}$$
$$M_{AT} = 10\% \times 85°F + 90\% \times 72°F$$
$$M_{AT} = 8.5 + 64.8$$
$$M_{AT} = \textbf{73.3°F}$$

Minimum-position relays are used to provide fresh air ventilation. See Figure 11-11. The pneumatic low-limit thermostat acts as a mixed air controller with an averaging element in the mixed air plenum. The output of the pneumatic low-limit thermostat is piped to the input of the minimum-position relay. The output of the relay is piped to the outside air damper. An outside air lockout causes the input signal at the minimum-position relay to fall to zero, causing the outside air damper to move to its minimum position. *Note:* This application does not include a complete control system.

Reversing Relays

A *reversing relay* is a relay designed to invert the output signal relative to the input signal. For example, as the input pressure from a controller to a reversing relay increases, the output pressure from the reversing relay to the controlled device decreases. Reversing relays are used with controlled devices that would otherwise be out of sequence. Reversing relays have an input port from the controller, output port to the controlled device, and main air input port.

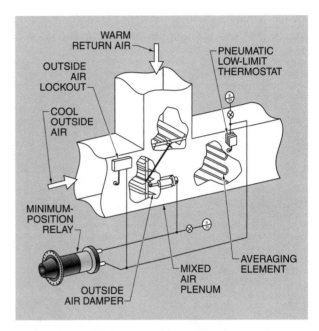

Figure 11-11. Minimum-position relays are used to provide fresh air ventilation.

A reversing relay can be used as a pneumatic inverter, volume amplifier, or inverse sequencer.

Reversing relays are used to invert a pneumatic signal. See Figure 11-12. In this application, a normally closed damper and a normally closed heating valve are controlled by the same direct-acting thermostat. Without a reversing relay, a rise in building space temperature causes a rise in pressure, opening the heating valve to allow additional hot water and opening the damper to allow additional cool air. A reversing relay is added to the damper piping so that, as the building space temperature increases, the heating valve closes and the damper opens.

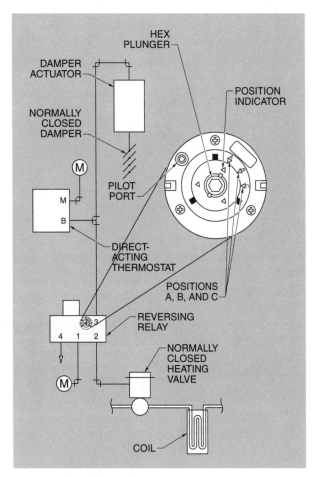

Figure 11-12. A reversing relay is used to invert a pneumatic signal.

Positioners

A *pneumatic positioner (pilot positioner)* is an auxiliary device mounted to a damper or valve actuator that ensures that the damper or actuator moves to a given extension. See Figure 11-13. Piping a certain air pressure does not guarantee that an actuator extends a

certain length. The extension of an actuator may be opposed by rust, corrosion, or external forces acting on the damper or valve (air flow striking damper blades). A positioner overcomes these forces and causes a damper to move a certain amount.

POSITIONERS

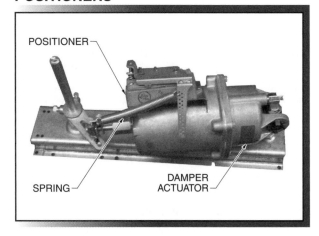

Figure 11-13. Pneumatic positioners ensure damper or valve actuators move to a given extension.

Pneumatic positioners are used to alter the effective spring range of an actuator. In some applications, it is necessary to closely sequence two or more damper or valve actuators. The actuator spring range must be closely adjusted to obtain the sequencing. Normally, the spring ranges of actuators are nonadjustable. The spring ranges can only be changed if the actuator is replaced. A positioner moves an actuator through its spring range when it receives a specific, adjustable pressure input. For example, an outside air damper actuator has a spring range of 3 psig to 11 psig. The actuator is required to move from open to closed when the air pressure input changes from 3 psig to 13 psig. When a positioner is calibrated properly, the pressure to the actuator changes from 3 psig to 11 psig when the input pressure to the positioner changes from 3 psig to 13 psig. Thus, the actuator appears to have a spring range of 3 psig to 13 psig.

Positioners are normally attached on the side of a damper or valve actuator and have an arm which connects to the actuator shaft through a spring assembly. A positioner has a starting point adjusting screw and an operating span adjustment. The operating span adjustment is made by placement of a spring in holes on the operating span lever arm.

See Figure 11-14. The three port connections on a positioner are the main air port, pilot port from the controller, and output port to the actuator.

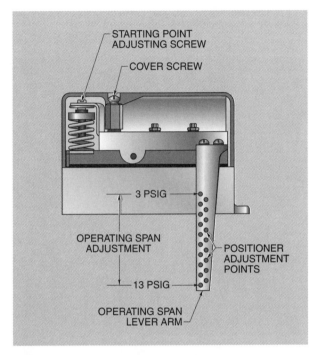

Figure 11-14. Positioner operating span adjustments are made by placing a spring in different holes in the operating span lever arm.

Positioners are also used to increase the main air capacity to multiple actuators. See Figure 11-15. In this application, the output pressure of a single-input receiver controller is piped to the pilot ports of two positioners. The positioners are attached to outside air damper actuators. The positioners ensure that the actuators extend smoothly and that any forces opposing their opening are overcome.

> *While normally among the easiest devices to set up, positioners are often the most confusing of all auxiliary devices. Regardless of the manufacturer, positioners have a start point and span adjustments, which must be calibrated properly. Many building automation systems use positioners. The sophisticated programming algorithms and features of a building automation system are ineffective if a damper or valve is not operating properly due to improper positioner calibration.*

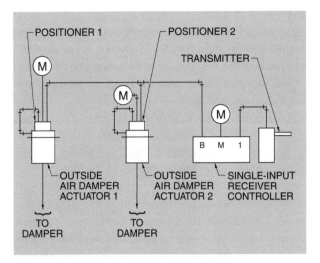

Figure 11-15. Positioners are used to increase the main air capacity to multiple actuators.

Positioners are also used to ensure proper sequencing of multiple actuators. See Figure 11-16. In this application, a normally open heating valve and a normally closed cooling valve are used to maintain a specific air supply temperature. The valves must be sequenced properly to prevent simultaneous heating and cooling. The heating valve is set up to operate from 3 psig to 11 psig and the cooling valve is set up to operate from 11 psig to 13 psig. The positioners enable easy setup and change of these sequence setpoints.

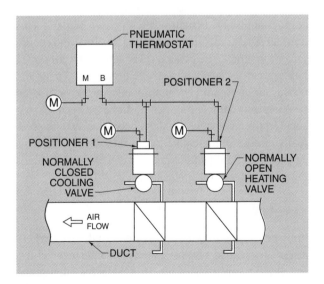

Figure 11-16. Positioners are used to ensure proper sequencing of multiple actuators.

Positioner Calibration. Proper pneumatic positioner calibration is required because pneumatic positioners are used by modern building automation systems and are responsible for the correct functioning of critical HVAC equipment. Positioners are calibrated by applying the procedure:

1. Determine the desired operational actuator range and start point. For example, an actuator is required to operate at pressures from 3 psig to 11 psig. The operational actuator range is 8 psig (11 psig – 3 psig = 8 psig). The start point is 3 psig.

2. Adjust the operational spring range to 8 psig. This is done differently for different positioners. Manufacturer literature should be consulted regarding operational spring range adjustments.

3. Input the start point pressure into the positioner pilot port (3 psig) using a squeeze bulb.

4. Adjust the output screw until the actuator begins to extend.

Biasing Relays

A *biasing (ratio) relay* is a relay used to perform a mathematical function on an input signal. Biasing relays are used to add, subtract, multiply, or divide. These operations are desired in HVAC control systems to accomplish control tasks.

Biasing relays enable multiple heating and cooling devices to be placed in the correct sequence. See Figure 11-17. In this application, a direct-acting, two-pipe thermostat is piped to a normally open heating valve, a normally closed cooling valve, and a normally closed outside air damper. The listed spring ranges for these damper and valve actuators are 3 psig to 13 psig. This normally causes overlapped heating, cooling, and economizer functions. The biasing relay at the normally open heating valve causes it to stroke between a 3 psig and 6 psig input signal. The biasing relay at the outside air (economizer) damper causes it to stroke between a 6 psig and 11 psig input signal, and the biasing relay at the normally closed cooling valve causes it to stroke between a 11 psig and 12 psig input signal. This close sequencing is due to the action of the biasing relays.

Electric/Pneumatic Switches

An *electric/pneumatic (EP) switch* is a device that enables an electric control system to interface with a pneumatic control system. In an electric/pneumatic switch, an electrical signal operates a pneumatic device. Electric/pneumatic switches are often referred

to as three-way, solenoid-operated air valves (SAVs). The solenoid has two wires that are connected to the appropriate point in the system. Solenoids of different voltages are available, with 24 VAC, 120 VAC, and 220 VAC the most common. Standard 60 Hz and 50 Hz models are available.

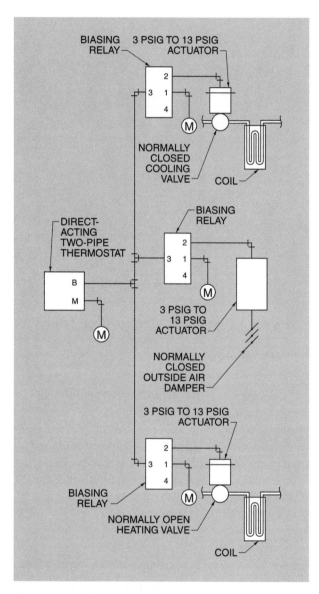

Figure 11-17. Biasing relays enable multiple heating and cooling devices to be placed in the correct sequence.

Electric/pneumatic switches have a normally open port, a normally closed port, and a common port. If no electrical power is applied to the solenoid, the normally open port and common port are connected internally. The normally closed port is disconnected from the other ports. If electric power is applied to the solenoid, the normally closed port and the common port are connected internally. The normally open port is disconnected from the other ports. The ports are labeled differently by each manufacturer. Manufacturer's designations include 1, 2, 3, and A, B, C. Some models include a manual bypass that can be used in case the solenoid burns out.

Electric/pneumatic switches are used to control a damper by the operation of a fan. See Figure 11-18. In this application, the electric/pneumatic switch solenoid is wired in parallel with the fan or fan starter. When the fan is ON, the electric/pneumatic switch is energized. When the fan is OFF, the electric/pneumatic switch is de-energized. Main air supply is connected to the normally closed (1) port. The common (3) port is connected to the damper actuator in the air handling unit. The normally open (2) port is vented to the atmosphere. When the fan starts, the electric/pneumatic switch is energized, the normally closed (NC) port and common (C) port are connected, and main air flows to the damper, causing it to open. When the fan stops, the electric/pneumatic switch is de-energized, the normally open (NO) port and common (C) port are connected, and the actuator air is vented to the atmosphere, causing the damper to close.

An electric/pneumatic switch is also used to close a pneumatic device when an electrical device is OFF. See Figure 11-19. In this application, the single-input receiver controller (mixed air control) has an averaging element transmitter located in the mixed air plenum. The controller output is piped to a minimum-position relay. The output of the minimum-position relay is piped to the normally closed (NC) port of the electric/pneumatic switch. The common (C) port is piped to a normally closed outside air damper. The normally open (NO) port of the electric/pneumatic switch is vented to the atmosphere.

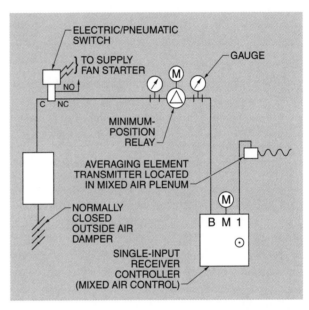

Figure 11-19. Electric/pneumatic switches are used to close a pneumatic device when an electrical device is OFF.

Without the electric/pneumatic switch, the outside air damper is always at its minimum position. This is not desirable because it can cause freezing problems. When the fan is shut OFF, the electric/pneumatic switch is de-energized and the air at the actuator is vented to the atmosphere, forcing the damper closed. When the fan is turned ON, the electric/pneumatic switch is energized, the normally closed (NC) port and common (C) port are connected, and the receiver controller controls the dampers to maintain the mixed air setpoint or minimum position.

Pneumatic/Electric Switches

A *pneumatic/electric (PE) switch* is a device that allows an air pressure signal to energize or de-energize an electrical device such as a fan, pump, compressor,

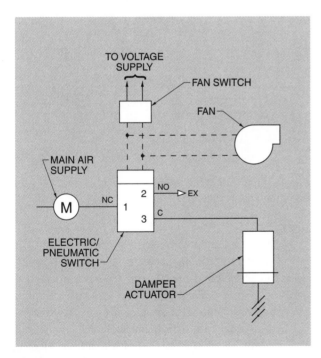

Figure 11-18. Electric/pneumatic switches enable a pneumatic device such as a damper to respond to the operation of an electrical device such as a fan.

or electric heating element. Pneumatic/electric switches perform a task that is the opposite of that of electric/pneumatic switches and cannot be interchanged with them. Pneumatic/electric switches have one inlet air port that activates an electric switch through a bellows mechanism. Pneumatic/electric switches have a normally open electrical terminal, a normally closed electrical terminal, and a common electrical terminal.

The normally closed and common terminals are connected electrically if the incoming air pressure is lower than setpoint. In this case, the normally open terminal is disconnected. The normally open and common terminals are connected electrically if the incoming air pressure is higher than setpoint. In this case, the normally closed terminal is disconnected.

> ⓘ *Pneumatic/electric switches are calibrated using a squeeze bulb and digital multimeter. A calibration simulator kit can be used if more accurate results are required.*

The pneumatic/electric switch is said to "make on a rise" of input pressure if the normally open and common terminals are used. The pneumatic/electric switch is said to "break on a rise" of input pressure if the normally closed and common terminals are used. *Note:* Electrical flow terminology is opposite of fluid flow terminology. A normally open valve or damper has 100% flow when no air is applied. A normally open electrical device has 0% flow when no electricity is applied.

A pneumatic/electric switch has an adjustable pressure setpoint and differential. These adjustments allow for the electrical device to be turned ON or OFF at different controller setpoints. The differential prevents short cycling due to small changes in the incoming air pressure. A differential of 1 psig means that the pressure must change a minimum of 1 psig from setpoint before the pneumatic/electric switch changes state. The adjustable setpoint allows staging or sequencing of different electrical devices from the same input signal.

Pneumatic/electric switches are used to energize and de-energize stages of heating or cooling. See Figure 11-20. In this application, two pneumatic/electric switches are used to energize two different stages of direct expansion (DX) cooling. Pneumatic/electric switch 1 controls the first stage of cooling while pneumatic/electric switch 2 controls the second stage. The cooling capacities of the stages are usually set up so

that the first stage of cooling has one-third the total cooling capacity, while the second stage of cooling has the remainder (two-thirds) of the total capacity. This provides good capacity control under light load conditions.

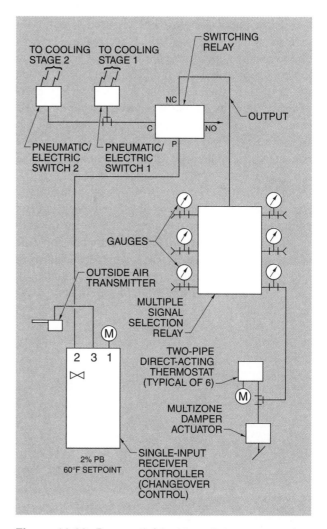

Figure 11-20. Pneumatic/electric switches are used to energize and de-energize stages of heating or cooling.

Both pneumatic/electric switches are piped to the output of a multiple signal selection relay. The multiple signal selection relay measures the output pressures of six zones. The highest pressure of the six zones is the warmest zone. This zone controls both stages of cooling in order to satisfy that zone. The setpoints of the cooling stages are staggered. Pneumatic/electric switch 1 energizes at a pressure of approximately 10 psig. On a call for additional cooling, the pressure to pneumatic/electric switch 1 continues to rise. Pneumatic/

electric switch 2 energizes at approximately 12 psig. The pneumatic/electric switch normally open and common terminals are used to "make on a rise" in pressure. Any loss of air pressure causes the cooling stages to be de-energized.

Pneumatic/electric switches are also used to allow the signal from an outside air transmitter to energize and de-energize a hot water pump. See Figure 11-21. In this application, a pneumatic/electric switch is piped to an outside air transmitter. The pneumatic/electric switch energizes a hot water pump when the outside air temperature is below 55°F. The normally closed terminal and common terminal are used to "break on a rise" in pressure. Any failure causes the hot water pump to continue running. The setpoint of the pneumatic/electric switch must correspond to the pressure output of the transmitter at 55°F. The pressure corresponding to 55°F is 9.9 psig for a –40°F to 160°F transmitter.

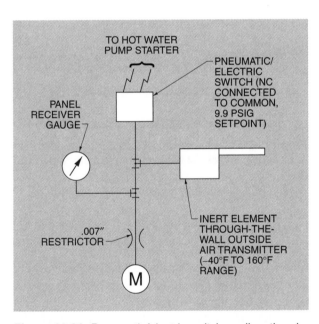

Figure 11-21. Pneumatic/electric switches allow the signal from an outside air transmitter to be used to energize and de-energize a hot water pump.

Transducers

A *transducer* is a device that changes one type of proportional control signal into another. Modern control systems must provide a method of interfacing with an electronic or automated control system. Transducers include electronic/pneumatic and pneumatic/electronic transducers.

Electronic/Pneumatic Transducers. An *electronic/pneumatic transducer (EPT)* is a device that converts an electronic input signal to an air pressure output signal. Electronic/pneumatic transducers are used to allow an electronic or automated system to drive pneumatic damper and valve actuators. See Figure 11-22. A 0 VDC to 10 VDC input signal causes a 0 psig to 20 psig output signal from the transducer. Electronic/pneumatic transducers have an output that is proportional to the input signal. For example, if the input signal to the transducer is 5 VDC the output is 10 psig. Models are available that have other voltage or current ranges.

ELECTRONIC/PNEUMATIC TRANSDUCERS

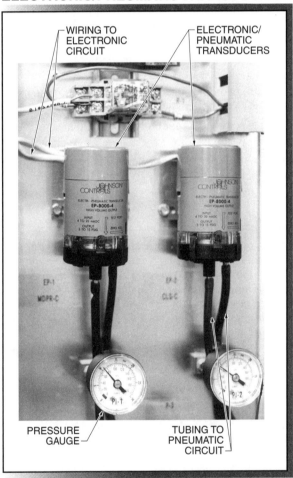

Figure 11-22. Transducers are used to allow a building automation system 0 VDC to 10 VDC signal to control a pneumatic damper or valve.

Pneumatic/Electronic Transducers. A *pneumatic/electronic transducer (PET)* is a device that converts an air pressure input signal to an electronic output signal. Pneumatic/electronic transducers are the opposite of electronic/pneumatic transducers and are not interchangeable with them. Pneumatic/electronic transducers are used to drive electronic actuators. The most common application of pneumatic/electronic transducers is to monitor air pressure in automated or electronic systems.

Low Temperature Cutout Controls

A *low temperature cutout control* is a device that protects against damage due to a low temperature condition. Low temperature cutout controls are commonly used in commercial air handling units. Low temperature cutout controls have capillary tube sensing elements and are usually located in the ductwork after the first heating coil in the system. See Figure 11-23. Low temperature cutout controls stop supply fan operation if the temperature drops near freezing, protecting water coils against bursting. Low temperature cutout controls are available in electric and pneumatic models. Pneumatic low temperature cutout controls must be used with pneumatic/electric switches to break the supply fan circuit. Low temperature cutout control setpoints are adjustable, with the most common being between 35°F and 40°F.

> *Low temperature cutout controls are normally located directly after the first heating or cooling coil in an air handling system. The sensing element must be installed to ensure adequate coverage of the duct. High-quality cutout controls are designed to open the contacts and stop the fan if any small length (½″) of the sensing element is below setpoint.*

TROUBLESHOOTING AUXILIARY DEVICES

Auxiliary devices require the correct setup for proper operation. Auxiliary devices must be calibrated and correctly connected in the circuit. In addition, auxiliary devices must be used for the application for which they are designed. Common problems associated with auxiliary devices include excessively hot and excessively cold building spaces, or equipment operation lockout. Tools needed to troubleshoot auxiliary devices include a thermometer, pressure gauge with fittings, and wrench or knife to connect or disconnect the air fittings.

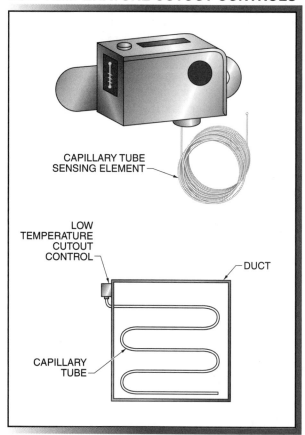

LOW TEMPERATURE CUTOUT CONTROLS

CAPILLARY TUBE SENSING ELEMENT

LOW TEMPERATURE CUTOUT CONTROL

DUCT

CAPILLARY TUBE

Figure 11-23. Low temperature cutout controls are used to reduce the possibility of damage from a frozen water coil.

Troubleshooting Application – Minimum-Position Out of Adjustment

A complaint of a building space being excessively hot may be caused by outside air dampers being open fully, admitting hot, humid air to the air handling unit. This overloads the cooling coil and causes the temperature to be excessively hot in the building space. See Figure 11-24. In this application, the minimum-position relay for the normally closed outside air damper is checked to ensure that it is adjusted correctly.

The input and output pressures of the minimum-position relay are checked using the pressure gauges. The output pressure should be at its minimum value if the inlet pressure is 0 psig. The minimum-position relay is out of adjustment if the output pressure is greater than its minimum value and the dampers are open more than their minimum amount. The adjustment knob at the temperature control panel is checked and adjusted. A minimum-position relay that is far out of adjustment may require removal of the adjustment knob and adjustment of the output pressure post.

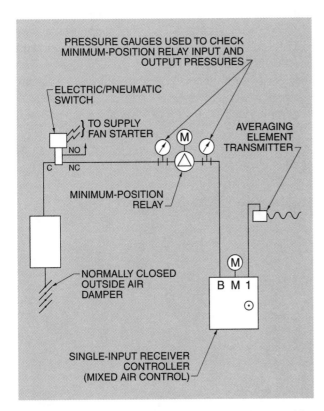

Figure 11-24. Pressure gauges are used when troubleshooting auxiliary devices to verify the pressure at various points in the system.

Troubleshooting Application – Switching Relay Out of Adjustment

A complaint of a building space being excessively hot may also be caused by a switching relay that is adjusted to the wrong pressure. In this application, the switching relay does not allow the outside air dampers to go to their minimum position, which allows excessive hot, humid air into the building space. See Figure 11-25. In this application, the input and output pressures of the switching relay are correct, but the normally open port and common port are connected instead of the normally closed port and common port. This is determined by checking the pilot, common, normally open port, and normally closed port pressures. In-line pressure gauges are installed to check these pressures. This problem may also be caused by a switching relay that requires replacement or an adjustment of its setpoint.

Troubleshooting Application – Hot Water Pump Not Operating

Electrical devices in commercial buildings, such as pumps, refrigeration compressors, and electric heat, are often interfaced with pneumatic controls. A lack of heat may be caused by a bad or incorrectly calibrated or connected pneumatic/electric switch. In this application, the hot water pump is not operating, so no heat is available. The hot water pump is turned ON and OFF by a pneumatic/electric switch that is connected in the outside air transmitter line. The pneumatic/electric switch is checked with a voltmeter to determine which terminals are electrically connected. A squeeze bulb is connected to the air pressure input to check and calibrate the pneumatic/electric switch. The electrical power can be disconnected and an ohmmeter used to check pneumatic/electric switch continuity. The setpoint should correspond to the transmitter pressure at the desired outside air temperature.

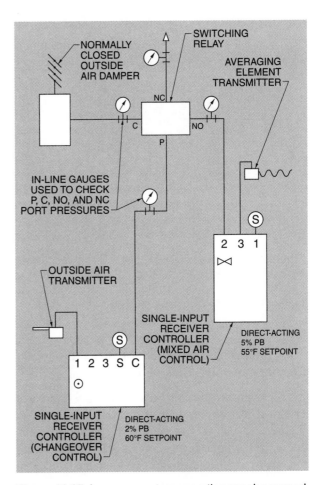

Figure 11-25. Improper system operation may be caused by an improper switching relay switch point pressure.

Troubleshooting Application – No Air on Controls

A complaint that a building space temperature is too hot may be caused by an electric/pneumatic solenoid that is burned out, preventing main air from passing through to the controllers. See Figure 11-26. In this application, the air handling unit is checked and the fan is found to be running. However, none of the air handling unit controls have air supplied to them. A quick check of the print confirms that when the electric/pneumatic switch is energized, it supplies main air to the controls. A check of the electric/pneumatic switch confirms that the proper power is supplied to the solenoid but the solenoid is not energized. The solenoid is burned out and must be replaced.

switches energize two different stages of cooling at different pressure setpoints. The multiple signal selection relay inputs are the output pressures of six direct-acting zone thermostats. Thermostats in signal selection systems must be calibrated at the same output pressure and have the same setpoints. If not, one thermostat always calls for full heat or cooling. In this application, one thermostat always needs full cooling and therefore has both stages of cooling energized continuously. A pressure gauge placed at each thermostat input is used to identify the thermostat that requires calibration. Occupants must be prevented from tampering with the thermostats, as this can cause the same problem. Lock boxes can be used at each zone thermostat if necessary.

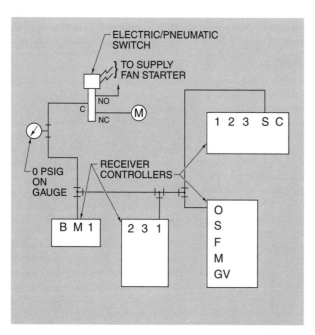

Figure 11-26. An excessively hot building space may be caused by an electric/pneumatic solenoid that is burned out, preventing main air from passing through to the controllers.

Troubleshooting Application – Two Stages of Cooling Operating Continuously

Improper pneumatic/electric switch points can cause electrical devices to work in an improper sequence. See Figure 11-27. In this application, the high signal output of the multiple signal selection relay energizes two pneumatic/electric switches. The pneumatic/electric

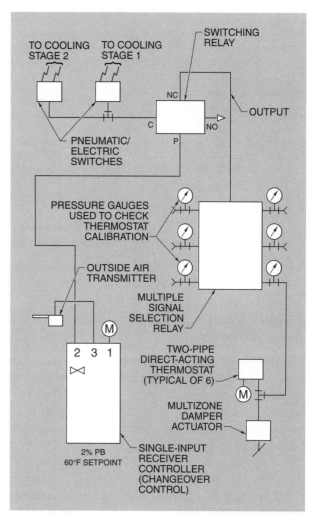

Figure 11-27. Improper pneumatic/electric switch points can cause electrical devices to work in an improper sequence.

Pneumatic control diagrams show the connection of pneumatic controls in a control system and are used for troubleshooting control system problems. A written sequence of operation describes, in basic language, the HVAC unit operation. Many control diagrams include a basic electrical schematic to show the interfacing of electrical devices.

PNEUMATIC CONTROL SYSTEM APPLICATIONS

Pneumatic control systems provide and maintain comfort in a building space. Pneumatic control systems require an understanding of individual component use in a particular HVAC system. Proper maintenance of a pneumatic control system requires the ability to read and understand pneumatic control diagrams.

> ⓘ *The ability to troubleshoot pneumatic control systems is one of the most sought after skills in the HVAC controls industry. Time should be taken to study the control diagrams before making any changes to the control system. This results in an efficient control system that keeps building occupants comfortable.*

Pneumatic Control Diagrams

A *pneumatic control diagram* is a pictorial and written representation of pneumatic controls and related equipment. Pneumatic control diagrams provide information regarding the design and implementation of each component in a pneumatic control system. Pneumatic control diagrams identify proper sequencing and control of equipment, as well as potential trouble spots and design flaws. Pneumatic control diagrams may be used

to recommend future control upgrades and to retrofit modern building automation systems to existing pneumatic control systems.

Pneumatic control diagrams are provided to a customer when a new mechanical system is installed or when an existing system is upgraded. The information included in pneumatic control diagrams is determined by the written specifications for a project. Customers are responsible for proper preservation and storage of pneumatic control diagrams. Problems that commonly occur when using pneumatic control diagrams include the following:

- Missing pneumatic control diagram components. Missing pneumatic control diagram components can make troubleshooting difficult.
- Inaccurate pneumatic control drawings. Existing systems that have been upgraded numerous times may have prints that are outdated or inaccurate. An *as-built* is a drawing provided for a new system on installation. Prints other than as-builts may be misleading when troubleshooting because changes may have occurred before the final system was installed.

Whether as-builts are used or a new set of prints are drawn, pneumatic control diagrams commonly consist of control drawings, a parts list, a written sequence of operation, a sequence chart, and an electrical interface diagram. See Figure 12-1.

PNEUMATIC CONTROL DIAGRAMS

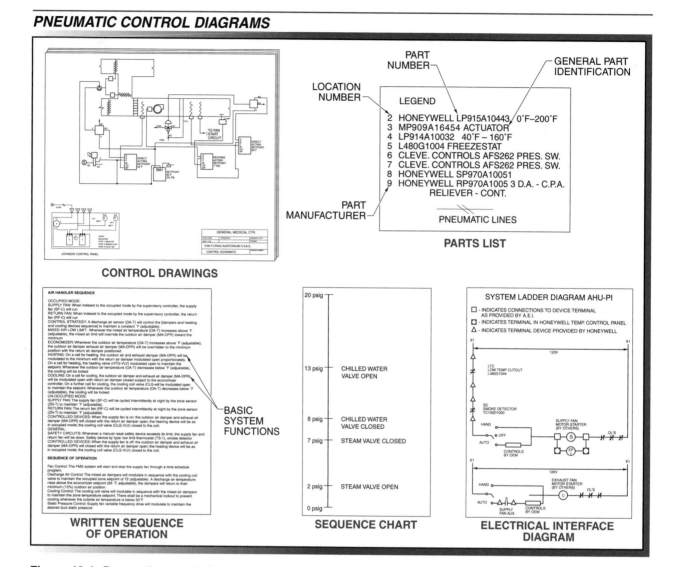

Figure 12-1. Pneumatic control diagrams commonly consist of control drawings, a parts list, a written sequence of operation, a sequence chart, and an electrical interface diagram.

Control Drawings. A *control drawing* is a drawing of a mechanical system that illustrates actual controls and piping between devices. A technician uses a control drawing to trace a control function for troubleshooting purposes. Control drawings are the most commonly used part of pneumatic control diagrams.

Parts List. A *parts list* is a reference list that indicates part description acronyms and actual manufacturer part names and numbers. A parts list is normally located on the control drawing and keys the part to a location number on the diagram, and gives part manufacturer, part number, and a general part identification. Parts lists are used when locating a part within a control drawing.

Major manufacturers provide cross-reference charts for easy part replacement. The entire part number is required to obtain the correct replacement part.

Written Sequence of Operation. A *written sequence of operation* is a written description of the operation of a control system. Recognizing individual parts or functions on a control diagram can be difficult. A written sequence of operation provides technicians with an understanding of circuit operation. The sequence of operation should be written in an easily understandable manner. Items such as part names, pressure or voltage setpoints, piping schemes, and port numbers should not be included. A written sequence of operation is broken down into paragraphs that describe the basic system functions. For example, a common air handling unit written sequence of operation may include the following:

- Fan Start: The fan moves warm or cool air into a building. If the fan is OFF no heating or cooling may flow into the building.

- Heating: A primary job of an air handling unit is to heat or cool the building space. A hot water, steam, or electric element or coil is installed in the air stream after the fan. A room thermostat or duct controller controls the application of heat.

In northern climates, the heating sequence is listed first. In southern climates, the cooling sequence is listed first. Cooling-only or heating-only air handling units are common, depending on the climate.

- Damper/Economizer Sequence: Air handling units also introduce the proper amount of fresh air into a building through outside air dampers. These dampers may also be used to satisfy the cooling requirements using cool outside air instead of energizing the mechanical cooling equipment. This is called the economizer mode. This sequence must also include the closing of the outside air dampers to the minimum position if the outside air conditions become unsuitable for cooling the building.

- Zone Temperature Control Sequence: Some air handling units perform building space temperature control. For example, variable air volume air handling units usually discharge 55°F air year-round. Zone thermostats then perform building space temperature control.

- Fire/Smoke Sequence: Various types of smoke removal and/or fire shutdown sequences may also be implemented.

- Humidification Sequence: Many air handling units provide humidity control. Humidity control can include humidification or dehumidification. The humidification sequence must include safety interlocks to ensure that the humidifier valve is forced closed. This prevents building damage due to high humidity levels.

> *A written sequence of operation should balance technical detail and readability. A sequence that is too technical may not be understood and a sequence that is easily understood may not have enough detail.*

Sequence Chart. A *sequence chart* is a chart that shows the numerical relationship between the different values in a pneumatic system. Many pneumatic control systems have numerical relationships that are difficult to understand. For example, a heating valve, outside air dampers, and a cooling valve may have different spring ranges that enable them to be sequenced properly. On a different system, a cooling tower may have three fans which are energized sequentially by pneumatic/electric switches with different setpoints. These applications may have sequence charts that show their numerical relationship.

Sequence charts may be used for installation and troubleshooting purposes. For example, a sequence chart may indicate that a cooling valve has a 9 psig to 13 psig spring range. A replacement cooling valve must have a 9 psig to 13 psig spring range. An incorrect spring range is determined if a cooling valve is indicated on the sequence chart as 9 psig to 13 psig but is tested as 5 psig to 8 psig.

Electrical Interface Diagram. An *electrical interface diagram* is a drawing showing the interconnection between the pneumatic components and electrical equipment in a system. Many pneumatic control systems must be interfaced with electrical equipment such as fans, pumps, lockouts, or individual stages of heating or cooling. An electrical interface diagram is commonly included in a pneumatic control diagram. Normally, only equipment interfaced directly to the pneumatic controls is included in a pneumatic control diagram. Additional electrical circuit connections are commonly included in the mechanical system or building electrical prints.

Single-Zone Air Handling Unit Control Application

Single-zone air handling units normally provide comfort for only one space within a commercial building. The building space may be as small as one office or as large as an entire floor of a building. As the building space becomes larger, the level of control decreases because sensing and control is normally performed from only one location. Also, as a building space becomes larger, greater temperature variations exist in the space. The temperature variations are not accounted for in the control of the air handling unit because the sensing is performed from only one location. Single-zone control is the most basic control method of any mechanical system. See Figure 12-2.

SINGLE-ZONE AIR HANDLING UNIT

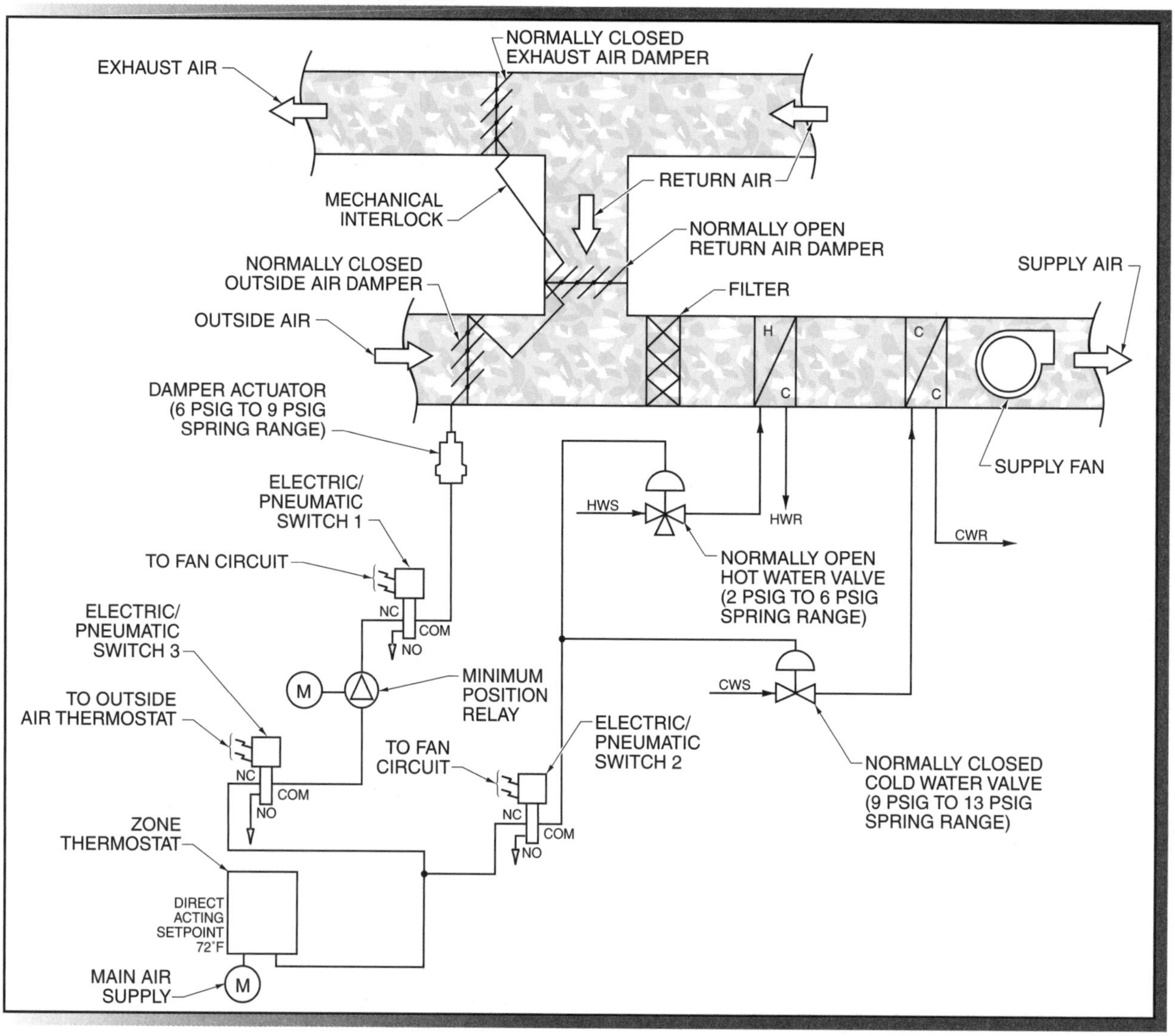

Figure 12-2. Single-zone air handling units provide temperature and humidity control for one building space.

In a single-zone air handling unit application, the supply fan is started by a time clock, facility management system, or manual switch. Some systems may run 24 hr per day. When the supply fan stops, electric/pneumatic switch 2 de-energizes, causing the hot water and cold water valves to return to their normal position. The zone thermostat modulates the outside air damper actuator, hot water valve, and cold water valve in sequence by means of electric/pneumatic switch 1, 2, and 3 to maintain its setpoint.

As the building space temperature increases, the zone thermostat opens the outside air dampers to provide cooling. As the outside air temperature increases, the outside air thermostat de-energizes electric/pneumatic switch 3, causing the outside air dampers to return to minimum position as determined by the setting of the minimum-position relay.

Single-Zone Air Handling Unit Optional Sequences. Additional devices or sequences can be added to a basic mechanical or control system to enhance system performance. Possible optional sequences for a single-zone air handling unit include the addition of humidification control. If building space humidification is desired, a humidifier with a normally closed valve can be inserted in the ductwork after the supply fan. The humidifier is controlled by a duct or zone humidistat. Humidity limit control would be included to prevent the discharge of excessive moisture or operation of the humidifier when the fan is OFF.

Direct expansion mechanical cooling may be required in warm climates. When chilled water is unavailable for cooling, the thermostat may energize mechanical cooling compressors through a pneumatic/electric switch. An interlock is included to prevent the compressors from operating if the supply fan is OFF.

There are multiple possible control interfaces with the fire and smoke safety controls of the building. All sequences must follow appropriate fire codes. Fire and smoke control sequences must be thoroughly tested by trained personnel. Some sequences provide for duct smoke detection and evacuation using the system fans or other dedicated exhaust fans. In most cases, the supply fan will be stopped and outside air dampers closed to prevent fresh air from feeding the fire.

Multizone Air Handling Unit Control Application

Multizone air handling units were common in buildings prior to the energy crisis of the 1970s. Multizone air handling units provide comfort for more than one building space. Some small multizone air handling units control only a limited number of building spaces. Large multizone air handling units may control 50 or more building spaces. Multizone air handling units are constant-volume air handling units. A *constant-volume air handling unit* is an air handling unit that moves a constant volume of air. For example, the fan in a constant-volume air handling unit always runs at 100% of its rated capacity. Constant-volume air handling units use a large amount of energy. See Figure 12-3.

Multizone air handling units are identified by the number of damper actuators mounted on the discharge end of the air handler. In a multizone air handling unit application, a time clock or facility management system starts the supply fan. If the fan is OFF, electric/pneumatic switch 1 causes the outside air damper to return to its normally closed position. Receiver controller 2 (hot deck control) modulates a three-way hot water (mixing) valve to maintain its setpoint, as reset by the outside air temperature.

In summer, as determined by receiver controller 5 (changeover control), the highest zone temperature energizes both stages of mechanical cooling in sequence. In addition, as determined by receiver controller 5, switching relay 2 changes position and causes the outside air damper to return to its minimum position, ensuring the proper amount of outside air. Zone thermostats 1 through 7 modulate the zone dampers in sequence to maintain their setpoints.

In winter, receiver controller 5 causes switching relay 3 to change position and disable the mechanical cooling. In addition, as determined by receiver controller 5, receiver controller 1 (mixed air control) modulates the outside air, return air, and exhaust air dampers to maintain its setpoint.

Usually, if a fire detection system is installed, the fire alarm system will stop the supply fan and cause the outside air dampers to close. The return air humidity controller modulates the normally closed (NC) steam humidifier valve to maintain its setpoint. If excessive moisture is discharged, receiver controller 4 (discharge high limit humidity control) causes the steam humidifier valve to close. In summer, as determined by receiver controller 5, switching relay 1 changes position, causing the steam humidifier valve to close. When the supply fan stops, electric/pneumatic switch 2 de-energizes, causing the steam humidifier valve to close. If the heating coil discharge temperature drops too low, the low temperature limit control stops the supply fan by de-energizing pneumatic/electric switch 3.

Multizone Air Handling Unit Optional Sequences. Optional sequences can be added to multizone air handling unit applications. For example, if chilled water is available, the direct expansion cooling could be replaced by a chilled water valve and coil. Also, the high signal selection relay could be replaced by a single-input receiver controller to control the cold deck temperature. Various types of smoke removal and/or fire shutdown sequences may also be implemented.

Variable Air Volume Air Handling Unit Control Application

Variable air volume air handling units are the most common mechanical HVAC systems installed today. A *variable air volume air handling unit* is an air handling unit that moves a variable volume of air. The volume of air is varied by a mechanism that throttles the supply fan. The supply fan throttling mechanism may consist of fan inlet vanes, variable speed drives (VSDs), or other methods. A static pressure controller maintains a constant static pressure in the duct. Most variable air volume air handling units maintain a constant static pressure of 1″ wc and a constant discharge air temperature of 55°F. Variable air volume terminal boxes are used to regulate the 55°F air flow into a building space to maintain its temperature setpoint. See Figure 12-4.

MULTIZONE AIR HANDLING UNIT

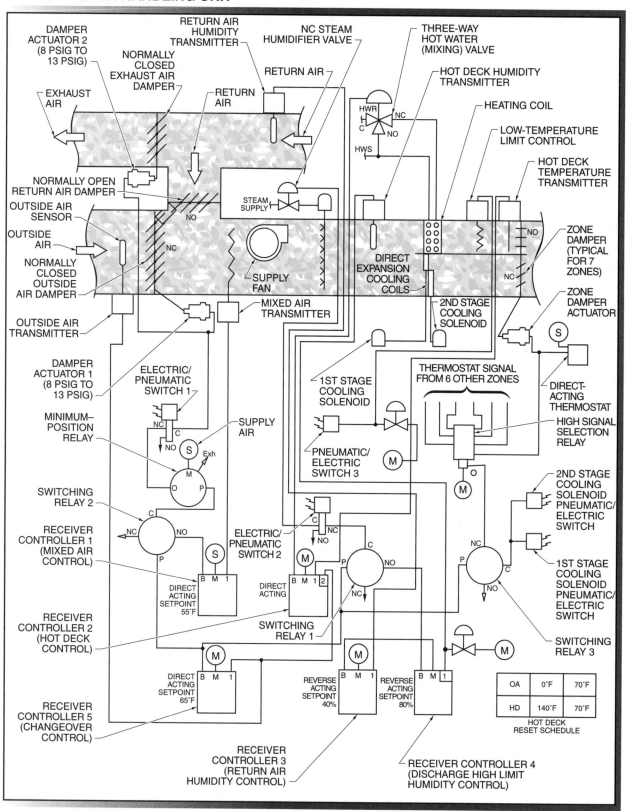

Figure 12-3. Multizone air handling units provide temperature control for many building spaces.

VARIABLE AIR VOLUME AIR HANDLING UNIT

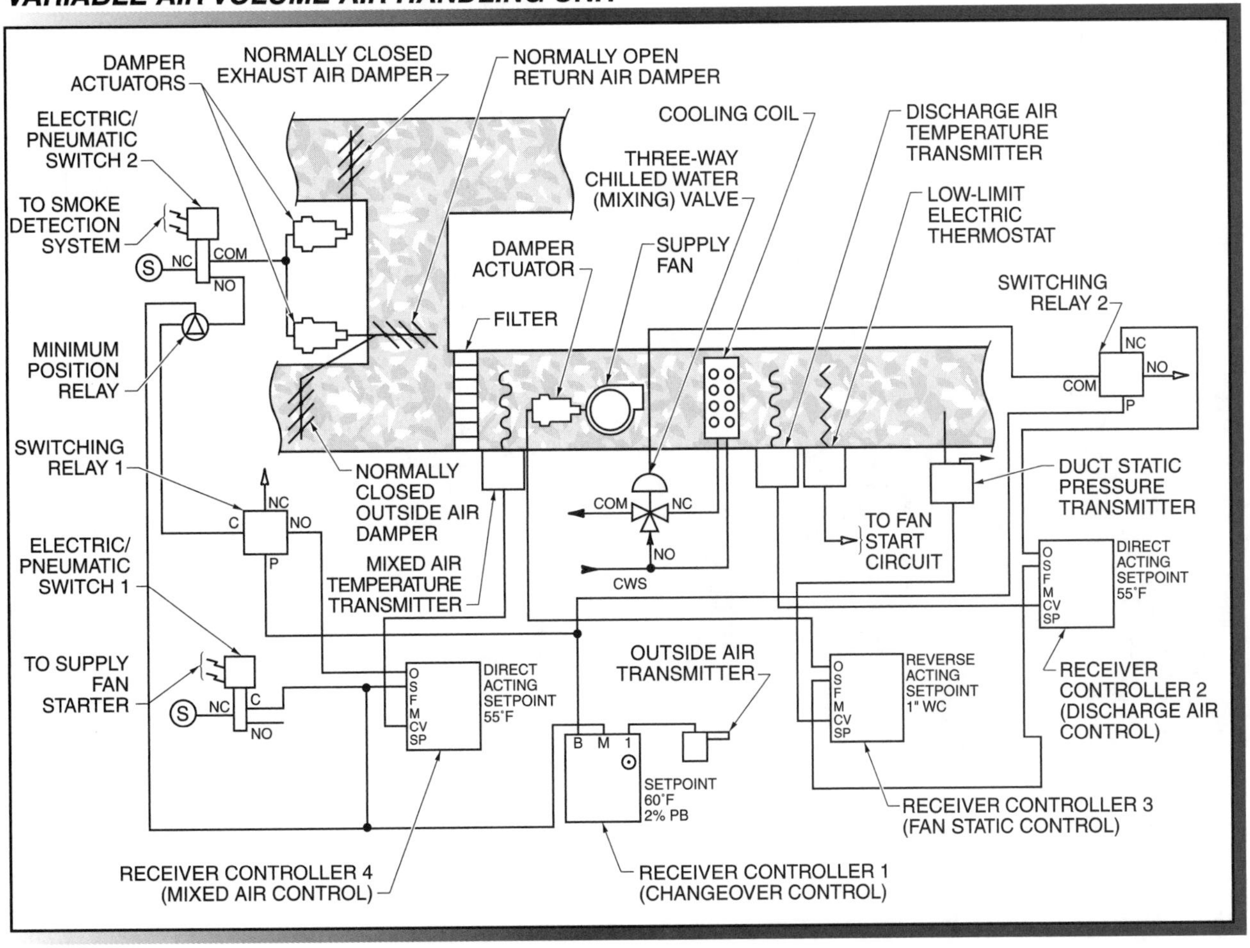

Figure 12-4. Variable air volume air handling units move a variable volume of air by a mechanism that throttles the supply fan.

In a variable air volume air handling unit application, on supply fan startup, electric/pneumatic switch 1 energizes, directing main air to receiver controllers 1, 2, 3, 4, and the minimum-position relay. Receiver controller 3 (fan static control) modulates the supply fan inlet vanes to maintain a pressure of 1″ wc in the supply duct. Receiver controller 2 (discharge air control) modulates the three-way chilled water (mixing) valve to maintain its setpoint.

Receiver controller 4 (mixed air control) modulates the outside air, return air, and exhaust air damper actuators in sequence to maintain its setpoint. As the outside air temperature increases, receiver controller 1 (changeover control) causes switching relay 1 to switch, returning the outside air dampers to their minimum position.

There are multiple possible control interfaces with the fire and smoke safety controls of the building. All sequences must follow appropriate fire codes. Fire and smoke control sequences must be thoroughly tested by trained personnel. In most cases the supply fan will be stopped and outside air dampers closed to prevent fresh air from feeding the fire. Often, electric/pneumatic switches are used to interface fan operation with damper and valve operation.

> *Care must be taken to prevent improper conditions at an air handling unit. For example, a low temperature limit shuts the supply fan OFF if the temperature in the duct falls excessively low. The low temperature limit controls must be checked and calibrated yearly for proper operation. If possible, the controls should be manually operated to ensure that the controlled device actually closes or stops.*

Variable Air Volume Air Handling Unit Optional Sequences. Optional sequences available for variable air volume air handling unit applications include return fan air volume control, outside air enthalpy changeover control, and heating coil control for the beginning of occupancy warm-up.

A pneumatic control system can be used to regulate the flow of hot water produced by a boiler throughout a commercial building.

Boiler Control Application

Many commercial buildings use hot water to provide heat for a building. Although small package boilers have their own integral electric control system, pneumatic controls are often used for maintaining the water loop temperature, including the starting and stopping of system pumps. See Figure 12-5.

In a boiler control application, as the outside temperature decreases, receiver controller 1 (changeover control) causes pneumatic/electric switch 1 to energize circulating pump 2. As the outside temperature decreases, receiver controller 1 causes pneumatic/electric switch 2 to start the boiler. The boiler and circulating pump 1 are controlled by the package boiler electric controls.

Receiver controller 2 (hot water control) modulates a three-way hot water valve to maintain its setpoint, as reset by the outside air temperature. Each zone thermostat modulates its zone hot water valve to maintain each zone setpoint. A hot water return transmitter indicates the return water temperature. System hot water differential pressure is maintained by a self-contained differential pressure valve.

Boiler Optional Sequences. Several optional sequences are available for use in a boiler control application. For example, if multiple pumps are used, operation of the pumps may be regulated through a lead/lag switch. If the boiler has multiple stages, a sequencer may be used to progressively bring on each successive stage. A variable speed drive may also be used to modulate the hot water pump water volume.

Variable Air Volume Terminal Box Control Application

Variable air volume air handling units produce a constant 55°F air temperature. The volume of air is modulated to maintain a constant static pressure of 1″ wc in the ductwork. Each zone in a building has a variable air volume terminal box, which supplies air to a building space or zone. Variable air volume terminal box control variations include multiple stages of heating, and fans in the terminal boxes to provide heat at night. Variable air volume terminal boxes may be pressure-dependent or pressure-independent systems.

In pressure-dependent variable air volume systems, a thermostat controls the damper actuator directly without reference to the volume of air flowing through the ductwork. Pressure-dependent variable air volume systems do not respond well to changes in the volume of air available in the ductwork.

In pressure-independent variable air volume systems, a reset flow controller is used. The reset flow controller measures air flow through the variable air volume terminal box. As the zone thermostat calls for more cooling, the reset flow controller maintains a specific volume of air flow through the terminal box that corresponds to that zone temperature. This air flow setpoint is maintained even if minor fluctuations in the volume of air available in the supply duct occur. The reset flow controller uses a reset schedule that includes building space temperature and air flow rate setpoint. For example, as building space temperature increases from 74°F to 78°F, the air flow rate setpoint changes from 300 cfm to 700 cfm. The reset flow controller can also be calibrated to allow a minimum and maximum air flow setting. The minimum setting provides fresh air ventilation and the maximum setting prevents excessive noise through the variable air volume terminal box. See Figure 12-6.

VAV terminal box reset schedules are normally located on the terminal box in the form of a chart or graph.

BOILER CONTROL

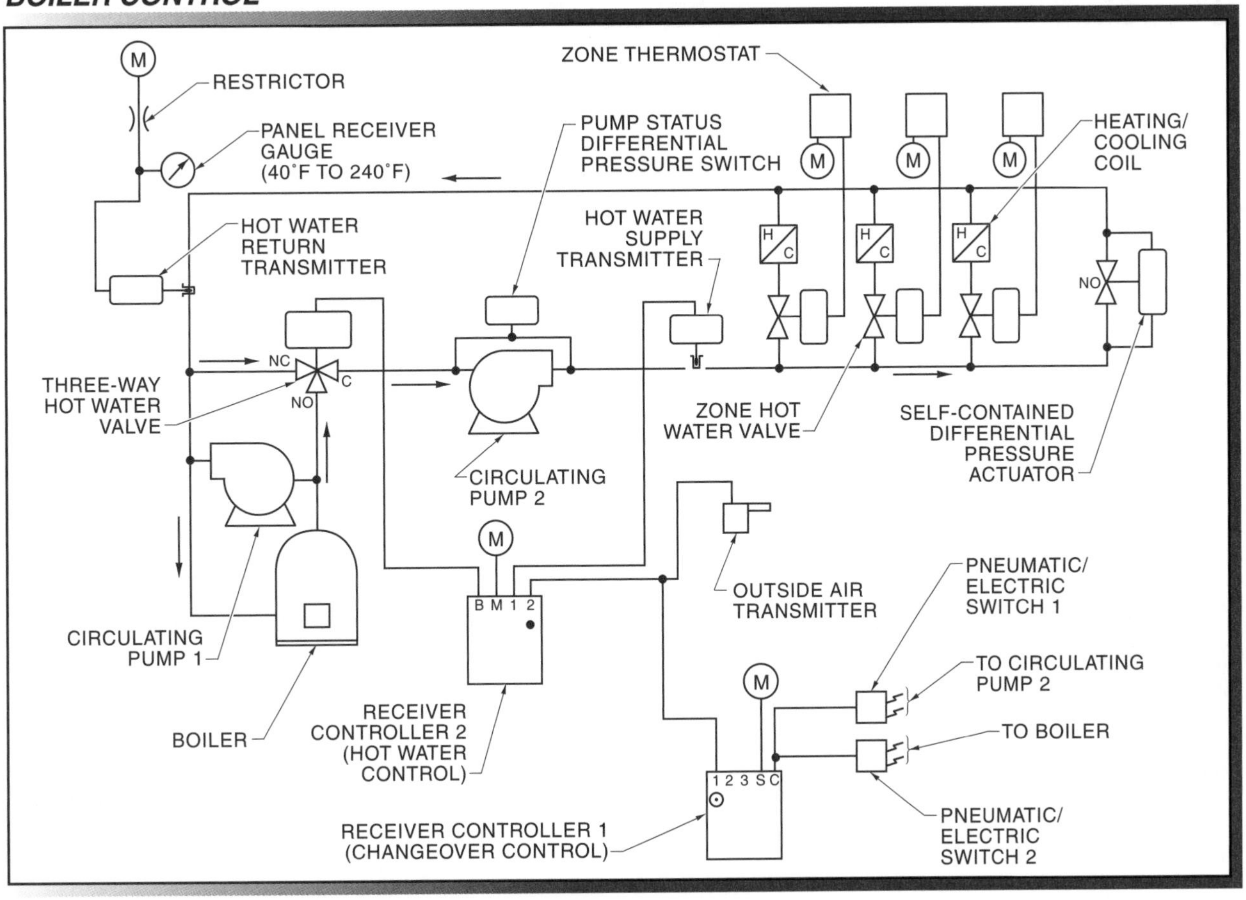

Figure 12-5. Many commercial buildings use pneumatic control systems to control hot water heating systems.

In a variable air volume terminal box control application, for cooling control, as the zone temperature increases, the zone thermostat causes the reset flow controller to increase its flow setpoint as determined by the terminal box reset schedule. When the zone temperature is satisfied, the reset flow controller maintains a minimum volume of air flow through the terminal box for ventilation.

Variable Air Volume Terminal Box Optional Sequences. Optional sequences for VAV terminal box control include hot water reheat valve control, single or multistage electric reheat control, baseboard hot water heat control, and dual-duct retrofit control.

Unit Ventilator Control Sequence Application

A *unit ventilator* is a small air handling unit mounted on the outside wall of each room in a building. Unit ventilators are used in schools, hotels, and areas that require individual space temperature control and ventilation. Most unit ventilators use steam, hot water, or electricity

for heating and outside air for cooling. The limited access inside unit ventilators can make repair and troubleshooting difficult. Some manufacturers use a slide-out chassis for easy replacement and maintenance. See Figure 12-7.

In a unit ventilator control application, occupants can control the fan operation by using an ON/OFF switch. The occupants also control the fan speed. When the fan is shut OFF, the electric/pneumatic switch is de-energized, causing the dampers and hot water valve to return to their normal positions.

When heating is required, the zone thermostat modulates the hot water valve, outside air damper, return air damper, discharge air damper, and bypass air dampers to maintain its setpoint. As the room temperature increases, the zone thermostat closes the hot water valve and opens the bypass air damper, causing air to bypass the hot water coil. The dual-spring range outside air/return air damper actuator opens to allow the outside air damper to go to minimum position.

VARIABLE AIR VOLUME TERMINAL BOX CONTROL

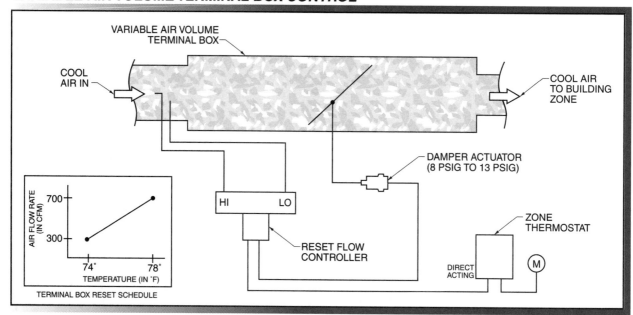

Figure 12-6. Pressure-independent variable air volume terminal boxes reset the flow of air as the building space temperature increases.

UNIT VENTILATOR CONTROL

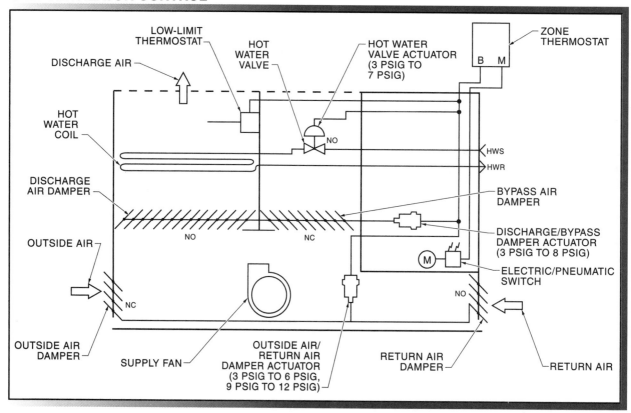

Figure 12-7. Unit ventilators are small air handling units used to provide heating, cooling, and ventilation for individual rooms in a building.

When the zone thermostat is satisfied, the outside air damper is at minimum position, and the discharge air and bypass air dampers modulate along with the hot water valve to maintain setpoint.

When cooling is required, the zone thermostat causes the dual-spring range outside air/return air damper actuator to open the outside air damper and close the return air damper. This allows a large amount of outside air to enter the building space to provide cooling. At the same time, the discharge air and bypass air dampers cause the air to bypass the heating coil.

The low-limit thermostat opens the hot water valve and closes the outside air damper if the discharge air temperature is excessively low. The discharge air and bypass air dampers force air across the hot water coil. This occurs regardless of whether the building space temperature is satisfied.

The American Society of Heating, Refrigeration, and Air-Conditioning Engineers (ASHRAE) has defined four distinct sequences of unit ventilators. Unit ventilator sequences are:

- ASHRAE Cycle I – The outside air dampers open to 100% almost immediately. The discharge air low-limit thermostat overrides the zone thermostat to ensure a minimum discharge air temperature (normally 60°F).
- ASHRAE Cycle II – A fixed amount of outside air (normally 10% to 20%) is brought in during the heating cycle. This percentage is gradually increased during the ventilation and cooling cycles. The discharge air low-limit thermostat opens the hot water valve and closes the outside air damper to ensure a minimum discharge air temperature.
- ASHRAE Cycle III – The discharge air low-limit thermostat is moved to become a mixed air thermostat in the mixed air section of the unit ventilator. The mixed air thermostat modulates the outside air and return air dampers to maintain its setpoint of 55°F. The zone thermostat controls the hot water valve to maintain its setpoint without interference from the mixed air thermostat.
- ASHRAE Cycle W – Cycle W is the same as cycle II except that the discharge air low-limit thermostat controls only the outside air and return air dampers. The zone thermostat controls the hot water valve without interference from the discharge air low-limit thermostat.

Unit Ventilator Optional Sequences. In addition to the ASHRAE cycles, optional sequences available in unit ventilator control applications include using chilled water or a compressor in the unit if mechanical cooling is needed. A day/night thermostat may also be substituted for the standard single-setpoint thermostat. Fan operation may also be accomplished through a facility management system controller instead of a manual switch.

Cooling Tower Control Application

Many commercial buildings use liquid chillers to provide cooling. A *liquid chiller* is a system that uses a liquid (normally water) to cool building spaces. The refrigeration effect is normally produced by absorption or mechanical compression.

In a cooling tower, water from a condenser is cooled by evaporation.

Liquid chillers normally reject heat to the atmosphere through a cooling tower. A *cooling tower* is an evaporative water cooler that uses natural evaporation to cool water. In a cooling tower, condenser water is sprayed across baffles, which break the spray into small, cascading droplets. The droplets are cooled by evaporation as air passes through louvers in the cooling tower. A fan is used to circulate additional air if necessary.

The cool water from the cooling tower is pumped to a condenser in the liquid chiller. A *condenser* is a heat exchanger that removes heat from high-pressure refrigerant vapor. The water circulating through the condenser contacts tubes filled with hot refrigerant vapor. The cool water absorbs heat from the refrigerant and is pumped to the cooling tower, where the cycle is repeated. Water is added as required because of the evaporation at the cooling tower. See Figure 12-8.

COOLING TOWER CONTROL

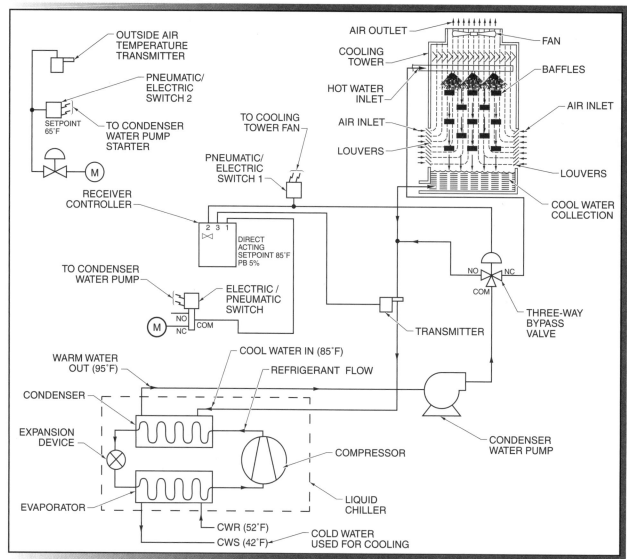

Figure 12-8. Cooling towers are used to reject heat from water that is used to cool commercial buildings.

In a cooling tower control application, as the outside air temperature drops, pneumatic/electric switch 2 energizes the condenser water pump. When condenser water flow is proved (verified), the electric/pneumatic switch energizes, switching on air to the receiver controller. The receiver controller modulates a three-way bypass valve to maintain its setpoint. If the condenser water temperature continues to rise, the receiver controller energizes the cooling tower fan through pneumatic/electric switch 1.

Cooling Tower Optional Sequences. Optional sequences in cooling tower control applications include staged multiple cooling tower fans or variable-speed drives on the cooling tower fans. An outside air tem-

perature lockout may be added which consists of an electric thermostat or a pneumatic controller piped to pneumatic/electric switch 2. This would turn off the condenser water pump if the outside air temperature was excessively low. A heater may also be included in the cooling tower sump to keep the temperature above freezing for winter operation. The heater could be energized by an additional pneumatic/electric switch.

Many basic control systems use a single setpoint. However, the ideal cooling tower setpoint is related to the outside air wet bulb temperature.

The most common control systems used in HVAC equipment are electrical control systems. Electrical control systems use 24 VAC or higher electricity to operate the devices in the system. Electrical control system controls include power, operating, and safety controls. The standard tool used in troubleshooting an electrical control system is a digital multimeter (DMM).

ELECTRICAL CONTROL SYSTEMS

An *electrical control system* is a control system that uses electricity (24 VAC or higher) to operate devices in the system. Electrical control systems are the most common control systems used in the HVAC industry. Electrical control system applications include residential furnaces, split systems, boilers, cooling towers, and water chillers. An *electromechanical control system* is a control system that uses electricity (24 VAC or higher) in combination with a mechanism such as a pivot, mechanical bellows, or other device.

Electricity

Electricity is the energy released by the flow of electrons in a conductor (wire). An *electron* is a particle that has a negative electrical charge of one unit. Electrons orbit the nucleus of an atom. The *nucleus* is the heavy, dense center of an atom. The nucleus of an atom contains protons and neutrons. A *proton* is a particle that has a positive electrical charge of one unit. A *neutron* is a particle that has no electrical charge. See Figure 13-1.

 To prevent serious injury or death, HVAC technicians must always use proper safety equipment and follow all applicable safety procedures.

ATOMIC STRUCTURE

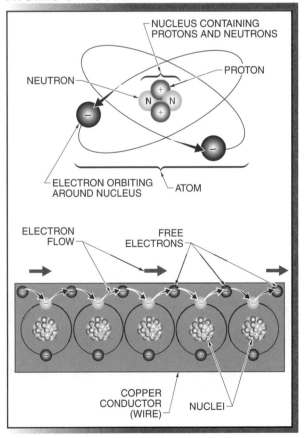

Figure 13-1. Atoms consist of electrons, protons, and neutrons.

Electrons flow from a power source that has an excess of electrons. Electrons flow easily through substances with little resistance (conductors). A *conductor* is a material that has little resistance and permits electrons to move through it easily. Conductors include materials such as copper and silver. Electrons do not flow easily through substances with high resistance (insulators). An *insulator* is a material that has a high resistance and resists the flow of electrons. Insulators include materials such as plastic, rubber, and glass.

Current is the amount of electrons flowing through a conductor. Current is measured in amperes (A). *Voltage* is the amount of electrical pressure in a conductor. Voltage is measured in volts (V). *Resistance* is the opposition to the flow of electrons. Resistance is measured in ohms (Ω). Current, voltage, and resistance are related mathematically through Ohm's law. *Ohm's law* expresses the relationship between voltage, current, and resistance in a circuit. Ohm's law states that current in a circuit is proportional to the voltage and inversely proportional to the resistance. Using Ohm's law, any value in this relationship can be found when the other two are known. See Figure 13-2.

OHM'S LAW

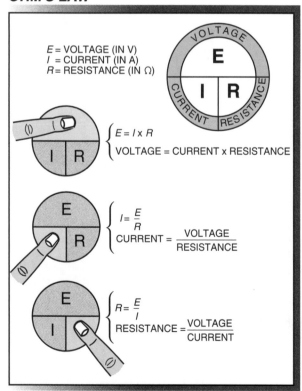

Figure 13-2. Ohm's law expresses the relationship between voltage, current, and resistance in a circuit.

Electrical Circuits

An *electrical circuit* is the interconnection of conductor(s) and electrical elements through which current is designed to flow. An electrical circuit consists of a power source, conductors, switch, and load (device that performs electrical work). See Figure 13-3. For a control circuit to operate properly, each control circuit element must perform its task. The control function experiences problems if any part of the circuit is not performing properly. An HVAC technician must be able to identify each component and its function in an electrical circuit.

ELECTRICAL CIRCUITS

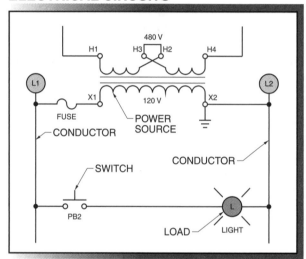

Figure 13-3. An electrical circuit consists of a power source, conductors, switch, and load (device that performs electrical work).

Control System Current

When current is applied to a control circuit, the current flow performs work. In general, the higher the current flow, the higher the amount of work performed by a circuit. Using control drawings, an HVAC technician must be able to trace current flow through the complete circuit, from its source to return. Ammeters are used to measure the current flow in a circuit. Current may be direct current or alternating current. See Figure 13-4.

Direct Current. *Direct current (DC)* is current that flows in one direction only. All DC voltage sources have positive and negative terminals, which establish polarity in a circuit. *Polarity* is the positive (+) or negative (–) state of an object. All points in a DC circuit have positive and negative polarity. Many electronic devices such as personal computers and cell phones operate using direct current.

CURRENT

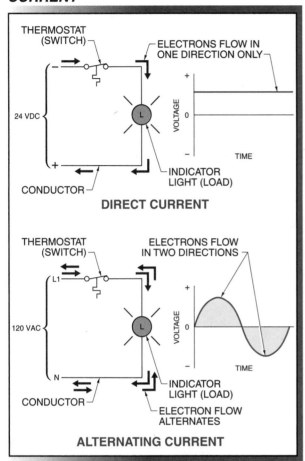

Figure 13-4. Direct current has electrons flowing in one direction only. Alternating current has electron flow changing direction at given intervals.

Alternating Current. *Alternating current (AC)* is current that reverses its direction at regular intervals. Most residential, commercial, and industrial HVAC systems use alternating current. In an AC system, the polarity alternates at regular intervals. The number of times (frequency) that a circuit alternates is measured in cycles per second. *Hertz (Hz)* is the international unit of frequency equal to one cycle per second. The United States, Canada, and Mexico use AC power having a frequency of 60 Hz.

Circuit Connections

HVAC electrical circuit components such as switches, conductors, and loads are connected in series, parallel, or series/parallel combinations. See Figure 13-5. A *series connection* is a connection that has two or more components connected so that there is one path for

current flow. In a series circuit, if any component is not connected, or if any switch is open, there is not a complete path for current flow. Without a complete path for current flow, the load, such as a motor or compressor, is not energized. Series connections are often used for devices that provide a safety function in a system. In this application, if any one safety device is open, the load does not operate.

CIRCUIT CONNECTIONS

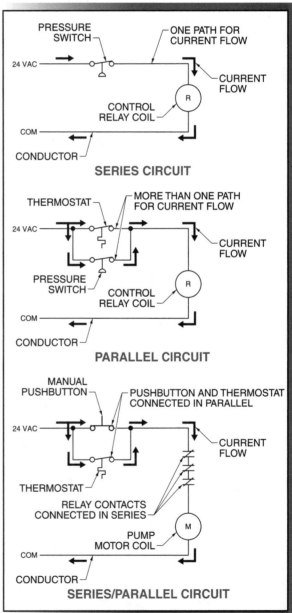

Figure 13-5. Electrical circuit components are connected in series, parallel, or series/parallel combinations.

A *parallel connection* is a connection that has two or more components connected so that there is more than one path for current flow. Parallel connections are commonly used in applications that contain multiple devices that are able to operate the same electrical load. A *series/parallel connection* is a combination of series- and parallel-connected components. Many HVAC control circuits use series- and parallel-connected components to operate a load.

Voltage Supply

AC voltage is constantly changing in polarity and magnitude. AC voltage supplied to an HVAC system may be low voltage, line voltage, or high voltage.

Low voltage is voltage at 30 VAC or less (commonly 24 VAC). Low-voltage control systems are the most common HVAC control system. In a low-voltage control system, high voltage is reduced to 24 VAC by a step-down transformer. At the low-voltage level, there is little danger of accidental shock or fire. Also, low-voltage control system conductors can be relatively small and do not require metal conduit. In a low-voltage control system, the low voltage is used to open and close relay contacts that switch line- or high-voltage loads in the system ON and OFF.

Line voltage is voltage at 120 VAC, up to 4160 VAC. Control systems at 120 VAC are the most common line-voltage control systems. Line-voltage control systems are used in industrial areas and systems where access can be restricted to authorized personnel. Specific measures are taken to reduce the possibility of shock and fire.

High voltage is voltage over 600 VAC. High voltage is used almost exclusively in commercial and industrial applications. In a high-voltage system, power is supplied from a utility at a high voltage level to power fan motors, compressors, and pumps. The high voltage allows HVAC units to come up to speed rapidly and generate large amounts of starting torque for large loads. High voltage is commonly supplied as 3ϕ power.

While high voltage is needed to start large loads, it is impractical in an HVAC control system because the amount of heat given off by the current flow affects sensitive temperature-sensing devices. High-voltage levels may also cause a spark when switched that could ignite flammable substances. High-voltage equipment is normally controlled by line- or low-voltage control circuits.

AC voltage is either single-phase (1ϕ) or three-phase (3ϕ). See Figure 13-6. Single-phase AC voltage contains one alternating voltage waveform. Single-phase

power at 120 VAC is produced with one hot wire and a grounded neutral (white wire). Single-phase power at 120 VAC uses a grounded neutral to create a potential difference between the hot and neutral wire. This potential difference causes electricity to flow. Single-phase power at 230 VAC is produced by using two hot wires. Single-phase circuits at 230 VAC do not require a neutral wire to create a potential difference. Potential difference exists between the two hot wires, causing electricity to flow in the circuit.

AC VOLTAGE

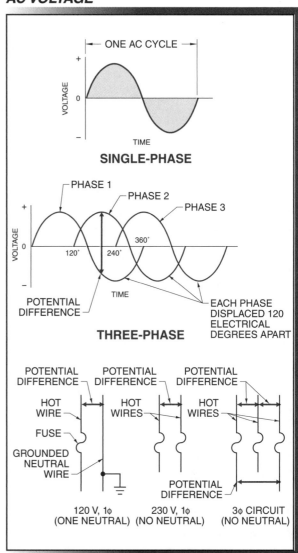

Figure 13-6. Single-phase power is used in residential and small commercial service. Three-phase power is used in large commercial and industrial applications.

Three-phase AC voltage is a combination of three alternating voltage waveforms, each displaced 120 electrical degrees apart. Three-phase AC voltage is produced by the three coils in commercial generators. The coils are out of phase with each other. Three-phase AC voltage is commonly required at industrial facilities. Common 3ϕ AC voltages are 230 V, and 460 V. Unlike 1ϕ circuits, 3ϕ circuits do not require a neutral wire. The potential difference between each phase causes current to flow in each phase. The voltage between each phase should be approximately equal. Three-phase voltage is widely used to supply large amounts of power and is ideal for operating large motors and compressors.

Power Supply

An electrical control system power supply is normally located between the building power supply and the HVAC unit. Most electrical control systems operate at 24 VAC. Because HVAC loads operate at line or high voltage, a transformer is used as a control system power supply to reduce the line voltage or high voltage to the required 24 VAC for use in the control circuit.

A *transformer* is a device that steps up or steps down alternating current. A *step-up transformer* is a transformer in which the secondary coil (high-voltage side) has more turns of wire than the primary coil (low-voltage side). Step-up transformers increase the voltage level from the primary side to the secondary side.

A *step-down transformer* is a transformer in which the secondary coil (low-voltage side) has fewer turns of wire than the primary coil (high-voltage side). Step-down transformers reduce the voltage level from the primary side to the secondary side. See Figure 13-7. A step-down transformer is often provided and prewired in packaged units such as rooftop units, split system air conditioning systems, residential furnaces, and heat pumps.

Disconnects

A *disconnect* is a switch that isolates electrical circuits from their voltage source to allow safe access for maintenance or repair. All HVAC units contain a disconnect. Disconnects have a lever on one side that opens or closes electrical contacts. The lever enables maintenance personnel to shut OFF the power to an HVAC unit so that maintenance tasks can be performed. See Figure 13-8.

TRANSFORMERS

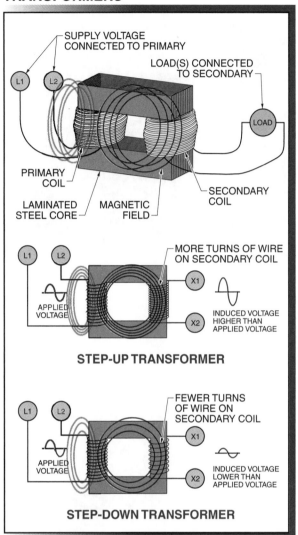

Figure 13-7. A transformer is a device that steps up or steps down alternating current.

The National Electrical Code® (NEC®) requires that a disconnect be provided within visual range of the equipment it disconnects. The position of the lever provides a visual indication of whether power is being supplied to the unit. The lever commonly has a tab that allows a lock or tag to be inserted when the power is disconnected. A *lockout* is the use of locks, chains, or other physical restraints to positively prevent the operation of equipment. A *tagout* is the process of attaching a danger tag to the source of power to indicate that the equipment may not be operated until the tag is removed. Lockouts and tagouts are removed only by authorized personnel.

DISCONNECTS

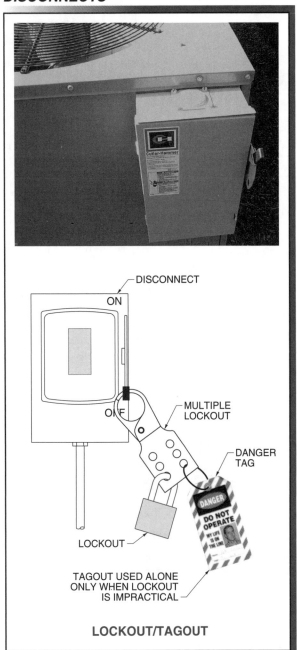

Figure 13-8. Disconnects are used to open or close electrical contacts and allow equipment to be serviced safely.

Know and follow the OSHA lockout/tagout procedure. Devices such as variable-speed motor drives can contain stored electrical energy. Each phase of multiple phase equipment must be tested before declaring that the equipment is locked or tagged out.

Overcurrent Protection

Electrical equipment must be protected from excessive current flow. *Overcurrent protection* is a device that shuts OFF the power supply when current flow is excessive. Overcurrent protection devices include fuses and circuit breakers. A *fuse* is an overcurrent protection device with a fusible link that melts and opens the circuit when an overload condition or short circuit occurs. A *circuit breaker* is an overcurrent protection device with a mechanism that automatically opens the circuit when an overload condition or short circuit occurs. Fuses and circuit breakers are located in an electrical panel that provides power to an HVAC unit. See Figure 13-9. Overcurrent protection devices sense the magnetic field caused by electricity passing through the conductors to an HVAC unit. The power is disconnected from the HVAC unit if the magnetic field is excessively high.

Figure 13-9. Circuit breakers provide overcurrent protection to a circuit.

Internal motor overcurrent protection can also be provided. With internal motor overload protection, a bimetallic thermodisc is embedded inside motor windings, where it senses the motor temperature directly. The bimetallic thermodisc warps and opens the circuit when the motor is overloaded.

Relays

Many HVAC control systems use low voltage such as 24 VAC. A *relay* is a device that uses a low voltage to switch a high voltage. Relays are required because most motors and compressors operate at line voltage. A relay consists of a coil, stationary contact, and movable contact. See Figure 13-10. When current passes through the coil, a magnetic field is developed which pulls the movable contact in contact with the stationary contact. This closes the load circuit, turning ON the device. Relays often consist of multiple contact configurations for different applications.

RELAYS

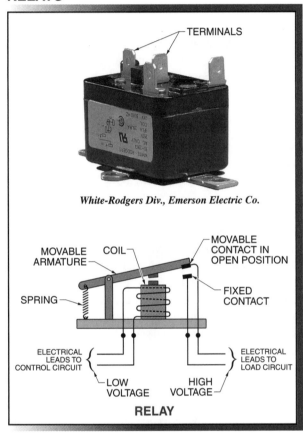

White-Rodgers Div., Emerson Electric Co.

Figure 13-10. Relays allow a low-voltage control circuit to start and stop line-voltage devices.

A normally open contact prevents current flow across the contacts if no power is applied to the coil. Normally open contacts close when power is applied to the coil. Normally closed contacts permit current flow across the contacts if no power is applied to the coil. Normally closed contacts open when power is applied to the coil.

Contactors and Motor Starters

A *contactor* is a control device that uses a small control current to energize or de-energize the load connected to it. Contactors are used in applications that have high current flow requirements. Contactors use the same principle as a relay but in a heavy-duty package. Contactors are used to switch lights and heating elements. A *motor starter* is an electrically operated switch (contactor) that includes motor overload protection. The overloads in a motor starter open if high or excess temperatures due to a high current draw are present. See Figure 13-11.

CONTACTORS AND MOTOR STARTERS

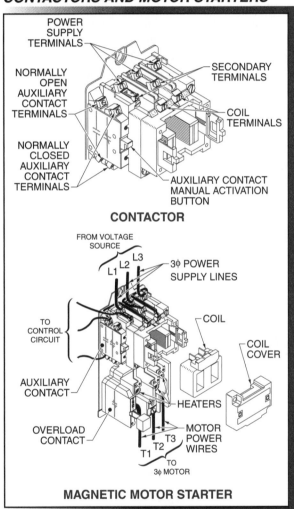

Figure 13-11. Contactors and motor starters are used to start and stop large loads such as heating elements and 3ϕ motors.

When testing a magnetic motor starter coil with a DMM, a reading of 50 Ω to 300 Ω indicates a good coil, a reading of 0 Ω to 2 Ω indicates a shorted coil, and a reading of OL indicates an open coil.

Operating Controls

Operating controls turn HVAC devices ON or OFF to meet specific conditions such as temperature, pressure, and humidity. HVAC unit operating controls are required to maintain the correct environment in commercial buildings. Operating controls include thermostats, pressure switches, and humidistats.

Thermostats. A *thermostat* is a temperature-actuated switch. Thermostats may be used to control the temperature in building spaces, piping, ducts, flues, or heat exchangers. Thermostats are commonly used in low-voltage applications but can also be used in line-voltage applications. Thermostats use bimetallic or remote bulb sensing elements to sense and respond to temperature changes. See Figure 13-12.

THERMOSTATS

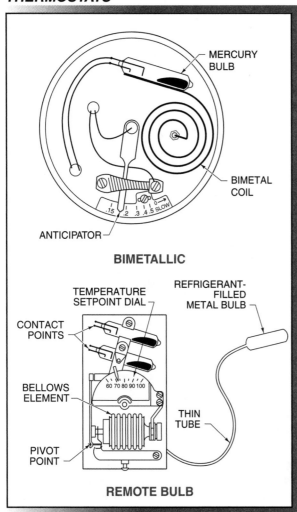

Figure 13-12. Thermostats use bimetallic or remote bulb sensing elements.

A bimetallic element consists of two metals with different rates of expansion and contraction that are joined together. The bimetallic element bends when the temperature changes. A bimetallic element can be attached to contacts or a mercury bulb that opens or closes an electrical control circuit when the proper temperature (setpoint) is reached. A duct thermostat often contains a helix-shaped bimetallic element that is inserted into a duct.

Remote bulb thermostats use a bulb filled with refrigerant. As the temperature at the bulb changes, the pressure inside the bulb changes. The pressure change opens or closes electrical contacts at the appropriate temperature (setpoint). Remote bulb thermostats have a bulb that may be mounted in a duct or pipe and a temperature setpoint adjustment on the front of the thermostat. Setpoint adjustments determine the temperature at which the thermostat electrical contacts are opened or closed.

Anticipators are often included with electrical thermostats. A *heating anticipator* is a small heater that causes an HVAC unit to stop heating before normal shutdown to avoid overshooting. *Overshooting* is the increasing of a controlled variable above the controller setpoint. Overshooting is normally caused by the control system operating the controlled device longer than required. Overshooting is common in electrical control systems. Cooling anticipators cause an HVAC unit to begin cooling before normal startup to prevent the building space temperature from rising excessively high before the unit responds.

Heating anticipators are wired in series with the heating device. Cooling anticipators are wired in parallel with the cooling device. While cooling anticipators are commonly fixed in value, heating anticipators can be adjusted to different values to reflect the heating system to which they are connected. Heating anticipators are commonly set to the value of the current draw of the heating device. When multiple stages of heating or cooling are needed, multiple elements with adjustable setpoints are included in the same thermostat package.

Most thermostats use a standard wire color-coding system. For example, red (R) is used for the wire connected to the transformer 24 VAC side, white (W) is used for heating, yellow (Y) is used for cooling, and green (G) is used for fan control. Most packaged units are prewired internally and provide a terminal block to which the thermostat wires are connected. Packaged unit thermostats also have mounted switches so occupants can run the system fan and also switch from the heating to cooling mode.

Pressure Switches. A *pressure switch* is an electric switch operated by pressure that acts on a diaphragm or bellows element. The diaphragm is made of neoprene or thin metal and the bellows is made of corrugated metal. A neoprene element is used for the low pressures found in ducts and building zones. A bellows element is often used for high-temperature and high-pressure applications such as steam and refrigeration systems. Pressure switches are available that measure small pressures measured in inches of water column (in. wc) or large pressures measured in pounds per square inch (psi). See Figure 13-13. Pressure switches normally contain a setpoint and differential adjustment similar to that found on a thermostat.

Humidistats. A *humidistat* is a device that senses the humidity level in the air. Humidistats are used because in many HVAC systems, there is a desire to control humidity in addition to temperature and pressure. Humidity-sensing elements normally consist of a hygroscopic element. A *hygroscopic element* is a device that changes its characteristics as the humidity changes. See Figure 13-14. The expansion of a hygroscopic element causes a mechanical contact to close or open. Care must be taken because dirt can cause problems with the operation of hygroscopic elements. The setpoint and differential calibration of a hygroscopic element should be checked with an accurate humidity-sensing instrument.

PRESSURE SWITCHES

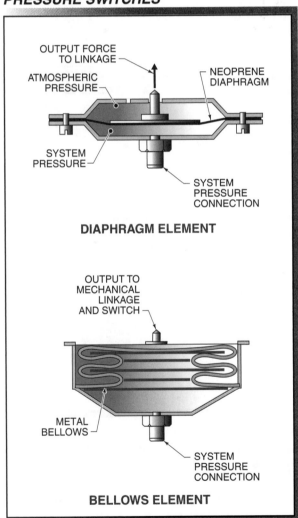

Figure 13-13. A pressure switch is an electric switch operated by the amount of pressure acting on a diaphragm or bellows element.

HUMIDISTATS

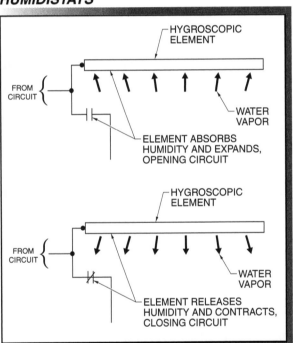

Figure 13-14. Humidity-sensing elements open or close a switch as the element expands or contracts because of changes in humidity level.

Safety Controls

In addition to controlling the temperature, pressure, and humidity of a building environment, safety controls are also required to provide safe operation of an HVAC system. High temperatures can cause equipment damage and are a fire hazard. Lack of water flow can cause a compressor to operate at higher than expected temperatures. Safety controls include limit switches, overloads, and timers.

Limit Switches. High-limit switches are electric switches that have an element which senses the temperature, pressure, or humidity of the surrounding fluid. High-limit switches are used as safety devices in HVAC control circuits. Thermostats, pressure switches, and humidistats can be used as safety devices. Safety devices normally open and stop equipment operation if abnormal conditions arise. For example, a high-limit temperature switch may be used to sense the plenum temperature in a heating application. See Figure 13-15. In this application, if the plenum temperature rises too high, the plenum high-limit temperature switch opens, shutting off power to the heating control circuit and de-energizing the heating circuit. Limit switches are generally connected in series with the operating controls to stop equipment operation if the limit condition is reached.

HIGH-LIMIT TEMPERATURE SWITCHES

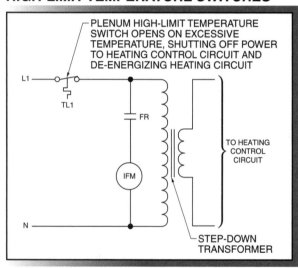

Figure 13-15. High-limit temperature switches shut down power to a device if the temperature is excessively high or low.

A common application of a high-limit temperature switch is maintaining safe operating temperatures in an HVAC system. High-limit temperature switches are commonly inserted into residential heat exchangers to stop heating equipment operation if the air plenum temperature becomes excessively high. Low-limit temperature switches are commonly used in air handling systems to protect the system against freezing if the temperature in the air stream becomes excessively low.

High-limit temperature switches can be reset when normal system operation resumes. The resetting operation can be automatic when normal operating values are achieved, or manual, which requires an operator to reset the switch by hand after normal operating conditions are achieved.

Pressure switches are used to ensure that system pressures are within safe limits. Pressure switches are used to lock out compressors if the discharge pressure becomes excessively high. Differential pressure switches can be used to ensure proper air and water flow by measuring the differential pressure across a fan or pump. When a fan or pump is operating properly, there is a specific difference in pressure generated between the inlet and discharge of the device. For example, a common differential pressure switch application is to verify air flow across a fan before heating elements are energized. See Figure 13-16. Electric heating elements may overheat and cause damage if a specific amount of air flow is not present across them. Similarly, water flow may need to be verified before a water chiller compressor operates.

DIFFERENTIAL PRESSURE SWITCHES

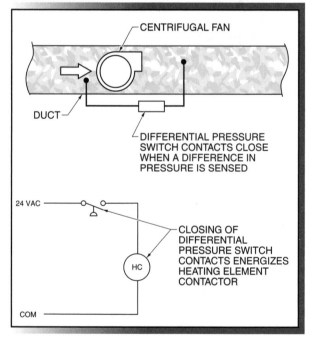

Figure 13-16. Differential pressure switches are used to verify air flow across a fan before heating elements are energized.

A humidifier that is stuck open and supplying excessive steam or water into a duct can cause water damage. Products such as paper, chocolate, and wood can be damaged by excessively high or low humidity. Most commercial humidifier control systems contain a humidity high limit inserted into the duct downstream of the humidifier element. The humidity high limit is set at a value such as 90% rh to 100% rh. A high value indicates the onset of condensation in the duct.

Overloads. An *overload* is a device that prevents overcurrent in an electrical circuit. Overloads may sense the electromagnetic field (magnetic overload) or high temperature (thermal overload) which indicates excessive current flow. For example, an air conditioning compressor working at full load normally draws a current of 10 A. If the compressor has a mechanical problem that prevents proper operation, it may draw 40 A or more. The overcurrent protection device opens and prevents excessive current flow, which could heat the power wiring and cause a fire.

Timers. Many HVAC control systems require close control of timing functions between stages and systems. Timers are often used to prevent equipment startup until a time period has elapsed, ensuring the proper conditions before equipment start. See Figure 13-17. Timers are classified as delay on make, delay on break, and defrost timers.

Figure 13-17. Timers are used to prevent equipment start until a time period has elapsed.

A *delay on make timer* is a timer that begins its timed operation when power is applied to a circuit. After the power is applied, a delay on make timer does not allow a device to operate until a specific length of time has elapsed. A *delay on break timer* is a timer that begins its operation when a circuit is de-energized after equipment is turned off. Delay on make and delay on break timers are used to ensure that motors and compressors do not short cycle, which may cause equipment damage.

Many refrigeration systems operate at temperatures below freezing. Ice can form on an evaporator coil and frost can accumulate. A *defrost timer* is a timer used to initiate a defrost cycle. A *defrost cycle* is a mechanical procedure that consists of reversing refrigerant flow in a system to melt frost or ice that builds up on the evaporator coil. See Figure 13-18. In a defrost cycle, the compressor runs and the hot gas that normally is piped to the condenser coil is diverted to the evaporator coil, melting the ice. HVAC technicians should not circumvent or jump a timer unless it is part of a recommended service procedure.

DEFROST TIMERS

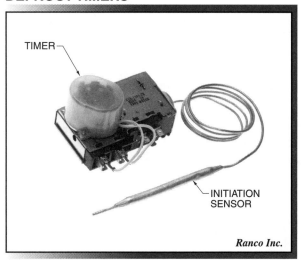

TIMER

INITIATION SENSOR

Ranco Inc.

Figure 13-18. Defrost timers are used to initiate defrost cycles in HVAC units.

Electrical Control System Applications

Electrical control systems are commonly used for heating control, cooling control, packaged unit control, refrigeration control, humidity control, hot water control, and heat pump control.

Heating Control. In a heating control application, a room thermostat energizes a gas valve coil which allows gas to flow to a burner. The burner ignites, producing heat for the building spaces. When the plenum temperature rises high enough, a fan switch turns on the fan motor. If a fan failure occurs, the plenum temperature rises excessively. In this case the high limit temperature switch de-energizes the gas valve solenoid coil. The pilot safety relay de-energizes the circuit if the pilot is lost. See Figure 13-19.

HEATING CONTROL

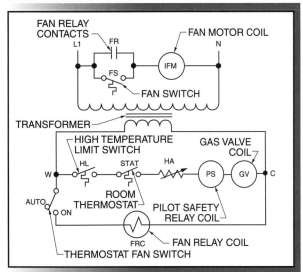

Figure 13-19. A heating control system includes a gas valve solenoid that is energized when the building space temperature drops and a high-limit temperature switch that shuts off the circuit if the plenum temperature becomes excessive.

Cooling Control. In a cooling control application, a thermostat energizes a compressor, condenser fan motor, and evaporator (indoor) fan motor when the thermostat is in the cool position and a call for cooling is received. See Figure 13-20. The high- and low-limit pressure switches open and stop the compressor and condenser fan motors if the suction pressure becomes excessively low or the discharge pressure becomes excessively high. The indoor fan continues to run. When the thermostat is in the fan position, the indoor fan is turned ON. When the thermostat is in the OFF position, the compressor, condenser fan, and indoor fan are OFF.

Packaged Unit Control. In a packaged unit control application, during a call for cooling, a room thermostat causes a compressor, outdoor fan motor, and indoor fan motor to run. See Figure 13-21. After the cooling need is satisfied and the compressor is OFF, the cooling anticipator adds false heat to the thermostat element, causing the compressor to energize slightly before setpoint.

COOLING CONTROL

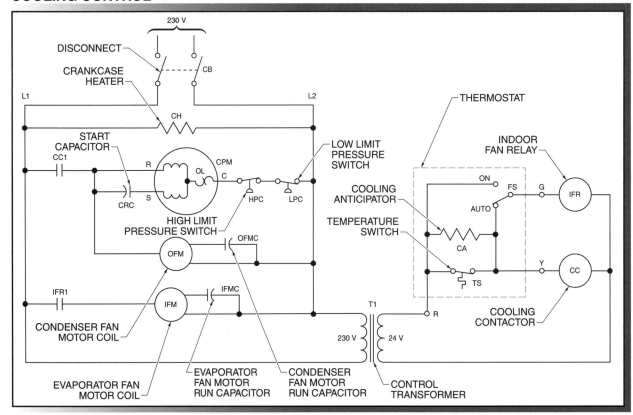

Figure 13-20. A cooling control system includes a thermostat that energizes a compressor, condenser fan motor, and evaporator (indoor) fan motor on a demand for cooling.

The indoor fan runs continuously if the thermostat is in the ON position. When heat is required, the thermostat energizes the electric heating element to provide heat to the building space.

PACKAGED UNIT CONTROL

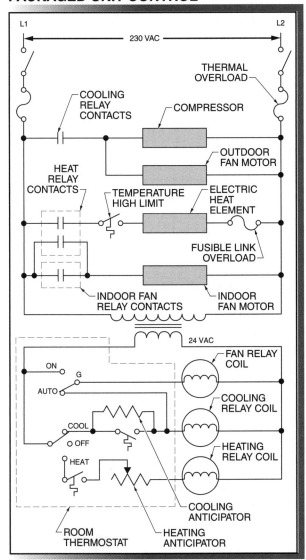

Figure 13-21. A packaged unit control system includes a thermostat that turns on a compressor, outdoor fan motor, and indoor fan motor on a demand for cooling.

Packaged units consist of a single unit that is normally pre-wired internally and ready for power and control connections. A terminal strip is provided to terminate the controller, and leads are provided for connection to the power supply.

Refrigeration Control. In a refrigeration control application, when power is applied to the HVAC unit, the compressor crankcase heater is energized. During a demand for cooling, the compressor, evaporator fan, and liquid line solenoid valve are energized. See Figure 13-22. When the temperature drops below setpoint, the liquid line solenoid valve closes, the evaporator pressure drops, and the low-pressure switch de-energizes the compressor. If a high-pressure condition occurs, the high-pressure limit switch opens, de-energizing the compressor.

REFRIGERATION CONTROL

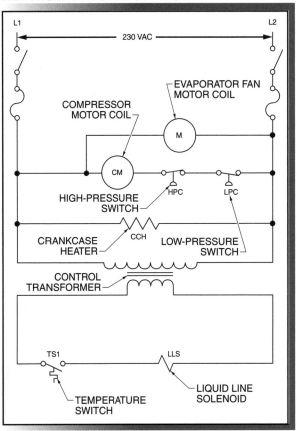

Figure 13-22. A refrigeration control system includes high- and low-pressure switches to shut off the compressor if system pressures become excessively high or low.

Humidity Control. In a humidity control application, a humidistat energizes a humidifier water solenoid valve on a drop in humidity. The humidifier adds moisture to the air flowing to the building space. If the discharge humidity is excessively high or air flow is lost, the humidifier water solenoid valve is de-energized by the high humidity switch or differential pressure switch. See Figure 13-23.

HUMIDITY CONTROL

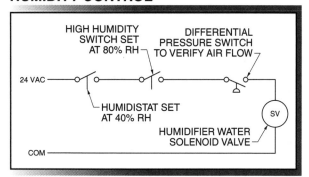

Figure 13-23. A humidity control system includes safety controls for high humidity and loss of air flow.

Hot Water Control. In a hot water control application, a hot water operating control (aquastat) closes when the supply water temperature drops below setpoint. This energizes the burner relay coil, starting the boiler. If the water temperature is excessively high, the high-limit temperature switch opens and stops the burner operation. If the gas pressure is excessively high or low, the appropriate switch opens and stops boiler operation. Additional safety control may be required to comply with all necessary codes. See Figure 13-24.

HOT WATER CONTROL

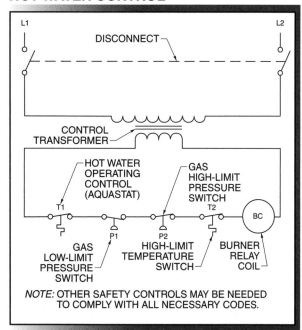

Figure 13-24. A hot water control system includes safety controls for high and low gas pressure as well as high water temperature.

Heat Pump Control. In a heat pump control application, on a demand for cooling, a thermostat energizes a compressor relay, an outdoor (condenser) fan relay, and an indoor fan relay, causing the compressor, outdoor fan, and indoor fan motors to operate. During a demand for heat, the compressor, outdoor fan, and reversing valve operate. If the air temperature continues to drop, an auxiliary heat relay is energized. If ice builds up on the outdoor coil, the reversing valve switches the refrigerant flow, defrosting the outdoor coil. See Figure 13-25.

Troubleshooting Electrical Control Systems

Electrical control systems are the most common control systems in the HVAC industry. An HVAC controls technician requires knowledge of basic control system troubleshooting tasks. The standard tool used when troubleshooting electrical control systems is a digital multimeter (DMM). A *digital multimeter* is a test tool used to measure two or more electrical values. The most common measurements taken when using a digital multimeter to troubleshoot an electrical control system are voltage, current, and resistance measurements.

Voltage is measured across an electrical control device by placing the leads in parallel across the device being tested. See Figure 13-26. Caution must be taken because voltage is measured with the circuit energized. The device (switch) is open if a voltage level is measured. The device (switch) is closed if no voltage is measured. By connecting a DMM from point to point in a circuit, an open device, which prevents a unit from operating, can be found. Voltage testing can also be performed at the power supply to check that the correct amount of voltage is being provided to the circuit.

Current measurements are valuable in a circuit because the reading indicates the amount of work being performed. Current can be measured in a variety of ways. For example, a DMM can be inserted in series with the circuit under test. A clamp-on meter may also be used. A clamp-on meter is a device designed to measure current in a circuit by measuring the strength of the magnetic field around a single conductor. A DMM can be used with a clamp-on adapter to measure current flow. A clamp-on meter and clamp-on adapter do not require that the circuit be opened to measure current. See Figure 13-27.

A common practice is to check DMM operation on a known circuit before troubleshooting a suspect circuit.

HEAT PUMP CONTROL

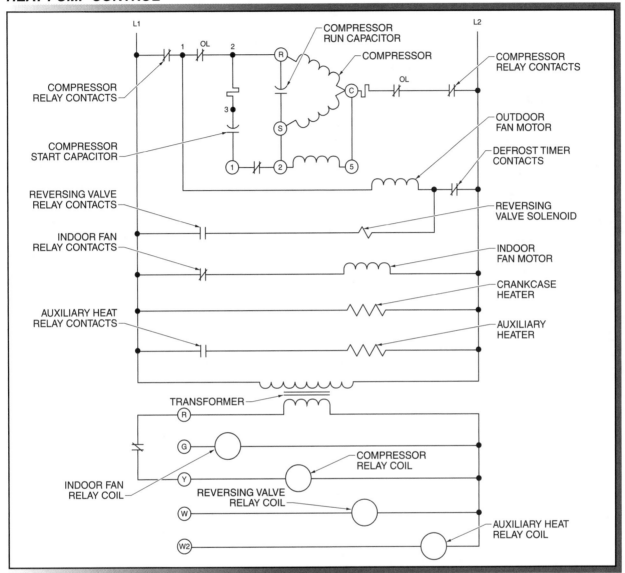

Figure 13-25. A heat pump control system includes a reversing valve to provide heating and cooling to a building space.

Motor windings are often measured for resistance to determine their condition. Resistance may also be checked by continuity. *Continuity* is the presence of a complete path for current flow. DMMs commonly offer a continuity mode. A DMM provides an audible signal if there is no resistance between two points in a circuit. This indicates a complete electrical path (electrical continuity) between the points tested. Continuity checking is helpful when performing large numbers of point-to-point tests to isolate a fault in a control panel. A technician can concentrate on correct placement of the leads and not be concerned with the meter readings.

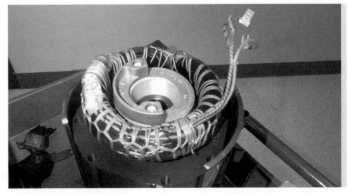

Motor windings can be checked with a DMM set to measure resistance if a motor problem is suspected.

VOLTAGE MEASUREMENT

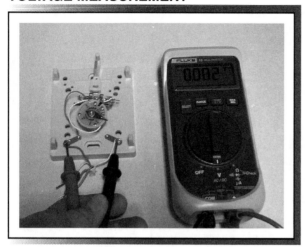

Figure 13-26. Voltage is measured across a device by connecting a DMM in parallel with the device.

Manufacturer recommendations must be followed for each type of HVAC system and test equipment involved. All applicable safety codes and regulations must be followed per the NEC® and/or authority having jurisdiction. Most control system problems fall into the categories of open or closed control device, short circuit, or improper voltage level.

Control devices may be normally closed or normally open. When a device that is normally closed is open, it prevents the unit from operating. When a device that is normally open is closed, it allows the unit to operate. This problem is determined by using a DMM, set to measure voltage, to check whether the devices function as designed open or closed. See Figure 13-28. For example, a unit may have an access door that prevents the unit from operating if the door is open.

CURRENT MEASUREMENT

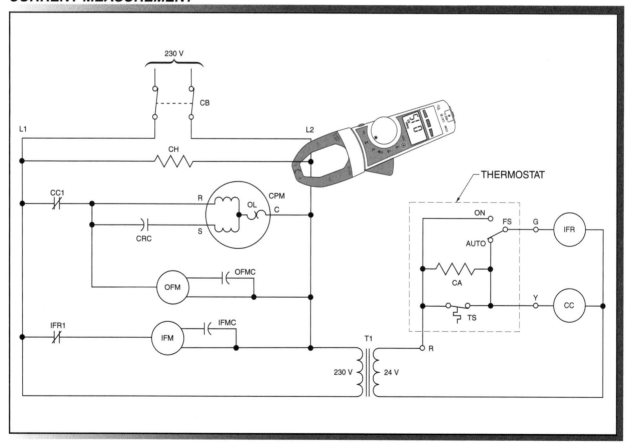

Figure 13-27. A clamp-on meter is used to measure current flow through a conductor by measuring the magnetic field around a conductor.

VOLTAGE TESTING

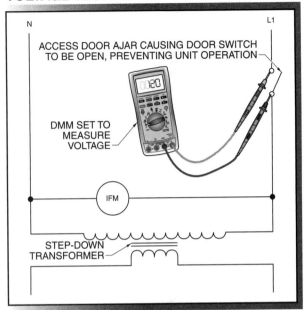

Figure 13-28. A DMM set to measure voltage can be used to determine if a switch is functioning correctly.

A *short circuit* is current that leaves the normal current-carrying path by going around the load and back to the power source or to ground. A control device is bypassed by the current flow if a short circuit is present. A *dead short* is a short circuit that opens the overcurrent protection device as soon as the circuit is energized or when the section of the circuit containing the short is energized. A *partial short* is a short circuit of only a section or several sections of a machine. A short circuit can be determined by measuring the resistance between a device and ground. For example, a reading of 0 Ω between a motor and ground indicates that a motor winding is touching the motor housing. See Figure 13-29.

If there is an improper voltage supply level, either excessively high or low, a control system may perform improperly. A power supply problem may be caused by utility power problems, electrical interference due to lightning, or other causes. DMMs often have a minimum/maximum recording mode that can be used to show fluctuations in the power supply.

SHORT CIRCUIT TESTING

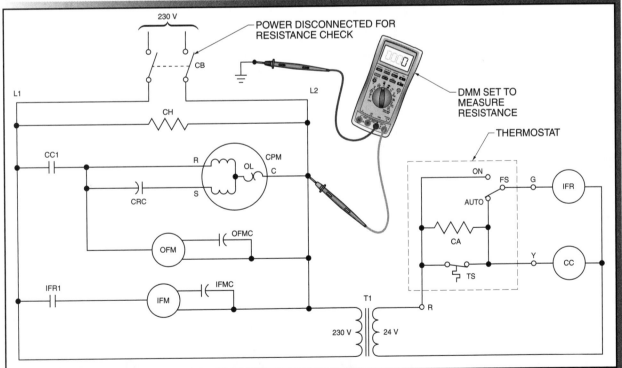

Figure 13-29. A short circuit can be determined by measuring the resistance between a motor and ground.

Low-pressure controls are used in many product refrigeration systems to start and stop the compressor, depending on refrigerant pressure.

Troubleshooting Application – Open Power Circuit.

In this application, a complaint is received that the building space is excessively hot. The technician notices that the rooftop unit fan and cooling system are not operating. The thermostat setpoint is 76°F but the actual building space temperature is 85°F. The technician attempts to turn the fan on by using the fan ON/AUTO switch on the thermostat. The fan does not run. The technician opens the unit control center door and uses a DMM to check for incoming power. See Figure 13-30. The reading is 0 VAC. The technician traces the power supply to a fused disconnect and notices that the disconnect is open. There is no lock or tag on the equipment. The technician checks with building personnel and finds that the disconnect was opened for cleaning purposes and was not turned on after completion of the cleaning.

Troubleshooting Application – Low-temperature Limit Switch Open.

In this application, a complaint is received that the building space is excessively cold. The technician notices that the unit is not producing heat and the air handling unit fan is not operating. The thermostat setpoint is 72°F but the building space temperature is 65°F. See Figure 13-31. The technician

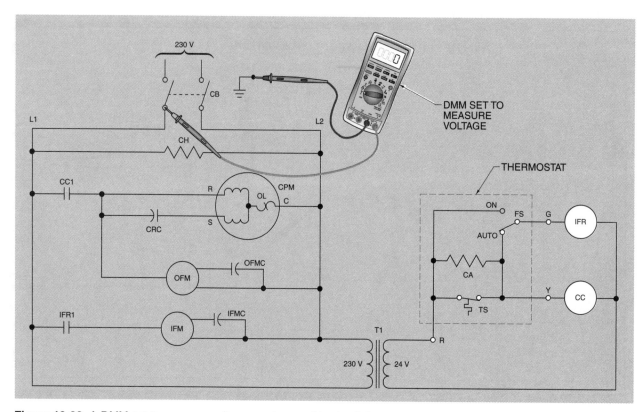

Figure 13-30. A DMM set to measure voltage can be used to check for an open power circuit.

attempts to turn the fan on by using the fan ON/AUTO switch on the thermostat. The fan does not run. The technician uses a DMM to test for incoming power. The DMM reading is line voltage, indicating that the unit has the proper electrical supply. The technician uses the unit ladder diagram and measures each safety switch with the DMM. The technician discovers that the low-limit temperature switch is open. The technician manually resets the low-limit temperature switch and the unit starts.

Troubleshooting Application – Improper Voltage Level. In this application, a complaint is received that the building space is excessively hot. The technician notices that the rooftop unit fan and cooling system are not operating. The thermostat setpoint is 76°F but the building space temperature is 85°F. See Figure 13-32. The technician attempts to turn the fan on by using the fan ON/AUTO switch on the thermostat. The fan does not run. The technician uses a DMM to test for incoming power. The reading is more than 20% less than the line voltage needed for the unit to operate. The technician or owner contacts the power utility to resolve the electrical supply problem.

Figure 13-31. A heat pump that does not operate may contain an open low-limit temperature switch.

Figure 13-32. An HVAC system may not operate because of a low supply voltage.

14

HVAC Control Systems
Electronic Control Systems

Electronic control systems use solid-state components that operate at low power and heat levels. Solid-state components include diodes, transistors, and thyristors. Integrated circuits consist of thousands of solid-state devices all contained in a single package or chip. Troubleshooting electronic circuits normally consists of finding and replacing the printed circuit (PC) board that has the problem.

ELECTRONIC CONTROL SYSTEMS

An *electronic control system* is a control system in which the power supply is 24 VDC or less. Electronic control systems are similar to electrical, pneumatic, and automated control systems. For example, electronic control systems use a power supply similar to electrical control systems. In addition, many pneumatic control system components such as thermostats, switching relays, and damper actuators have electronic control system counterparts. Also, the solid-state components used in electronic control systems are also used to construct automated control system controllers.

Electronic control systems use solid-state electronic components to control HVAC equipment. Electronic control systems use sensors to measure the temperature, humidity, or pressure in an HVAC system. These variables are represented by resistance values of 0 Ω to over 1000 Ω, voltage values of 0 VDC to 10 VDC, or current values of 0 mA to 20 mA. The electronic controller compares the input signal from the sensor to an internal setpoint. The controller also has an adjustment to determine the amount that the controller output changes when the measured variable changes. The output of the electronic controller is sent to a motor-driven damper or valve actuator. Electronic control systems may also include switching, averaging, or other electronic controls that complete the sequence of operation for the HVAC equipment.

Electronic control systems are often referred to as analog electronic systems because they provide variable (analog) switching. With the advent of modern automated systems, analog electronic control systems have lost some of their popularity. Early electronic devices used vacuum tubes to switch or amplify a signal. Modern electronic devices are made of semiconductor materials.

Vacuum Tubes

In the mid 1940s, the most common electronic device was the AM radio. Immediately after World War II, the military began building the first electronic computers. However, the development of computers was hindered because the main component used for electrical switching was the vacuum tube. A *vacuum tube* is a device that switches or amplifies electronic signals. Vacuum tubes perform these functions by allowing electrons to flow to plates that are located inside a glass tube. See Figure 14-1. The glass tube contains plates that are in a vacuum. A vacuum is required because air is an insulator and interferes with the function of the plates. The anode and cathode (main plates) become electrically charged and attract or repel charges at the other plate. The control grid placed between the two main plates modifies the signal between the two main plates. The control grid enables a small control signal to control a large signal between the main plates.

VACUUM TUBE CONSTRUCTION

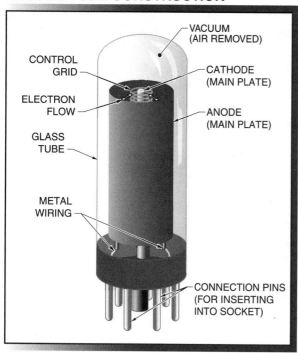

VACUUM
(AIR REMOVED)

CONTROL
GRID

CATHODE
(MAIN PLATE)

ELECTRON
FLOW

ANODE
(MAIN PLATE)

GLASS
TUBE

METAL
WIRING

CONNECTION PINS
(FOR INSERTING
INTO SOCKET)

Figure 14-1. Vacuum tubes perform switching functions by allowing electrons to flow to plates located inside a glass tube.

Vacuum tubes consume large amounts of power, produce excessive amounts of heat, and have high failure rates. These characteristics limited the usefulness of vacuum tubes. The search for smaller, more efficient, and more reliable electrical switching and amplifying components led to the development of semiconductor devices.

Semiconductors

A *semiconductor* is a material in which electrical conductivity is between that of a conductor (high conductivity) and that of an insulator (low conductivity). The application of a voltage to the semiconductor material causes the material to change electrical characteristics. Devices constructed from semiconductor material eliminate or reduce many of the problems associated with vacuum tubes. For example, semiconductors use small amounts of power and produce little heat. The small amount of heat produced increases the reliability of the devices. The systems used for component cooling can also be reduced or eliminated. Semiconductors (solid-state devices) are continually being manufactured in smaller packages, allowing increased capability in a smaller package.

The most common material used for semiconductor devices is silicon. Silicon is one of the most common naturally occurring elements, which makes devices made from silicon inexpensive. The main costs involved are in the engineering and design of a solid-state device, not in the raw materials. Other elements and/or materials are added to silicon to create semiconductor materials with various characteristics. The various materials cause the silicon to act differently under various voltage and current flow conditions.

Doping is the addition of material to a base element to alter the crystal structure of the element. The addition of material to the crystal structure of silicon creates N-type material or P-type material. *N-type material* is material created by doping a region of a crystal with atoms from an element that has more electrons in its outer shell than the crystal. N-type material has extra electrons in its crystal structure available for current flow. *P-type material* is material created by doping a region of a crystal with atoms from an element that has fewer electrons in its outer shell than the crystal. P-type material has empty spaces (holes) in its crystal structure into which electrons can be placed. See Figure 14-2. Most semiconductor devices are constructed from different arrangements of P-type and N-type material. Semiconductor devices include diodes, transistors, thyristors, and integrated circuits.

Diodes. A *diode* is a semiconductor device that allows current to flow in one direction only. Diodes are the most common semiconductor devices. A diode is constructed by placing one layer of P-type material and one layer of N-type material back to back. The area where the materials contact is referred to as the PN junction (PN region). See Figure 14-3.

The P-type material of a diode has holes that are available to receive electrons, and the N-type material has extra electrons available. When a voltage source of .6 VDC or more is connected to the N-type material, the extra electrons of the N-type material jump across the PN junction (current flow) and fill the holes of the P-type material. The .6 VDC is used to forward bias the diode (allow current flow). The .6 VDC required is a function of the doping process performed to the silicon.

Modern devices such as personal computers, DVD players, pagers, and cell phones were made possible by the development of semiconductor devices.

DOPING SILICON

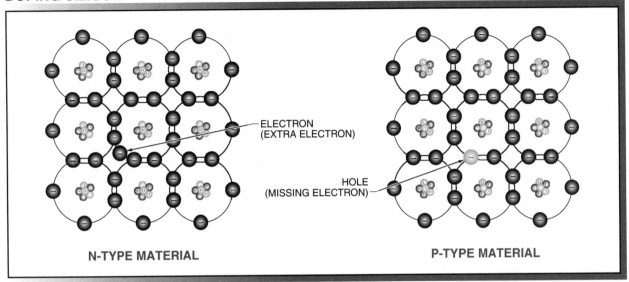

Figure 14-2. N-type material has extra electrons in its crystal structure available for current flow. P-type material has empty spaces (holes) in its crystal structure into which electrons can be placed.

DIODES

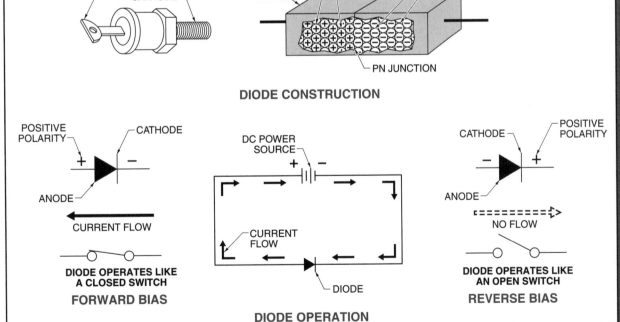

Figure 14-3. A diode is an electronic device that allows current to flow in one direction only.

A voltage supply connected to the P-type material is used to reverse bias a diode. When a diode is reverse biased, electrons are drawn away from the PN junction, preventing current flow. A high voltage connected to the P-type material of a diode makes the diode conduct current temporarily. After a couple of milliseconds, the diode melts or explodes. The low voltage (.6 VDC) normally used by semiconductor devices allows low current flow and creates low amounts of heat. Small power supplies can also be used because of the low current flow.

Diodes are used as electronic check valves and as rectifiers. A *rectifier* is a device that changes AC voltage into DC voltage. Diodes are used in power supplies to change 120 VAC supply voltage into rectified low-voltage DC for an electronic circuit. Rectifiers include half-wave, full-wave, and bridge rectifiers. See Figure 14-4.

DIODE RECTIFIERS

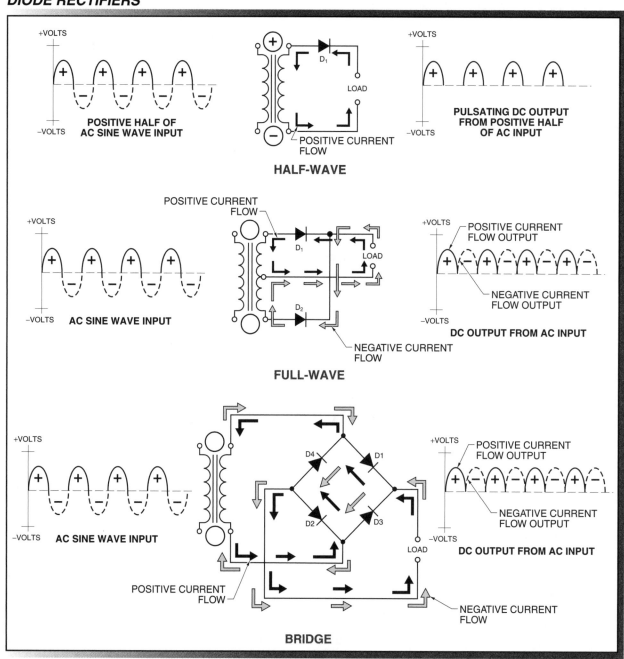

Figure 14-4. Rectifiers are used to change AC voltage into DC voltage and include half-wave, full-wave, and bridge rectifiers.

A *half-wave rectifier* is a circuit containing one diode that allows only half of the input AC sine wave to pass. Half-wave rectification is accomplished because current is allowed to flow only when the anode terminal has a positive polarity with respect to the cathode. Current is not allowed to flow through the rectifier when the cathode has a positive polarity with respect to the anode. Half-wave rectification is inefficient for most applications because one-half of the input sine wave is not used.

A *full-wave rectifier* is a circuit containing two diodes and a center-tapped transformer that permits both halves of the input AC sine wave to pass. Full-wave rectification is accomplished by one diode passing the positive half of the AC sine wave and the second diode passing the negative half of the AC sine wave. A full-wave rectifier is more efficient than a half-wave rectifier because both halves of the input sine wave are used.

A *bridge rectifier* is a circuit containing four diodes that permits both halves of the input AC sine wave to pass. A bridge rectifier is more efficient than a half-wave or full-wave rectifier and is the most common rectifier circuit used in rectification circuits. The output of a bridge rectifier is a pulsating DC voltage that must be filtered (smoothed) before it can be used in most electronic equipment.

Diodes are used for rectification and to block current flow. Some diodes are designed with certain characteristics and are used to perform specific tasks. These diodes include zener diodes, light emitting diodes, and photodiodes.

A *zener diode* is a diode designed to operate in a reverse-biased mode without being damaged. See Figure 14-5. Different types of zener diodes are available that use various voltages to trigger (turn ON) the diode. Zener diodes are commonly used as a voltage shunt or electronic safety valve. When a power supply voltage to an electronic circuit increases due to a malfunction, a zener diode passes the unwanted voltage to ground, protecting the sensitive electronic circuit from overvoltage.

Zener diodes are often specified and installed on electronic controls used in dangerous environments. Zener diodes are used because they shunt any arc current to ground before the arc can occur and cause an explosion.

ZENER DIODES

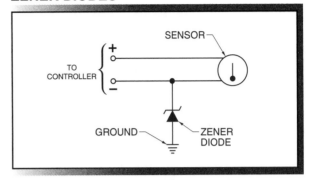

Figure 14-5. Zener diodes are designed to operate in the reverse-biased mode at a certain voltage level.

A *light emitting diode* (LED) is a diode designed to produce light when forward biased. The light emitted from most diodes is very small and not visible, but LEDs are specifically designed to create the maximum amount of light. An LED consists of a small diode chip enclosed in an epoxy housing that contains light diffusing particles. When an LED is forward biased, light is released and diffused in different directions by the epoxy housing. The amount of light produced is directly proportional to the current flow through the LED. See Figure 14-6. To prevent LEDs from being damaged by overcurrent, a resistor is normally placed in series with the LED to limit current flow. The light wavelength determines the color of the LED, with the most common colors being red and green. The wavelength is determined by the forward-bias voltage of differently doped LEDs.

LEDs are commonly used to visually indicate conditions of equipment. For example, LEDs are used to indicate equipment conditions such as power being supplied, that a device is operating properly, or communication with other devices is taking place. Many devices use LEDs that flash in various patterns to indicate to an operator or technician a particular problem with a machine or piece of equipment when the pattern is explained in the technical manual of the machine.

A *photodiode* is an electronic device that changes resistance or switches ON when exposed to light. In many ways, photodiodes are the opposite of LEDs. See Figure 14-7. An adequate light source striking the surface of the PN junction of a photodiode chip causes electrons to jump the PN junction and current to flow. Photodiodes can be designed to switch only when exposed to specific amounts of light or wavelengths of light, allowing photodiodes to be used as light detectors.

LIGHT EMITTING DIODES

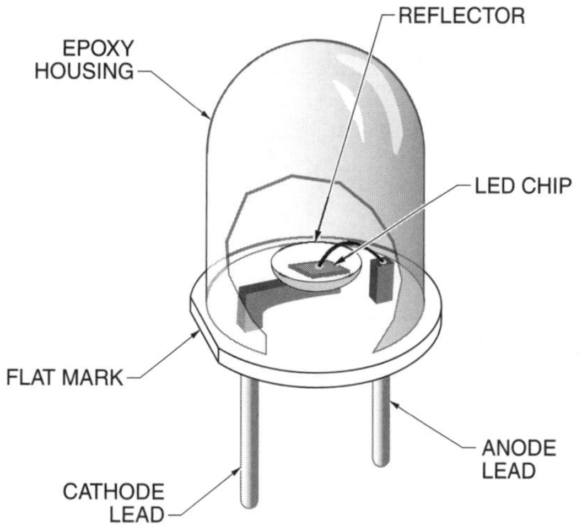

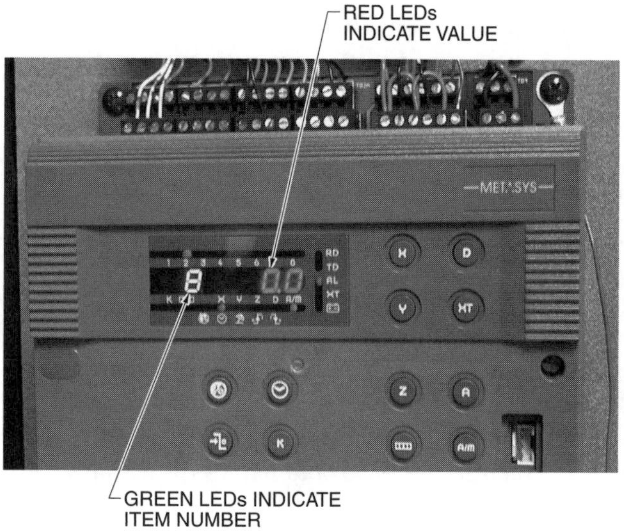

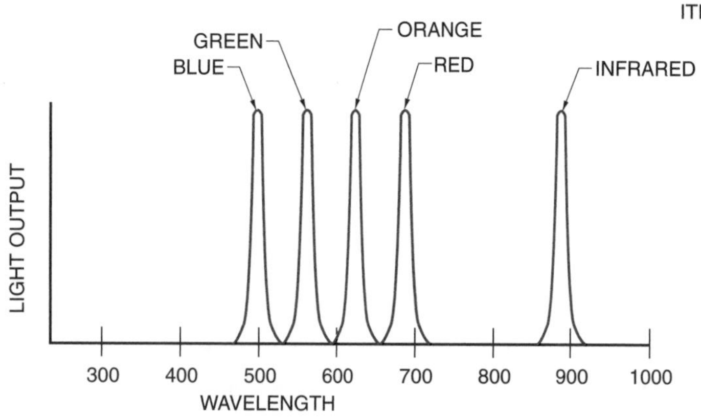

WAVELENGTH	VOLTAGE
490 (BLUE)	2.1 – 2.6
565 (GREEN)	2.2 – 3.0
590 (YELLOW)	2.2 – 3.0
615 (ORANGE)	1.8 – 2.7
640 (RED)	1.6 – 2.0
690 (RED)	2.2 – 3.0
880 (INFRARED)	2.0 – 2.5
900 (INFRARED)	1.2 – 1.6
940 (INFRARED)	1.3 – 1.7

Figure 14-6. Light emitting diodes are designed to produce light when forward biased.

PHOTODIODES

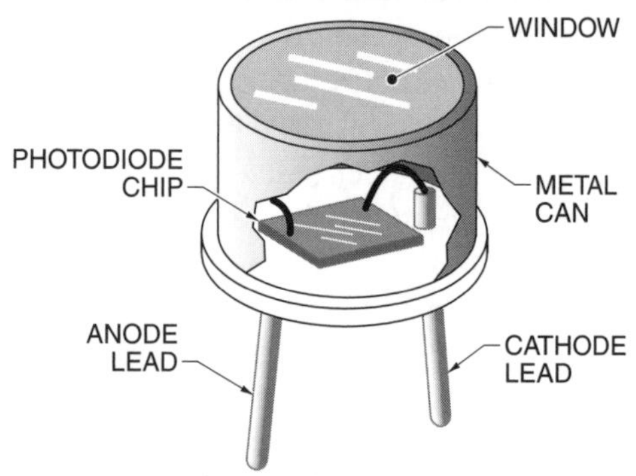

Figure 14-7. A photodiode changes resistance or switches ON when exposed to light.

Specific models of photodiodes are designed to switch when exposed to infrared (heat) energy. Infrared photodiodes are used as motion detectors for alarm systems. In many cases, photodiodes have the P-type material and N-type material formed into flat segments to expose the maximum amount of material to the light source.

Transistors. A *transistor* is a three-terminal semiconductor device that controls current according to the amount of voltage applied to the base. Transistors are used as amplifiers or DC switches, with some transistors able to be used as either one. Transistors include bipolar junction, field-effect, photo, and photodarlington transistors.

Bipolar junction transistors were the first transistors developed. A *bipolar junction transistor (BJT)* is a transistor that controls the flow of current through the emitter (E) and collector (C) with a properly biased base (B). Bipolar junction transistors have three separate layers of P- and N-type material and can be layered in a PNP configuration or an NPN configuration. See Figure 14-8. The layers of a bipolar junction transistor are in physical contact with each other and have terminals connected to the layers. The three terminals are the emitter, collector, and base.

The emitter is the common contact for the base and collector circuits. The collector is the connection that is connected to the load. The base connection is a low-powered connection that functions as a valve (switch). The base layer is very thin with a low amount of doping material. The base layer is used to control the flow of electrons between the emitter and collector. A very small flow of electrons between the base and emitter can cause a much larger flow of electrons between the emitter and collector, thus functioning as an amplifier. When a variable resistor is placed in the base circuit, the collector output signal can be changed. See Figure 14-9. Bipolar junction transistors are commonly used to amplify small control signals into large signals that are used to drive large-capacity devices such as electronically controlled valve and damper actuators.

A *field-effect transistor (FET)* is a transistor that controls the flow of current through the drain (D) and source (S) with a properly biased gate (G). Field-effect transistors include junction field-effect transistors (JFETs) and metal-oxide semiconductor field-effect transistors (MOSFETs). Junction field-effect transistors were developed as a better switching device than bipolar junction transistors. Junction field-effect transistors are named because they are formed from a single piece of N-type or P-type material. A layer (channel) of opposite material is formed around the center section of the P-type or N-type material. See Figure 14-10.

Junction field-effect transistors may be N-channel or P-channel JFETs. When a voltage is applied to the material that surrounds the channel, the effect of the electrons being present causes the channel to have various states of conductivity. The channel acts like a valve or switch. The action takes place with a very low current and voltage at the gate. An advantage of JFETs over bipolar junction transistors is that current and voltage requirements are much lower in JFETs. The low power consumption of JFETs makes them ideal for use in a wide variety of circuits.

BIPOLAR JUNCTION TRANSISTOR CONSTRUCTION

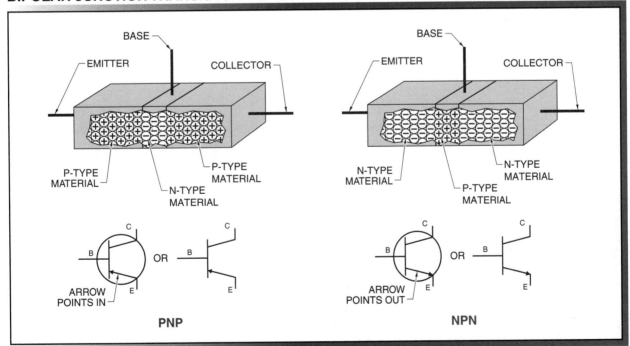

Figure 14-8. A bipolar junction transistor controls the flow of current through the emitter and collector with a properly biased base.

BIPOLAR JUNCTION TRANSISTOR OPERATION

LARGE CURRENT FLOW THROUGH LOAD

HIGH VOLTAGE

HIGH VOLTAGE

SWITCH CLOSED

LOW VOLTAGE

APPROXIMATELY 95%

APPROXIMATELY 5%

VARIABLE RESISTOR

CURRENT FLOWS THROUGH BASE AND EMITTER – TRANSISTOR ON

LOW VOLTAGE

CURRENT FLOW

LOAD

PNP CIRCUIT

LARGE CURRENT FLOW THROUGH LOAD

HIGH VOLTAGE

SWITCH CLOSED

LOW VOLTAGE

APPROXIMATELY 95%

APPROXIMATELY 5%

VARIABLE RESISTOR

CURRENT FLOWS THROUGH BASE AND EMITTER – TRANSISTOR ON

LOW VOLTAGE

CURRENT FLOW

LOAD

NPN CIRCUIT

Figure 14-9. Bipolar junction transistors are commonly used to amplify small control signals into large signals that are used to drive large-capacity devices.

Modern building automation system controllers use different transistors for circuit operation.

The most common transistors used today are metal-oxide semiconductor field-effect transistors (MOSFETs). Metal-oxide semiconductor field-effect transistors are constructed from two separate sections of either P-type or N-type material at opposite ends of the transistor. Between the two sections is a gate lead that is connected to the transistor by a metal-oxide material and insulator. The insulator allows no physical contact between the gate and the rest of the transistor. A voltage present at the gate of a MOSFET causes conductivity between the source and drain leads. See Figure 14-11.

JUNCTION FIELD-EFFECT TRANSISTOR CONSTRUCTION

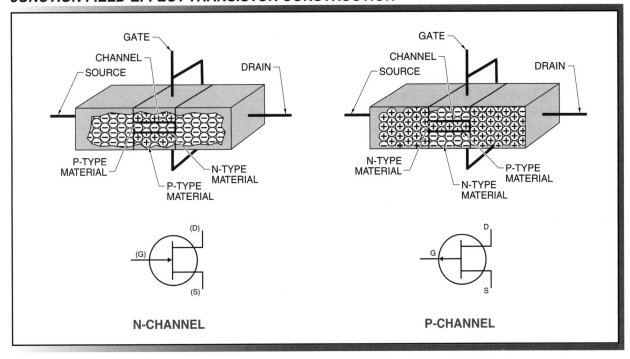

Figure 14-10. Junction field-effect transistors use a small voltage on the gate to control current flow between the source and drain.

METAL-OXIDE SEMICONDUCTOR FIELD-EFFECT TRANSISTOR CONSTRUCTION

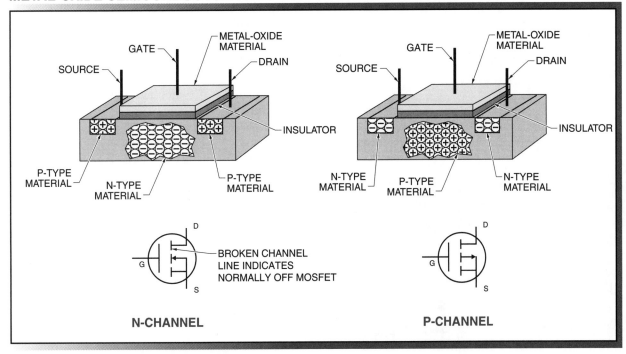

Figure 14-11. Metal-oxide semiconductor field-effect transistors use a small amount of current flow to turn ON and are the basis for computer chips.

Metal-oxide semiconductor field-effect transistors have almost infinite resistance between the gate and either the source or drain leads because of the insulator. Because of the high resistance, a tiny current flow at the gate can cause the device to switch ON. Metal-oxide semiconductor field-effect transistors are used as switches and represent ON/OFF signals in large numbers of computer chips. Static electricity or overcurrent can cause the insulator of a MOSFET to be destroyed.

A *phototransistor* is an NPN transistor that has a large, thin base region that is switched ON when exposed to a light source. When light strikes the base region, current flow is induced that causes the emitter-collector current to switch ON. See Figure 14-12. The drawback of a phototransistor is that the large, thin base region causes the current-carrying ability of the transistor to be reduced. In many cases, a device must have a large current-carrying ability to function properly in a circuit.

A *photodarlington transistor* is a transistor that consists of a phototransistor and a standard NPN transistor in a single package. In a photodarlington transistor, the sensitive phototransistor is used to switch the standard NPN transistor. See Figure 14-13. The phototransistor provides a high sensitivity while the standard NPN transistor provides the high current flow necessary for specific applications such as energizing relays.

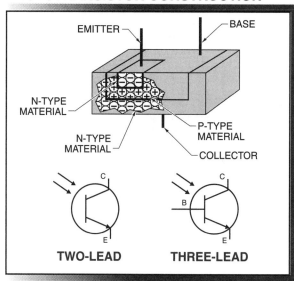

PHOTOTRANSISTOR CONSTRUCTION

Figure 14-12. Phototransistors are designed to allow current flow when exposed to a light source.

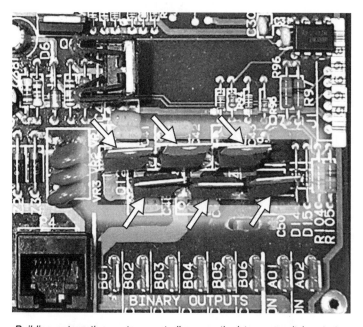

Building automation system controllers use thyristors as switches to turn equipment ON or OFF.

Thyristors. Another major category of semiconductor devices is thyristors. A *thyristor* is a solid-state switching device that switches current ON by a quick pulse of control current. Thyristors act as solid-state relays that can switch devices such as heaters, compressors, motors, and relays ON and OFF. The two major categories of thyristors are silicon-controlled rectifiers (SCRs) and triacs.

A *silicon-controlled rectifier (SCR)* is a thyristor that is capable of switching direct current. Silicon-controlled rectifiers have four layers of P-type and N-type material. See Figure 14-14. One of the layers is referred to as the gate. When a small voltage is introduced at the gate, the middle layer switches ON, and the SCR allows current flow in one direction only. The SCR continues to conduct current even when the gate voltage is removed. To stop an SCR from conducting, system power must be disconnected.

Silicon-controlled rectifiers are used in HVAC applications such as electric heating element control and in variable-frequency drives (VFDs) used for motor speed control. Silicon-controlled rectifiers have a high current-carrying capability that causes them to create high amounts of heat. The heat must be removed by heat sinks or fans. The disadvantage of silicon-controlled rectifiers is that they can only be used with DC.

PHOTODARLINGTON TRANSISTORS

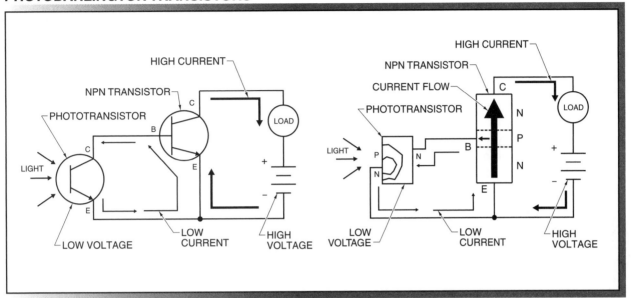

Figure 14-13. Photodarlington transistors are used in applications where high sensitivity to light is required with high current flow.

SILICON-CONTROLLED RECTIFIERS (SCRs)

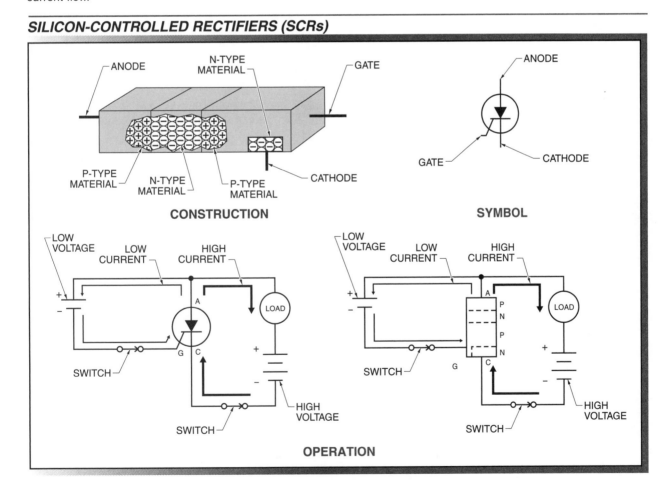

Figure 14-14. Silicon-controlled rectifiers (SCRs) are semiconductor devices used to switch direct current.

A *triac* is a solid-state switching device used to switch alternating current. A triac is triggered into conduction in either direction by a small voltage to its gate. A triac consists of two SCRs connected in a reverse-parallel configuration. See Figure 14-15. One SCR conducts current in one direction, while the other SCR conducts current in the opposite direction. Triacs are used as solid-state relays. Unlike electromechanical relays, triacs have no moving parts and are capable of millions of ON/OFF cycles without failure. Triacs are used in automated systems to turn electrical devices ON and OFF and are limited only by the amount of current they can carry.

Integrated Circuits. An *integrated circuit* is an electronic device in which all components (transistors, diodes, and resistors) are contained in a single package or chip. Integrated circuits were developed because as electronic devices were being used to perform more and more functions, it was discovered that many transistors, diodes, and other devices could be deposited microscopically on a small chip of silicon. See Figure 14-16.

Microscopically depositing many semiconductor devices on a small chip of silicon enabled many functions to be performed by a single semiconductor chip. One function is to represent ON/OFF (digital) decisions. The ON/OFF decision devices are referred to as logic gates and follow a set of logic rules known as Boolean logic. Logic gates include AND gates, OR gates, NOT gates, NOR gates, etc. Combining logic gates onto a single chip in the minimum amount of space so that they use the minimum amount of power enables a ¼″ square chip of silicon to contain hundreds of thousands or millions of individual semiconductor devices.

TRIACS

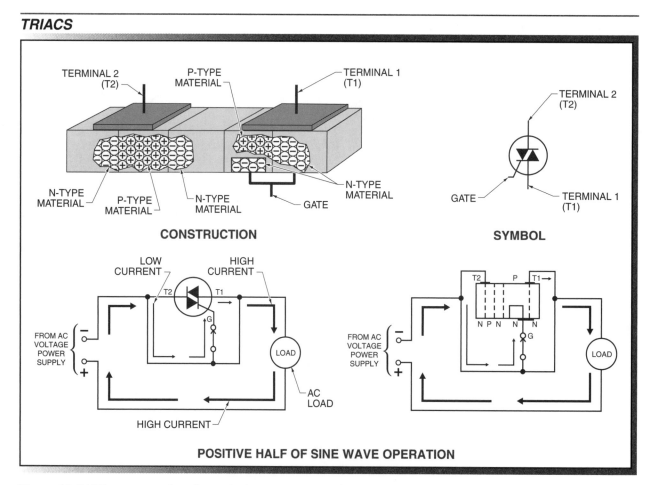

Figure 14-15. Triacs are semiconductor devices used to switch alternating current.

INTEGRATED CIRCUITS

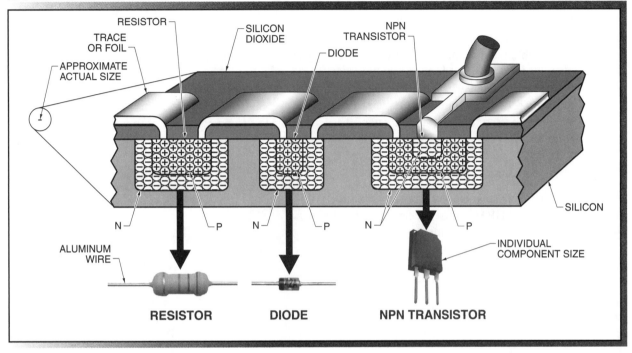

Figure 14-16. Layers of P-type and N-type materials are deposited on silicon to form integrated circuits.

These devices are deposited by different processes such as photolithography (etching) or by burning with a laser. A chip is attached to a chip holder that protects the chip from damage and attaches the chip to wire leads that carry the electrical signals to and from the other devices in the circuit. Chip holders have the logo of the manufacturer, date code, and index mark imprinted on them. The logo and date code help identify the chip, while the index mark indicates the correct orientation of the chip when it requires replacement.

Some chip holders have windows which enable the chip inside to be erased and reprogrammed—for example by light. Many manufacturers also use stickers on the chips with version numbers for easy identification. Care must be taken when replacing chips in control circuits. Chips must never be forced or have excessive pressure placed on the fragile pins. A chip removal tool is used to aid in chip removal. Integrated circuits are used in many HVAC applications such as chiller control, boiler control, and building automation systems.

Troubleshooting Semiconductor Devices

HVAC technicians require an overview of electronic component troubleshooting procedures and methods. Before troubleshooting any semiconductor device,

always verify that the power is disconnected. If power is connected, serious injury, death, or equipment damage can result. Even low-voltage power can damage electronic circuit components.

Always verify adequate personal grounding. An inexpensive wrist strap can be connected to ground to protect against static electricity discharge to electronic components. Electronic components are damaged or destroyed by even small static electricity discharges. When the cover is removed from an electronic device, the device is unprotected and vulnerable to static electricity discharges.

Always use quality test meters. The use of a poor test meter can cause faulty or inaccurate readings. Digital multimeters measure small voltages and currents properly and display the readings in an easy-to-read format. Many digital multimeters have an autoranging function that automatically displays the correct value. The digital readout helps to eliminate reading guesswork. Analog meters are used on certain circuits where circuit loading is required to obtain accurate readings.

The most common electronic circuit troubleshooting situation is to find the correct printed circuit (PC) board, determine if the board has a problem, and replace the board if necessary. A failed PC board is normally returned to the manufacturer for replacement or repair. PC board troubleshooting requires that a technician find the defective semiconductor device on the board. A close visual inspection often reveals burn marks on components or foils (traces), indicating a failed component. It is possible that a semiconductor device has been completely destroyed and is missing. In many cases, an overheated component causes a burning smell. The sense of touch (a short time after power is removed) can often be used to detect a component that is operating excessively hot.

Electronic Control System Applications

Electronic control systems use various semiconductor devices to control HVAC units. Electronic control systems include the sensors, controllers, switching components, and output devices such as dampers and valve actuators. These components then provide temperature, pressure, and humidity control in the building. Electronic control system components include sensors and electronic thermostats that are used in applications such as multizone unit control, boiler control, and chiller control.

Sensors. Many temperature sensors use electronic components. See Figure 14-17. Temperature sensors use semiconductor materials that change resistance characteristics as the temperature around the sensor changes. Humidity sensors change resistance as humidity is absorbed or adsorbed by electronic elements. Pressure sensors may use a piezoelectric crystal, which changes output voltage as pressure is exerted on the crystal.

ELECTRONIC TEMPERATURE SENSORS

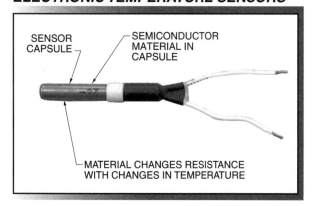

Figure 14-17. Electronic sensors use variables such as temperature, humidity, and pressure to change resistance or to create current flow.

Electronic Thermostats. Many commercial and residential thermostats are electronic. See Figure 14-18. Applications such as rooftop heat pumps and other packaged equipment commonly use electronic thermostats. Electronic thermostats use electronic sensors to sample the building space temperature and use triacs to energize and de-energize heating, cooling, and fan functions. Some electronic thermostats include LEDs to indicate system operation or trouble conditions.

Figure 14-18. Electronic thermostats are used to control residential and commercial HVAC units such as split system air conditioning and heat pump systems.

Multizone Unit Control. In multizone unit control, each zone damper is controlled by a zone thermostat while the air handling unit is controlled by several electronic controllers. In this application, the supply fan runs continuously. The hot deck controller controls the hot deck electronic valve to maintain the hot deck setpoint. The setpoint of the hot deck controller is reset from outside air. As the outside air temperature varies from −10°F to 65°F, the hot deck setpoint varies from 120°F to 55°F. The cold deck controller maintains a constant temperature of approximately 55°F. See Figure 14-19.

Boiler Control. Electronic controls are used to control boilers and their associated equipment. In this application, the electronic controller changes the hot water setpoint from 100°F to 190°F as the outside air temperature changes from 65°F to −10°F. The electronic controller also modulates a three-way mixing valve to obtain the correct water temperature. The electronic controller also turns the boiler ON below a specific outside air temperature to begin to supply heat. See Figure 14-20.

MULTIZONE UNIT CONTROL

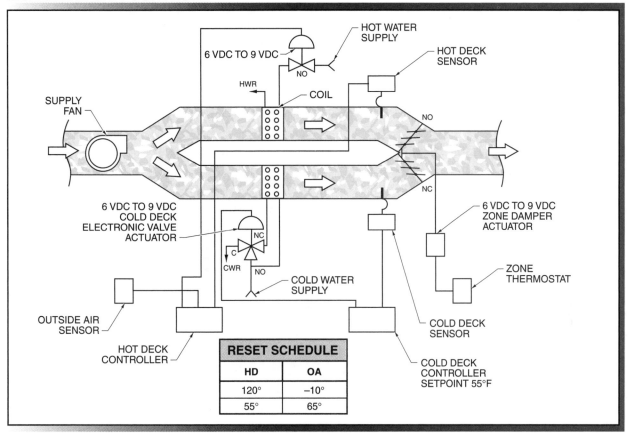

Figure 14-19. Electronic controls can be used to control multizone units.

RESET SCHEDULE

HD	OA
120°	−10°
55°	65°

BOILER CONTROL

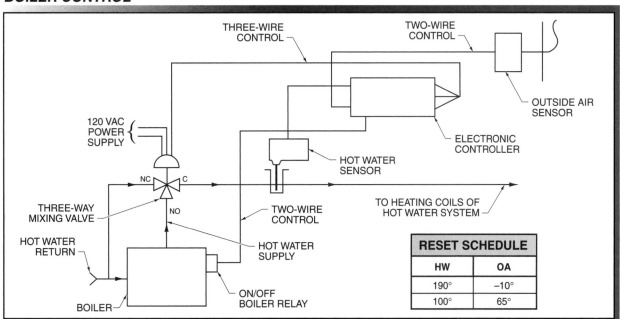

Figure 14-20. Electronic controls can be used to control boilers and their associated equipment.

RESET SCHEDULE

HW	OA
190°	−10°
100°	65°

Chiller Control. One or more water chillers can be controlled by electronic control systems. In this application, the electronic controller starts and stops the chiller when required. The controller also opens or closes the chiller inlet vane electronic actuator to maintain the correct chilled water supply or return temperature. Electronic outside air, chilled water return, and chilled water supply temperature sensors measure the temperature and send a signal back to the controller. Low water and low oil pressure safety control circuits ensure that the chiller does not operate if damage to the unit may occur. See Figure 14-21.

CHILLER CONTROL

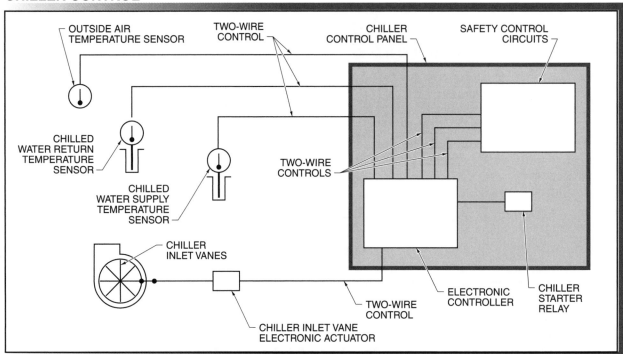

Figure 14-21. Electronic controls can be used to control chiller operation.

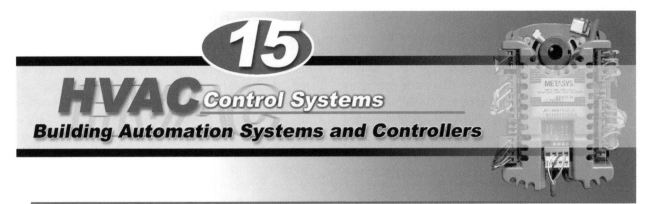

Building automation systems have evolved from central supervisory, to central- direct digital control, to distributed direct digital control systems. Central supervisory systems were the first facility management systems. Central-direct digital systems provide management and control functions. Distributed direct digital systems have system intelligence located throughout the system. Building automation system controllers include application-specific and universal input-output controllers, and network communication modules.

BUILDING AUTOMATION SYSTEMS

A *building automation system* is a system that uses microprocessors (computer chips) to control the energy-using devices in a building. Building automation systems are the most common control system installed in commercial buildings today. Building automation systems control HVAC equipment, lighting, security, and other essential functions in a building. Building automation systems also indicate abnormal conditions (alarms) and provide information regarding energy consumption and equipment maintenance. Building automation systems include central supervisory control systems, central-direct digital control systems, and distributed direct digital control systems.

Central Supervisory Control Systems

The earliest building automation system was the central supervisory control system. A *central supervisory control system* is a control system in which the decision-making equipment is located in one place and the system enables/disables local (primary) controllers. See Figure 15-1. Central supervisory control systems were developed in the early 1970s and were popular in the mid to late 1970s. Large central pneumatic control panels were popular before the development of central supervisory control systems. Some commercial buildings still use central supervisory control systems.

CENTRAL SUPERVISORY CONTROL SYSTEMS

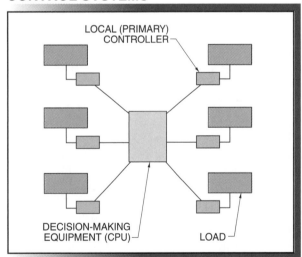

Figure 15-1. Central supervisory control systems have decision-making equipment located in one place and can only enable/disable local (primary) controllers.

In a central supervisory control system, all of the control decisions are made by a central processing unit (CPU) centrally located in the building or complex. The results of the control decisions are sent over a pair of wires to an interface device such as a relay. Temperature sensors are also wired to the CPU to indicate conditions in the building or system.

Central supervisory control systems required a large amount of wire to connect all of the equipment. Wiring cost was often 40% to 50% of the entire job cost. The high cost of the equipment and wiring made central supervisory control systems unaffordable for many small and medium-sized commercial buildings. Also, the long wire runs caused problems if one of the wires fails or was installed improperly.

The central location of the decision-making equipment makes central supervisory control systems vulnerable to any problem that affects the CPU. A system-wide failure occurs if the CPU of a central supervisory control system fails. This was a major problem with central supervisory control systems. Manual overrides were required to alleviate system-wide failures. In addition, the individual system components were not as reliable or capable as modern components. Thus, the failure rate of central supervisory control systems was high, causing customer dissatisfaction. Building automation systems of this time period had a poor reputation in the HVAC industry. Some manufacturers provide gateway interface modules that allow central supervisory control systems to communicate with modern building automation system networks. See Figure 15-2.

Central Supervisory Control System Functions. Central supervisory control systems perform a limited number of basic control functions. Control functions include electrical demand control, duty cycling, and time clock functions. Prior to the development of central supervisory control systems, these control functions were not performed or in some cases were performed by different controllers. For example, prior to central supervisory control systems, electromechanical time clocks were commonly used to turn electrical devices ON and OFF. A large building or complex might have had dozens of electromechanical time clocks. Changing the clocks for a holiday was time-consuming. The installation of a central supervisory control system allowed one technician at a central location to change dozens of time clocks for a holiday in a few hours.

Central supervisory control systems do not control a load directly, but enable or disable existing controllers. A central supervisory control system does not replace existing local controllers, but determines if they are allowed to operate. Local controllers include pneumatic receiver controllers, electric room thermostats, and light switches.

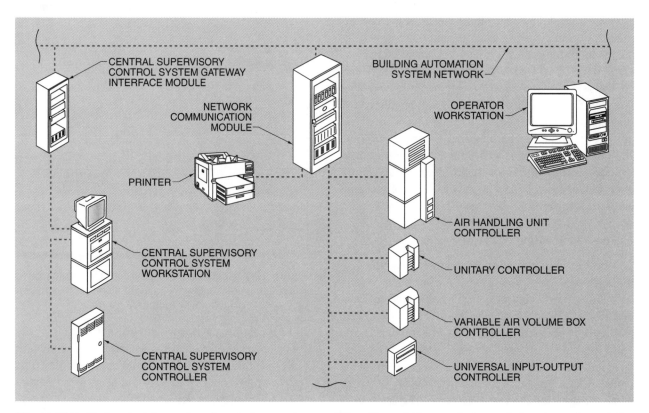

Figure 15-2. Modern building automation systems are compatible with central supervisory control systems through the use of gateway interface modules.

A disadvantage of central supervisory control systems is that they do not correct problems with existing local controls, they only enable or disable improperly operating controls. This is unacceptable in many building control applications. In addition, central supervisory control systems have a high failure rate, are expensive, and have limited control functions. Today, central supervisory control systems are obsolete and spare parts are difficult or impossible to find.

Central-Direct Digital Control Systems

As the reliability and functionality of solid-state electronics improved, additional capabilities were added to building control systems. A *central-direct digital control system* is a control system in which all decisions are made in one location and which provides closed loop control. See Figure 15-3. *Closed loop control* is control in which feedback occurs between the controller sensor and controlled device. A building automation system controller replaces the thermostat or other controller in the system. The building automation system controller has a sensor input and sends an output signal to the valve, damper, or other controlled device. Central-direct digital control systems were used in the late 1970s and early 1980s. Many central-direct digital control systems are still in use today.

Central-Direct Digital Control System Architecture. The decisions in central-direct digital control systems are made at one central processing unit. For this reason, central-direct digital control systems are similar to central supervisory control systems. However, central-direct digital control systems control loads such as dampers, valves, and compressors through field interface devices, without the need for local (primary) controllers. Central-direct digital control systems consist of a centrally located CPU connected to field interface devices by a communication network. The communication network normally consists of a twisted, shielded pair of wires.

Field Interface Devices. A *field interface device (FID)* is an electronic device that follows commands sent to it from the CPU of a central-direct digital control system. Field interface devices cannot perform control on their own. The commands sent by the central processing unit may be to turn a fan ON or OFF, change the position of a pneumatic valve, or report the value of a temperature sensor. The field interface devices may be referred to as data gathering panels (DGPs), data acquisition panels (DAPs), slave panels, or submaster panels.

Field interface devices may have limited or no decision-making ability. For this reason, they are often referred to as dumb panels or dumb input/outputs (I/Os).

CENTRAL-DIRECT DIGITAL CONTROL SYSTEMS

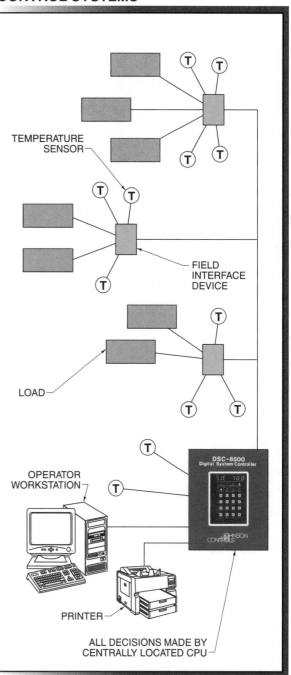

Figure 15-3. In a central-direct digital control system, the central CPU communicates all decisions to field interface devices that control the loads.

Recent field interface devices have limited decision-making ability and can provide limited control response in a system failure or loss of communication with the CPU. These field interface devices are located at or near the controlled devices. All sensors, relays, and transducers are wired to the field interface devices.

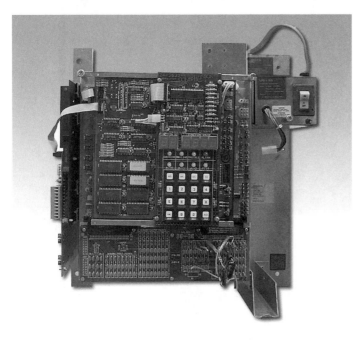

In central-direct digital control systems, all decisions are made by a single central processing unit.

Central-Direct Digital Control System Reliability. Central-direct digital control systems enable closed loop control in an automated control system. However, a control system must have the reliability to perform control functions time after time. A central-direct digital control system relies on a communication network to carry commands from the centrally located CPU to the field interface devices. The system fails if the centrally located CPU or communication network fails. In addition, if the centrally located CPU is down, the entire building automation system may fail. In many applications, the full capabilities of central-direct digital control systems are not realized because of reliability and critical failure issues. Many central-direct digital control systems are used for supervisory control instead of closed loop control. In this way, the field interface devices can be configured to fail-safe and allow existing local controllers to remain in control.

Central-Direct Digital Control System Functions. Central-direct digital control systems perform all of the control functions of central supervisory control systems. Functions include duty cycling, electrical demand control, time clock, and closed loop control functions. In addition, central-direct digital control systems also include extensive maintenance management functions. Maintenance management functions include data logging, alarms, and preventive maintenance reporting.

Data logging is the recording of information such as temperature and equipment ON/OFF status at regular time intervals. This information is viewed in tabular form. Data logging is used for evaluating equipment operation and predicting maintenance. An *alarm* is a notification of improper temperature or other conditions existing in a building. Alarms are displayed on a monitor, printed on paper, or dialed out to an off-site personal computer or pager. *Preventive maintenance reporting* is the generating of forms to notify maintenance personnel of routine maintenance procedures. Routine maintenance procedures include filter replacement and equipment lubrication. These additional functions increase the ability of a building automation system to manage the energy use in a building.

Central-direct digital control systems have greater control capabilities than central supervisory control systems, and are more reliable due to improved electronic components. In addition, the wiring cost of central-direct digital control systems is lower because a simple communication network connects the centrally-located CPU to the field interface devices. The field interface devices are located close to the equipment they control, making the wire runs short. Also, the field interface devices are inexpensive because they do not have CPU chips. The major drawback of central-direct digital control systems is that they are not very reliable because of possible CPU or communication problems.

Distributed Direct Digital Control Systems

A *distributed direct digital control system* is a control system that has multiple CPUs at the controller level. See Figure 15-4. Distributed direct digital control systems are the building automation systems currently used. In distributed direct digital control systems, decisions are made by each controller. Distributed direct digital control systems began replacing central-direct digital control systems in the mid 1980s.

DISTRIBUTED DIRECT DIGITAL CONTROL SYSTEMS

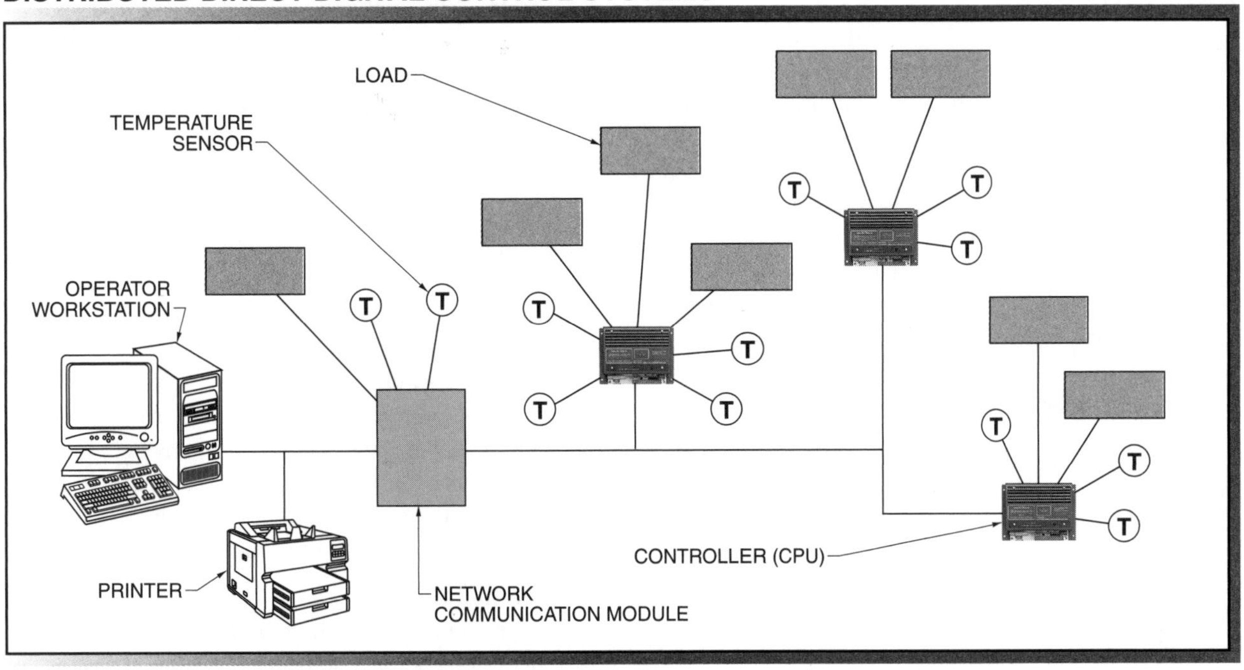

Figure 15-4. Distributed direct digital control systems have multiple CPUs at the controller level.

Distributed Direct Digital Control System Architecture. The difference between central-direct digital control systems and distributed direct digital control systems is that each controller in a distributed direct digital control system has the ability to make its own decisions. The ability is spread out (distributed) throughout the control system. Distributed direct digital control systems normally have a supervisory (master) controller. The supervisory controller provides communication support, a convenient place to connect a PC or modem for monitoring purposes, and connections for global data. *Global data* is data needed by all controllers in a network. Global data includes outside air temperature and electrical demand.

Stand-alone control is the ability of a controller to function on its own. Stand-alone control is less risky than other types of control because a controller failure is local and has minimal effect on the entire system. Because stand-alone control enables a system to function without the operation of a failed controller, it has increased the reliability of building automation systems.

Controllers in a network commonly share information through a communication network. A controller that needs information from a failed controller in the network is unable to access that information. The first controller functions, but at a reduced level or efficiency, until the failed controller is repaired or replaced.

Distributed Direct Digital Control System Functions. Distributed direct digital control systems have the same control functions as central-direct digital control systems, which include duty cycling, electric demand control, and time clock functions. Distributed direct digital control systems also can perform precise closed loop temperature control, humidity control, and pressure control.

Advantages of distributed direct digital control systems include improved reliability and increased capacity over previous systems. The stand-alone feature makes distributed direct digital control systems better than any previous system. Also, distributed direct digital control systems are modular and easily expandable. A building automation system can be started with a minimum number of controllers and expanded as required.

> *In general, the greater the amount of information that is shared across the network of a building automation system, the less the amount of stand-alone functionality. The control system of a critical system that must operate perfectly in the event of a partial system failure must be designed with minimum reliance on network information.*

BUILDING AUTOMATION SYSTEM CONTROLLERS

Building automation systems such as distributed direct digital control systems require a number of controllers. When connected by a communications network, the controllers provide comprehensive control and monitoring of the HVAC equipment in a building. Controllers are often referred to as control modules, modules, and panels. Building automation system controllers include application-specific controllers, universal input-output controllers, and network communication modules.

Application-Specific Controllers

An *application-specific controller (ASC)* is a controller designed to control only one type of HVAC system. See Figure 15-5. In most applications, application-specific controllers are more cost-effective than other types of controllers. Also, costs are reduced because most of the software programming is done by the manufacturer, reducing setup time.

APPLICATION-SPECIFIC CONTROLLERS

Figure 15-5. Application-specific controllers are designed to control only one type of HVAC system.

Application-Specific Controller Hardware. *Hardware* is the physical parts that make up a device. Application-specific controller hardware normally consists of an electronic circuit board, a wiring baseplate, and any external devices such as temperature sensors or flow switches. The external devices used vary widely depending on the application. The electronic circuit board contains an electronically erasable, programmable, read-only memory (EEPROM) chip. An *EEPROM* is an integrated-circuit

memory chip that has an internal switch to permit the user to erase the contents and write new contents by means of electrical signals. The EEPROM chip is factory-programmed to match the application-specific controller application. For example, a package unit application-specific controller has a rooftop-unit program installed at the factory.

Many manufacturers use two-piece construction for application-specific controllers. In two-piece construction, the baseplate and electronic circuit board are separate packages. This enables the baseplate to be installed, wired, and checked before the valuable electronic components are installed. This reduces the risk of theft of the electronic components from the construction site. This also reduces the possibility of damage to the electronic components from improper baseplate wiring. The baseplate is wired and voltages measured before the electronic components are mounted.

Application-Specific Controller Software. *Software* is the program that enables a controller to function. Application-specific controller software specifies the control functions performed. The EEPROM chip contains basic control functions that pertain to a specific HVAC system. Basic control functions include generic setpoints and sequences. In most cases, the setpoints and sequences must be altered to match a specific application. For example, the EEPROM chip setpoints on an application-specific controller from the factory may be 72°F for cooling and 70°F for heating. These setpoints may work fine for most applications but may need to be changed to 74°F for cooling and 68°F for heating for a particular application. In addition, an EEPROM chip may be factory programmed to recognize only one stage of cooling. A particular HVAC system may have two stages of cooling. In both cases, the EEPROM program must be modified.

Manufacturers provide controller software that can be loaded into a laptop PC. The existing controller program in the EEPROM can be changed or replaced when the laptop is connected to the application-specific controller. See Figure 15-6. The new controller program (database) is transferred (downloaded) from the laptop to the application-specific controller. Only basic changes can be made to the software, such as rooftop unit setpoints and functions. The application, such as rooftop controller, cannot be changed.

Controller software also has the ability to change the values and override valves and dampers to check their operation. This check is referred to as commissioning the controller. Another advantage of using

EEPROM-based equipment is that no battery backup is needed. The program on an EEPROM chip is nonvolatile (not lost if power is disconnected). Application-specific controllers include unitary controllers, air handling unit controllers, and variable air volume box controllers.

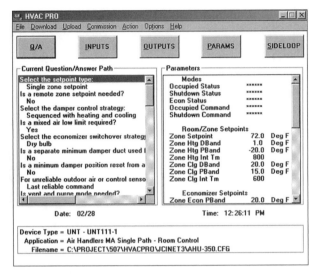

Figure 15-6. Building automation system software manufacturers provide software that is used to program application-specific controllers.

Unitary Controllers. A *unitary controller* is a controller designed for basic zone control using a standard wall-mount temperature sensor. See Figure 15-7. Unitary controllers are designed to control packaged HVAC equipment such as rooftop units, heat pumps, and fan coil units. Unitary controllers are normally compact to allow mounting at the packaged HVAC equipment. Waterproof and heated enclosures are often used which allow operation of multiple stages of heating and cooling, economizer dampers, heat pump reversing valves, and supply fans. Some unitary controllers provide supply air sensing capabilities, air flow switches, and dirty-filter-condition switches.

Unitary controllers are commonly used in rooftop unit applications. In this application, the unitary controller is used in place of an electromechanical low-voltage thermostat. See Figure 15-8. A room temperature sensor located in the building space is wired to the unitary controller. An air flow switch detects air flow status. The unitary controller cycles the supply fan, gas heating valve (heating stage), and cooling compressor (cooling stage) to maintain the programmed day and night building space temperature. Other devices such as a temporary occupancy switch, supply air temperature sensor, outside air temperature sensor, and filter status can also be wired to the unitary controller.

Unitary controllers are also used in water source heat pump applications. Many commercial buildings use heat pump systems. This application is similar to the rooftop unit application. The major difference is that the heat pump reversing valve is cycled by the unitary controller.

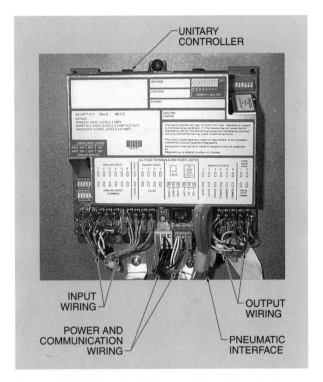

Figure 15-7. Unitary controllers are designed for simple zone control using a standard wall-mount temperature sensor.

Air Handling Unit Controllers. An *air handling unit (AHU) controller* is a controller that contains inputs and outputs required to operate large central-station air handling units. Air handling unit controllers control humidification, static pressure, and indoor air quality. See Figure 15-9. Many central-station air handling units have complex control sequences. Dual-duct and multizone air handling units have needs that cannot be met by a unitary controller. Air handling unit controllers are larger and more complex than unitary controllers, and air handling unit controller software is more complex than unitary controller software to allow the handling of the complex control sequences.

UNITARY CONTROLLER ROOFTOP UNIT APPLICATION

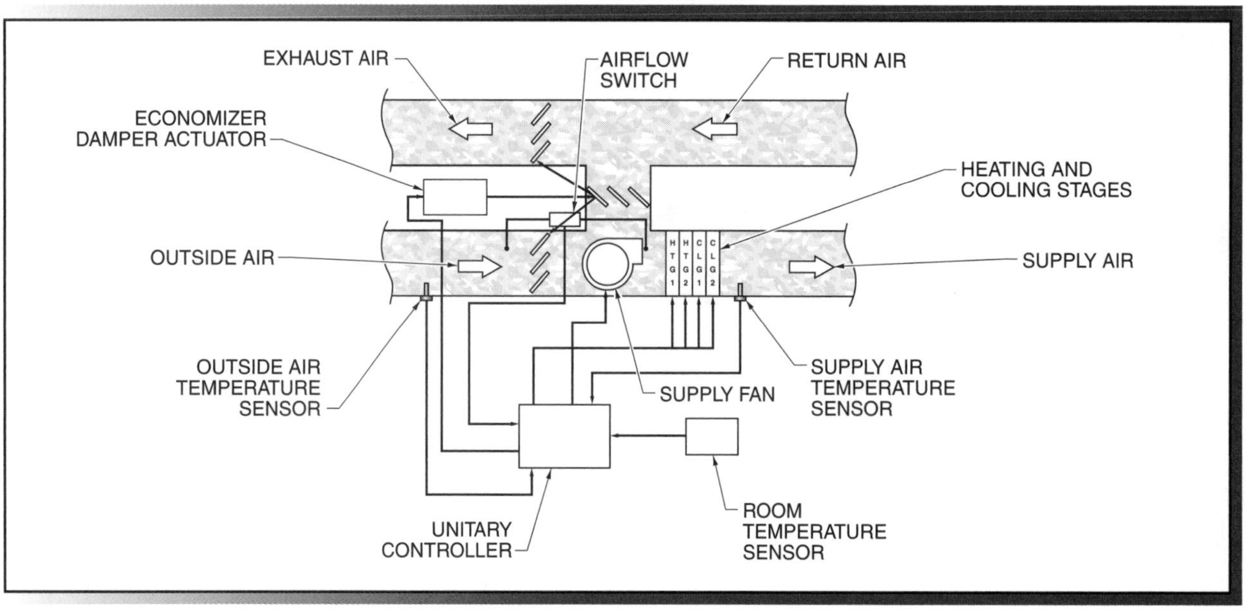

EXHAUST AIR

AIRFLOW SWITCH

RETURN AIR

ECONOMIZER DAMPER ACTUATOR

HEATING AND COOLING STAGES

OUTSIDE AIR

SUPPLY AIR

OUTSIDE AIR TEMPERATURE SENSOR

SUPPLY AIR TEMPERATURE SENSOR

SUPPLY FAN

UNITARY CONTROLLER

ROOM TEMPERATURE SENSOR

Figure 15-8. Unitary controllers are used to control rooftop unit fans, heating stages, and cooling stages.

AIR HANDLING UNIT CONTROLLER WIRING BASEPLATE

COMMUNICATION BUS ACTIVITY LIGHTS

Figure 15-9. Air handling unit controllers are used when sophisticated control of large central station air handling units is required.

Air handling unit controllers are used to control single-zone air handling units with building space temperature controls. In this application, a large air handling unit controls the temperature in a large building space. A unitary controller cannot be used because of specific control needs such as building space humidity, chilled water valve control, and reheat hot water valve control. This application requires a greater number of inputs and outputs than rooftop and heat pump applications.

A common air handling unit controller application is the control of a variable air volume air handling unit to maintain a minimum static pressure in the supply duct. See Figure 15-10. In this application, heating and cooling control is performed to maintain a constant discharge air temperature of 55°F. An output (static pressure sensor) from the air handling unit controller is provided to increase the volume of air provided by the supply fan. The air volume increase is produced by controlling the speed of the fan motor using a variable-frequency drive (VFD) or by opening dampers on the fan to admit a greater amount of air.

Variable Air Volume Box Controllers. A *variable air volume box controller* is a controller that modulates the damper inside a variable air volume (VAV) terminal box to maintain a specific building space temperature. Variable air volume box controllers are similar in appearance to unitary controllers. The difference is in the factory programming of the EEPROM chip. See Figure 15-11.

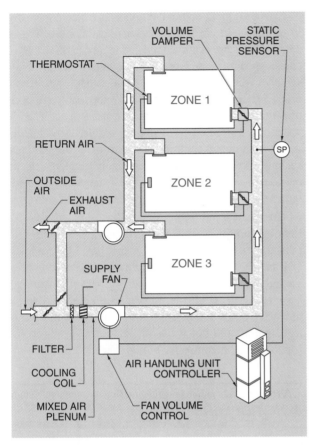

Figure 15-10. Air handling unit controllers are used to control variable air volume air handling units in order to maintain a minimum static pressure in the supply duct.

Variable air volume box controllers modulate the primary damper inside the VAV terminal box. Reheat valves can also be controlled. A flow input to the controller can be set up to indicate air flow (in cubic feet per minute). The controller turns the fan inside the VAV terminal box ON and OFF based on the input. The software of many variable air volume box controllers allows VAV terminal boxes to be checked for proper air balancing.

Variable air volume box controllers are used to control pressure-dependent VAV terminal boxes. Pressure-dependent VAV terminal boxes are basic VAV terminal boxes that control building space temperature only and do not measure air volume. In this application, the VAV terminal box damper is controlled to maintain desired building space temperature. The inlet air flow (in cubic feet per minute) is not measured or controlled.

Variable air volume box controllers are also used in pressure-independent VAV terminal box with reheat applications. In this application, the variable air volume box controller is used to control building space

temperature and measure and control the amount of air flow. See Figure 15-12. In addition to controlling building space temperature, a flow sensor is installed at the inlet to the VAV terminal box. The flow sensor is connected to a differential sensor, which is wired to the variable air volume box controller to measure and control air flow. An output connection from the controller is provided to open the reheat coil valve on a call for heat.

Unitary controllers, air handling unit controllers, and variable air volume box controllers are the most common application-specific controllers. Additional application-specific controllers are available for computer room air conditioning (CRAC), reciprocating or centrifugal chillers, and indoor air-quality calculations. Most manufacturers install application-specific controllers on the equipment at the factory. This is referred to as original equipment manufacturer (OEM) installation. Field installation and retrofit problems are minimized by purchasing HVAC equipment with the application-specific controllers installed at the factory.

VARIABLE AIR VOLUME TERMINAL BOX CONTROLLERS

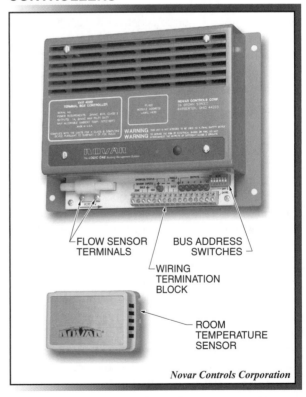

Novar Controls Corporation

Figure 15-11. Variable air volume box controllers are used to control building space temperature and air volume in VAV air handling systems.

PRESSURE-INDEPENDENT VAV TERMINAL BOX WITH REHEAT COIL APPLICATION

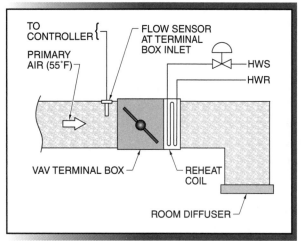

Figure 15-12. Pressure-independent VAV terminal box applications contain flow sensors to measure and control the amount of air flow.

Universal Input-Output Controllers

A *universal input-output controller (UIOC)* is a controller designed to control most HVAC equipment. See Figure 15-13. Many universal input-output controllers have the same two-piece construction as application-specific controllers. Universal input-output controllers have a fixed number of inputs and outputs, with eight and 16 being the most common. Most universal input-output controllers use a 24 VAC power supply.

Universal Input-Output Controller Software. The software of universal input-output controllers is different from the software of application-specific controllers. Application-specific controllers are factory programmed. Universal input-output controllers are programmed in the field. An application-specific controller can control only the equipment for which it is designed. Universal input-output controllers can control most HVAC equipment.

Universal input-output controllers are not programmed when shipped from the factory. Universal input-output controllers must be custom programmed in the field. Documentation and training are essential because the custom programming can be difficult to understand if the original programmer is no longer available. Custom programming is also more difficult to create. Universal input-output controller outputs can be used to control cooling towers and turn lighting loads ON and OFF. Universal input-output controllers are more expensive than application-specific controllers.

Universal input-output controllers are commonly used in central-plant hot water loop applications. See Figure 15-14. In this application, the hot water pumps are started and stopped by the universal input-output controller. Flow switches wired to the universal input-output controller are used to sense water flow. The hot water supply temperature and outside air temperature are also sensed. The hot water bypass (three-way) valve is modulated by the universal input-output controller to maintain the programmed setpoint based on the hot water supply temperature and outside air temperature. Additional features may include outside air reset and lead/lag hot water pump control.

Universal input-output controllers are also used for lighting control. In this application, the controller outputs are wired to lighting relays that turn lighting loads ON and OFF. A time schedule may be used to control the lighting functions. An outside light level sensor may be wired to the controller to control the lights. A single universal input-output controller may be used to control lighting and hot water applications as long as the capacity of the controller is not exceeded.

UNIVERSAL INPUT-OUTPUT CONTROLLERS

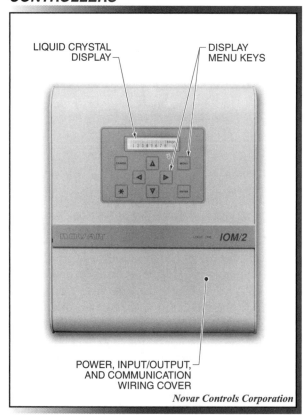

Novar Controls Corporation

Figure 15-13. Universal input-output controllers are fully programmable and can be used in most HVAC applications.

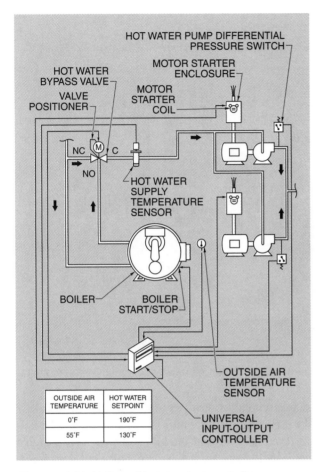

OUTSIDE AIR TEMPERATURE	HOT WATER SETPOINT
0°F	190°F
55°F	130°F

Figure 15-14. Universal input-output controllers are commonly used to control hot water loop applications.

Network Communication Modules

A *network communication module (NCM)* is a controller that coordinates communication from controller to controller on a network and provides a location for operator interface. See Figure 15-15. In most building automation systems, individual controllers do not communicate directly with each other. Generally, a network communication module obtains information from individual controllers and passes it along to other controllers in the network. One network communication module is normally required in a small- to medium-size building having up to 200 controllers. Large installations may have several network communication modules which are networked together.

Network communication modules are often able to use the same memory chips and network communication cards as personal computers.

NETWORK COMMUNICATION MODULES

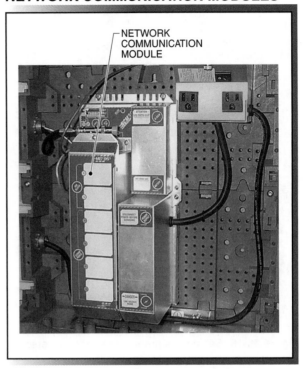

Figure 15-15. Network communication modules coordinate communication from controller to controller on a network and provide a location for operator interface.

Network Communication Module Hardware Connections. Network communication modules have connections for a power supply, communication network, operator interface devices, personal computers, and modems. The power supply is normally a 120 VAC to 24 VAC transformer. The communication network line connects the network communication module to the other controllers in the network. The communication network normally consists of a single pair of twisted, shielded wire. The wire is twisted to cancel out electromagnetic interference. The shielding is designed to prevent interference from reaching the wire. The shield is commonly terminated at ground to drain the interference to earth.

Network communication modules normally have a number of ports for connecting operator interface devices to the system. Serial ports are used to directly connect personal computers and modems to the network communication module. See Figure 15-16. A modem provides telephone communication to a remote personal computer. *Serial information* is information that is transmitted sequentially one bit (0 or 1) at a time. Network communication modules may contain more than one serial port. Parallel ports are provided to connect a local printer to the system. Parallel ports transmit information in multiple bits (0s or 1s) simultaneously.

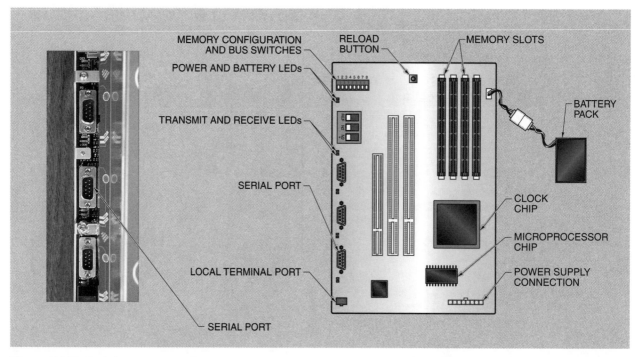

Figure 15-16. Serial ports are used to connect personal computers and modems to a network communication module.

Network Communication Module Software and Setup. Network communication module software provides communication capabilities instead of controlling the energy-using devices in a building. Network communication modules control communication between the controllers in a network. This enables information such as outside air temperature, occupied/unoccupied status, electrical demand, and outside light level to be shared with all controllers in a network. Global data is data required by all controllers in a network. The network communication module software allows global data sharing and alarms that are dialed out to a remote personal computer to be programmed. Telephone numbers and communication parameters such as data transmission speed and printer parameters are also set up in network communication module software.

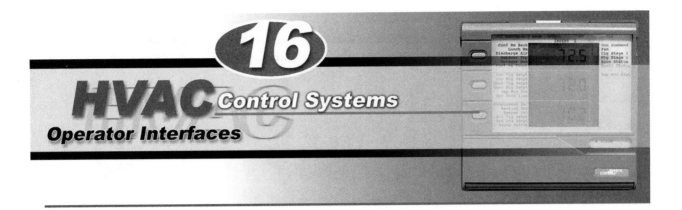

16

HVAC Control Systems
Operator Interfaces

Operator interfaces are means by which individuals can access and use a building automation system. Operator interfaces include desktop PCs, alarm printers, notebook PCs, operator terminals, keypad displays, dumb terminals, voice phone interface, pagers, fax machines, and cellular communication. Operator interfaces require proper training for long-term success with building automation system operation.

OPERATOR INTERFACES

A variety of operator interfaces are available by which individuals can access information and use a building automation system. An *operator interface* is a device that allows an individual to access and respond to building automation system information. Operator interfaces include the use of on-site and off-site devices. Choosing the correct operator interface device(s) and using them properly are required for the proper performance of the system.

On-Site Operator Interfaces

In medium- and large-sized buildings such as hospitals, colleges, or office buildings, maintenance or management personnel commonly access and troubleshoot a building automation system as part of their normal work responsibilities. On-site operator interfaces used in large buildings to access and troubleshoot a building automation system include desktop PCs, alarm printers, notebook PCs, portable operator terminals, keypad displays, and dumb terminals.

On-Site Desktop PCs. The operator interface method most commonly used in commercial buildings is a desktop personal computer (PC) connected to the building automation system network. See Figure 16-1. Most building automation systems use Ethernet as the building automation system network. *Ethernet* is a local area

network (LAN) architecture that can connect up to 1024 nodes and supports data transfer rates of 10 Megabits per second (Mbps). A *local area network (LAN)* is a communication network that spans a relatively small area. Most LANs are confined to a single building or group of buildings. A *node* is a device, such as a computer or a printer, that has a unique address and is attached to a network.

Maintenance personnel and supervisors can use on-site desktop PCs to check alarms and data logs, check or change setpoints and time schedules, and override equipment ON/OFF functions. On-site desktop PCs are normally located in the office of the maintenance engineer. The primary desktop PC used to communicate with a building automation system is commonly referred to as the front end. Most building automation systems (90%) have only one primary desktop PC, and therefore one front end. However, in large building automation systems that have multiple desktop PCs, all PCs are equal, and all are considered front ends.

Building automation system changes may be temporary (ending at a set time such as midnight) or permanent (lasting until changed). It may take a higher operator access level to override a permanent change. Depending on the manufacturer, some building automation systems allow operators to define new points, delete old points, or reconfigure the system based on the operator access level.

BUILDING AUTOMATION SYSTEM NETWORK (ETHERNET)

BUILDING AUTOMATION SYSTEM NETWORK (ETHERNET)

DESKTOP PC

NETWORK COMMUNICATION MODULE

NETWORK COMMUNICATION MODULES

UNITARY CONTROLLER

VARIABLE AIR VOLUME BOX CONTROLLER

LIGHTING CONTROLLER

UNIVERSAL INPUT-OUTPUT CONTROLLER

ALARM PRINTER

VARIABLE AIR VOLUME BOX CONTROLLER

SECURITY ACCESS CONTROLLER

AIR HANDLING UNIT CONTROLLER

Figure 16-1. Desktop PCs are the most common type of on-site building automation system operator interface.

Alarm Printers. An *alarm printer* is a printer used with a building automation system to produce hard copies of alarms (indications of improper system operation), preventive maintenance messages, and data trends. An alarm printer may be connected to a desktop PC or directly to a network communication module. See Figure 16-2. Alarm printers are one-way operator interfaces with the building automation system. Alarm printers are sent information and print it without any intervention by the operator. In some installations, the alarm printer is located at a 24-hour guard station or boiler room office. The alarm or message may include the name and phone number of an individual to be contacted to fix the alarm condition. This individual has the appropriate access level and has an operator interface method available.

ℹ️ *Not all building automation systems are compatible with all printers. Obtain a list of approved printers from the building automation system manufacturer.*

ALARM PRINTERS

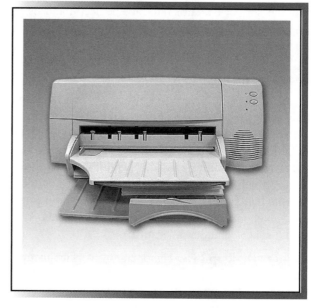

Figure 16-2. An alarm printer is a standard office printer that provides an operator with hard copies of alarms, preventive maintenance messages, and data trends.

On-Site Notebook PCs. In some installations, notebook PCs have replaced desktop PCs as the operator interface method. See Figure 16-3. Advances in the speed and capacity of notebook PCs have resulted in a portable operator interface method comparable to desktop PCs in flexibility. When loaded with the appropriate software, a notebook PC can function as a portable service/maintenance tool that can be used to change setpoints and time schedules, or check a controller for proper operation. Most modern building automation systems provide a phone jack connection at the temperature sensor for checking system operation.

NOTEBOOK PC OPERATOR INTERFACE

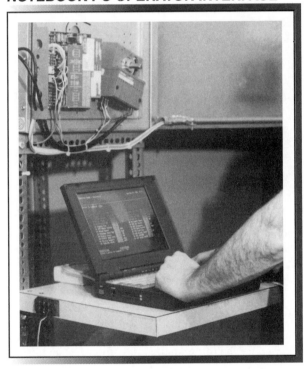

Figure 16-3. Notebook PCs have become common building automation system operator interfaces due to their speed, capacity, and portability.

On-Site Portable Operator Terminals. *A portable operator terminal (POT)* is a small, lightweight, handheld device that allows access to basic building automation system functions from various controllers throughout a building. A portable operator terminal is used in place of a notebook PC when the expense and setup effort of a notebook PC are inappropriate. See Figure 16-4. A portable operator terminal can display only two to four lines on its liquid crystal

display (LCD) and can change only times and setpoints. Portable operator terminals can receive controller alarms and are access code protected. Portable operator terminals normally have a phone jack connection similar to a notebook PC.

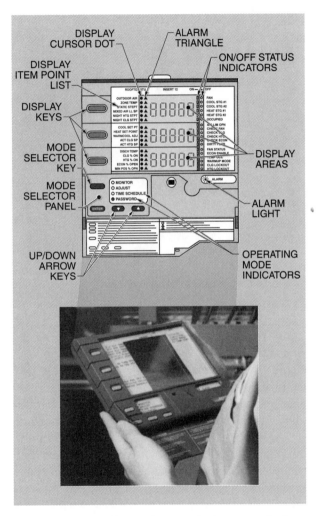

Figure 16-4. Portable operator terminals are used where the expense and setup of a notebook PC is inappropriate.

Keypad Displays. A *keypad display* is a controller-mounted device that consists of a small number of keys and a small display. Keypad displays are used if a limited method of operator interface is desired from a fixed location. See Figure 16-5. Keypad displays are commonly mounted in the front panel of a building automation system network communication module. Keypad displays can display two to four lines of information and have limited capability such as changing times and setpoints. Some manufacturers provide a device that can be used as a permanent keypad display or carried as a portable operator terminal.

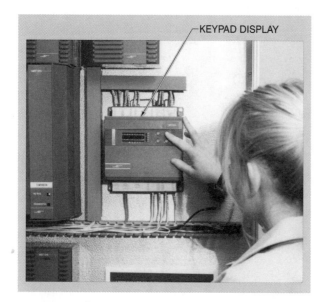

Figure 16-5. Keypad displays are used if a limited operator interface at a fixed location is desired.

Dumb Terminals. A *dumb terminal* is a display monitor and keyboard, with no processing capabilities. See Figure 16-6. A dumb terminal is used if the amount of information displayed on a portable operator terminal or keypad display is too limited. A dumb terminal has similar capabilities to a portable operator terminal or keypad display. The difference is that a dumb terminal can display 20 to 60 lines of information. This makes a dumb terminal easier to use than a portable operator terminal or a keypad display. A portable operator terminal or keypad display must be scrolled two lines at a time to view the same information.

A personal computer can be used as a dumb terminal. In this application, the PC simulates (emulates) another type of device. Many popular software packages provide this function. Dumb terminal emulation has standards that can be changed in the software to correspond to particular building automation system settings.

Off-Site Operator Interfaces

Off-site operator interfaces are often provided in addition to on-site operator interfaces used by maintenance and management personnel. The widespread use of personal computers and modems enables an alarm that occurs when a building is unoccupied to be reported to on-call maintenance personnel. In addition, off-site operator interfaces are used by the controls contractor to provide system monitoring and adjustment. Off-site operator interfaces include off-site desktop PCs, off-site portable operator terminals, off-site notebook PCs, voice phone interface, pagers, fax machines, and cellular communication.

DUMB TERMINALS

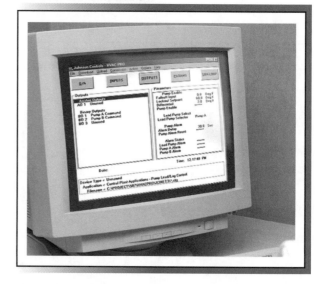

Figure 16-6. A dumb terminal is a display monitor and keyboard with no processing capabilities.

Off-Site Desktop PCs. The most common off-site operator interface is an off-site desktop PC. For small commercial facilities, such as fast food restaurants or convenience stores, a desktop PC at the office of the building contractor may be the only operator interface method used. An off-site desktop PC uses a modem to dial in to the building automation system or allows the building automation system to dial in to the off-site desktop PC. See Figure 16-7. Off-site desktop PCs can be used to perform the following:

- create and download control software to the network communication modules in a building
- change the software at network communication modules to update, expand, or fine-tune the control sequences
- receive alarms initiated by the network communication modules in a building

- retrieve data logs and preventive maintenance messages
- retrieve information used by third-party software packages to create graphs and pie charts that indicate system performance, mechanical equipment condition, and energy consumption

In addition to communication between a building automation system and an off-site desktop PC in the office of the building contractor, communication can also be performed between a building automation system and desktop PCs at the residence of a technician on call. At night or on weekends, when the contractor's office is closed, alarms and other information can be sent to the desktop PC at the residence of a technician on call. The software of most building automation systems can be programmed to dial out to more than one phone number. The software chooses the phone number of a person who is on call at the time the alarm occurs, enabling 24-hour coverage.

Off-Site Portable Operator Terminals. A portable operator terminal with a modem located at the residence of a technician on call can perform the same tasks as an on-site portable operator terminal. Functions of an off-site portable operator terminal commonly include setpoint adjustments, time schedule changes, and timed override of equipment.

Off-Site Notebook PCs. Notebook PCs used on site as a service/maintenance tool may also be used off site in a remote location at night or on weekends as a mobile monitoring tool. With the addition of a docking station, notebook PCs have the same monitoring and printing capabilities as desktop PCs. A *docking station* is a base station for a notebook computer that includes a power supply, expansion slots, and monitor and keyboard connectors.

Voice Phone Interface. *Voice phone interface* is an automated phone service in which a voice on a computer chip prompts the caller to press various numbers

OFF-SITE DESKTOP PCs

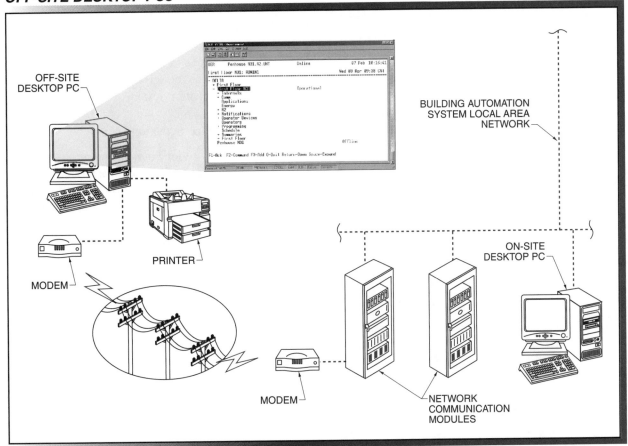

Figure 16-7. Modems allow off-site communication with a building automation system.

for different functions when dialing in on a standard touch-tone phone. For example, a system may require that the number 1 be pressed for a setpoint change, or the number 2 be pressed for a time schedule change. Voice phone interface is password-protected, inexpensive, and easy to use. Some manufacturers provide durable, pocket size cards that contain the operator access code and procedure for accessing information such as load numbers, setpoints, schedules, and alarms. Voice phone interface is effective for busy facility managers who are responsible for a number of facilities located over a large area.

Pagers. A *pager (beeper)* is a small portable electronic device that vibrates, emits a beeping signal, or displays a text message when the individual carrying it is paged. Pagers use the dial-out capabilities of a building automation system for remote notification of alarms. See Figure 16-8. The building automation system calls the pager through a paging service and places a numeric or alphanumeric message on the digital display. The message alerts the on-call technician that a problem has occurred and an alarm has been activated. The on-call technician must use an operator interface such as an off-site desktop PC to modify system parameters. Pagers have a low initial cost and enable 24-hour coverage.

PAGERS

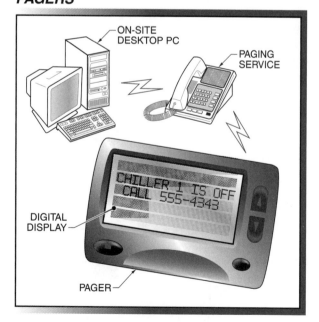

Figure 16-8. Building automation systems can be integrated with paging services to alert maintenance personnel to alarms.

Fax Machines. A building automation system can send alarms and logs to fax machines in addition to off-site PCs. This removes the need for a dedicated printer for the building automation system. However, care must be taken that the amount of information reported to the fax machine is not excessive because this ties up the fax machine and makes it difficult to use for normal business purposes. Conversely, if a fax machine is heavily used for normal business purposes, the information from the building automation system may not get through to its intended recipient.

> *When integrating pagers with building automation systems, ensure that the correct information is sent to the appropriate pager. All of the maintenance personnel in a facility should not be inconvenienced by low-priority alarms and messages being routed to their pagers. In addition, multiple pagers may be required for different times of the day and different days of the week.*

Cellular Communication. Cellular communication can be used as a building automation system operator interface. A cellular phone may use a voice phone interface to enable changes to the setpoints and time schedules of a building automation system.

Operator Interface Selection

The operator interface for a building automation system is selected according to the level of interface required. The controls contractor, building owner, and manager require an operator interface method to enable system parameter modification. Guidelines for selecting an operator interface include the following:

- An off-site desktop PC, notebook PC, or pager is required if 24-hour, 7 day per week coverage is desired.
- An on-site desktop PC or notebook PC is required if only a few individuals need access to the building automation system. A dumb terminal or portable operator terminal is required if a large number of people need access to the building automation system.
- A basic operator interface, such as a dumb terminal or keypad display, is required if the maintenance staff has poor computer skills or training is not provided.

Operator Interface Training

One of the primary reasons for poor customer satisfaction with a building automation system is the lack of effective training. Routine tasks are not performed as expected if maintenance personnel are unable to properly use the selected operator interface. The content of a training course should be examined closely and a training needs analysis survey performed on each new building automation system installation. See Figure 16-9. The training needs analysis helps ensure the following:

- tasks the personnel need to perform are presented
- a method of performance measurement exists
- training covers the tasks needed in a particular building
- training is not over the heads of the building maintenance personnel
- the computer literacy level of the training is matched to the level of the personnel

Training Needs Analysis

TRAINING REQUISITION FORM

Originator Name ____________ Address ____________
Job Title ____________ City ____________
Department ____________ Telephone ____________

COURSE INFORMATION
Course Topic
Course Description
(please attach final course outline or describe suggested training)

JOB INFORMATION
Target Audience

What tasks can employees not perform that the course will train them to do?

What previous training have the employees received on these tasks?

What specific productivity improvements do you expect from this proposed course?

Figure 16-9. A training needs analysis identifies tasks and skills required by building automation system personnel.

Training Availability and Scheduling. A training package should be included as part of the building automation system specification and purchase process. As part of this package, training dates and times should be scheduled in advance to enable attendance by the required personnel. See Figure 16-10. Operator training should be provided simultaneously with the building automation system startup. Also, the location, times, and dates of the training should be agreed upon in advance to minimize problems with schedules, travel arrangements, and shift coverage. All major building automation system manufacturers have a training center and staff available. If training is performed on site, the times should be specified to minimize overtime costs, enable second and third shift personnel to easily attend, and reduce interference with normal building operations.

ABC CONTROLS COMPANY - Training Schedule
Subject to change; courses will be on an as-needed basis.

Course #	*Course Name*	*Course Start*	*Course End*	*Location of Training*
120	Facility Operators	January 21	January 22	Anderson College
130	Controller Engineering/ Programming	February 2	February 8	Anderson College
121	ASC Engineering	February 18	February 22	Anderson College
120	Facility Operators	March 18	March 22	Anderson College
127	Hardware Troubleshooting	April 8	April 12	Anderson College
167	Graphics Development	April 23	May 1	Anderson College
100	HVAC Basics	May 3	May 11	Anderson College
134	VAV Operators	May 21	May 22	Anderson College
120	Facility Operators	June 1	June 7	Anderson College
164	Control Engineering	June 17	June 21	Anderson College
100	HVAC Basics	July 5	July 11	Anderson College
124	HVAC Maintenance	July 22	July 26	Anderson College
122	LAN Engineering	August 2	August 3	Anderson College
120	Facility Operators	September 3	September 11	Anderson College
132	Software Update	September 24	September 28	Anderson College
121	ASC Engineering	October 14	October 18	Anderson College
100	HVAC Basics	November 2	November 7	Anderson College
131	Graphics Engineering	November 18	November 22	Anderson College

To enroll, please call 1-800-555-4343.

Figure 16-10. Building automation system training must be scheduled in advance to enable attendance by the required personnel.

Ongoing Training. In addition to an adequate training program initiated on building automation system startup, ongoing training is also required. If a needs analysis is performed, a long-term comprehensive training plan can be established. The comprehensive training plan can include building automation system as well as HVAC system training, and should be integrated with the human resources department of a facility. The human resources department is responsible for setting up policies and procedures regarding professional development goals, incentives for improvement, staffing needs, and long-term management goals. Acquisition of new employees, updating of software, or expansion of the building automation system may require additional training which should be part of the comprehensive training plan.

Training Support Materials. To be effective, training programs require the proper training support materials. Manuals and other literature must be updated and available at all times. Lists of points, alarms, and logs should be given to all maintenance personnel who use the system. Some manufacturers provide waterproof pocket-size cards that contain point and access information.

With the widespread availability of VCRs, training tapes and manuals are provided by many manufacturers to allow individual study at home. CD-ROMs with built-in tests that indicate proficiency in specific areas can also be used. Training tapes, manuals, and CD-ROMs should be provided on a rotating system, and individuals should be encouraged to use them. In addition, training simulators may be purchased from building automation system manufacturers for use by the maintenance staff. See Figure 16-11.

TRAINING SIMULATORS

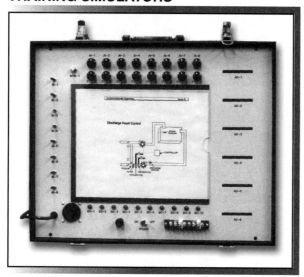

Figure 16-11. Training simulators are used by maintenance staff to gain controller knowledge and programming experience.

Access Levels and Codes. An essential part of operator interface training is the assigning of access levels and codes (passwords). An access level is the group of items and functions that an operator is permitted to perform. The access level assigned to operators or other personnel is determined by the individual responsible for the building automation system project. Access levels include supervisor, engineering, lead technician, and technician levels. See Figure 16-12.

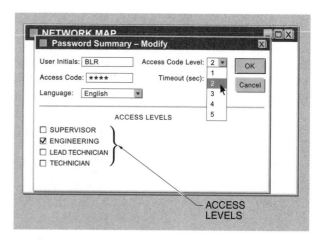

Figure 16-12. Access levels are used to limit access to information and data.

The supervisor level allows the modification of any parameter in the system. The supervisor level also allows the assigning of access levels and codes to other individuals. The supervisor level requires the highest amount of training because any parameter may be changed. The engineering level allows the creation or modification of any parameter in the system but does not permit assigning access levels and codes. The engineering level is normally used to add or delete items from the database. The lead technician level allows modification of setpoints and time schedules but does not allow the creation or changing of building automation system parameters. The lead technician level also allows the checking of alarms, point status, and data logs. The changes may be permanent or temporary. The technician level allows the viewing of setpoints, time schedules, alarms, data logs, and point status. The technician level does not allow changes.

When an access level is assigned, the operator is given an alphanumeric access code that is entered into the building automation system database. The access code activates the operator access level. Most manufacturers use three or four digits that appear as asterisks on a monitor to prevent unauthorized misuse of the access code. Some manufacturers use a 20 sec timeout between access attempts to prevent quick multiple guesses. A further safeguard immobilizes the access code system for 10 min, 20 min, or 30 min if three wrong access codes are entered. A special log is often set aside to record the names, times, and dates of operator interface so that action can be taken in case of improper changes.

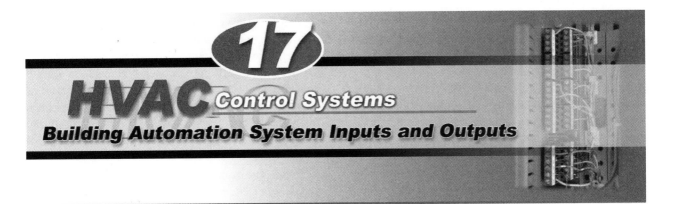

Building automation systems include inputs (devices that indicate building conditions to a controller) and outputs (devices that change the state of a controlled device). Inputs and outputs may be analog or digital. Analog inputs send a range of continuous values to a controller. Digital inputs produce an ON or OFF signal. Analog outputs produce a continuous signal between two values. Digital outputs accept an ON or OFF signal.

BUILDING AUTOMATION SYSTEM INPUTS

A *building automation system input* is a device that senses and sends building condition information to a controller. The central processing unit (CPU) of a building automation system controller makes decisions based on the input information. The decisions are used to change the state of output devices. Building automation system inputs may be analog inputs or digital inputs.

Analog Inputs

An *analog input (AI)* is a device that senses a variable such as temperature, pressure, or humidity and causes a proportional electrical signal change at the building automation system controller. See Figure 17-1. Analog inputs enable HVAC technicians to obtain a readout of the variable on a personal computer (PC) or other operator interface device. Analog inputs are normally more expensive than digital inputs.

The proper analog input must be selected for a particular task in a commercial building. Factors considered include price and reliability of the input device, its proposed location, and the environment (humidity, dirt, etc.). Once the input device is selected, proper installation procedures must be followed to ensure proper

device operation. Analog input devices include temperature sensors, humidity sensors, pressure sensors, light sensors, and specialized analog input devices.

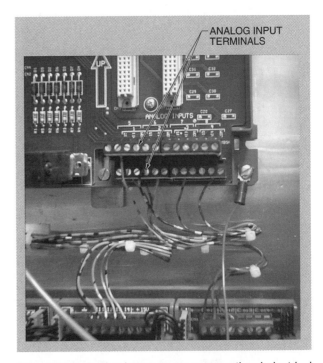

Figure 17-1. Analog inputs cause a proportional electrical signal change at building automation system controllers.

Temperature Sensors. A *temperature sensor* is a device which measures the temperature in a duct, pipe, or room and sends a signal to a controller. Temperature sensors are the most common analog inputs used in building automation systems. Temperature sensors are available in a variety of packages and mounting configurations. Common temperature sensors include wall-mount, duct-mount, well-mount, and averaging temperature sensors. See Figure 17-2. Wall-mount temperature sensors are the most common temperature sensors used in building automation systems and are used to sense air temperatures in building spaces. Duct-mount temperature sensors are used to sense air temperatures inside ductwork. Well-mount temperature sensors are used to sense water temperature in piping systems. Averaging temperature sensors are used in large duct systems to obtain an accurate temperature reading. Temperature sensors that are designed to sense outside air temperatures are also available.

Temperature sensors are commonly referred to as positive-temperature-coefficient (PTC) or negative-temperature-coefficient (NTC) sensors. A positive-temperature-coefficient sensor increases its output resistance as the temperature increases, and decreases its output resistance as the temperature decreases. A negative-temperature-coefficient sensor decreases its output resistance as the temperature increases, and increases its output resistance as the temperature decreases.

Temperature sensor data sheets are available that provide resistance/temperature curves and charts for a particular sensor. See Figure 17-3. Sensor characteristics vary based on the material from which they are constructed. Care should be taken to choose a sensor that has a relatively linear resistance/temperature curve. A linear resistance/temperature curve ensures accuracy across the entire temperature range of the sensor. Temperature sensor sensing elements may be thermistors or resistance temperature detectors.

TEMPERATURE SENSORS

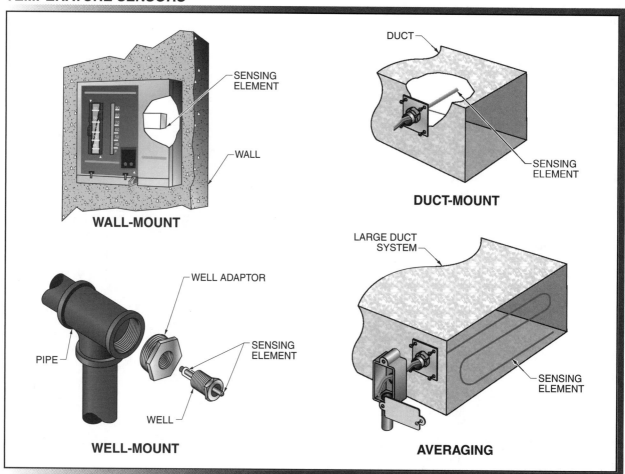

Figure 17-2. Temperature sensors are available in a variety of packages and mounting configurations.

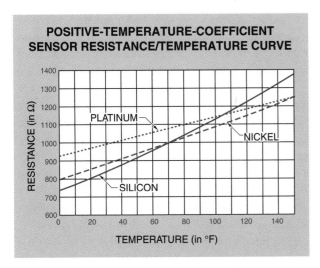

Figure 17-3. A positive-temperature-coefficient sensor increases its output resistance as the temperature increases and decreases its output resistance as the temperature decreases.

A *thermistor* is a resistor made of semiconductor material in which electrical resistance varies based on changes in temperature. Thermistors are inexpensive but susceptible to inaccuracy if long wire runs are used. Long wire runs add resistance to a circuit, which leads to sensor inaccuracy because of the sensor resistance/temperature relationship. Many building automation systems allow the changing of sensor values in the system software to compensate for inaccuracies due to long wire runs. See Figure 17-4.

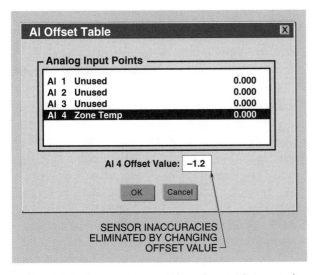

Figure 17-4. Temperature-sensing element inaccuracies can be compensated for by changing the sensor offset value in the building automation system software.

A *resistance temperature detector (RTD)* is a resistor made of a material for which the electrical resistance is a known function of the temperature. Resistance temperature detectors are constructed from long pieces of nickel, silicon, or platinum. Resistance temperature detectors are accurate but expensive. Resistance temperature detectors can be adjusted in software like thermistors or may be factory-calibrated by burning off part of the precious metal. Factory-calibrated resistance temperature detectors are referred to as laser-trimmed sensors.

All temperature sensors must be placed in the proper location for measurement accuracy. For example, outside air temperature sensors must be located away from exposure to direct sunlight. All manufacturers provide sunshields to prevent direct sunlight from affecting outside air temperature sensor operation. See Figure 17-5.

TEMPERATURE SENSOR SUNSHIELD

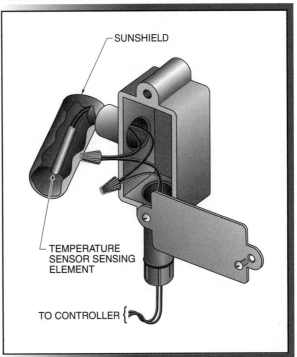

Figure 17-5. A sunshield protects outside air temperature sensor sensing elements from direct solar radiation.

Averaging sensors must be installed properly by stringing the flexible sensing element across the duct. Most temperature sensors are wired to a controller using two wires and are not polarity-sensitive. A minimum of 18-gauge wire should be used because light-gauge wire is more easily damaged and kinked. Light-gauge wire also has more resistance per foot and may cause greater inaccuracy at the sensor. A single continuous run of wire should be used because wire nuts and other connectors are subject to poor connections and failure, especially if equipment vibration is present. Also, care must be taken to prevent accidental damage to the element from service procedures. Many temperature sensors have built-in electronic circuits that produce a sensor output signal of 4 mA to 20 mA or 0 VDC to 10 VDC.

Humidity Sensors. Human comfort, product integrity, and corrosion prevention require control of the humidity level in a commercial building. A *humidity sensor* is a device which senses the amount of moisture in the air and sends a signal to a controller. The most common humidity sensors measure percent relative humidity (% rh), while other sensors measure dewpoint or absolute humidity. Most humidity sensors use a hygroscopic element. A *hygroscopic element* is a device that changes its characteristics as the humidity changes. See Figure 17-6.

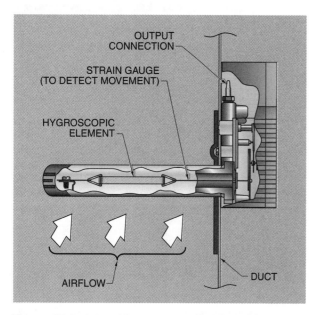

Figure 17-6. A humidity sensor contains a hygroscopic element that changes physical characteristics as the humidity changes.

Accurate humidity sensing and control is more difficult than temperature sensing and control. Humidity sensors are susceptible to dirt on the element. Many humidity sensors drift over time, with inaccuracies of 1% per year common. Also, if certain gases or solvents are present in the air, they may damage the sensing element or prevent it from functioning properly. Proper sensor function is essential in process or product storage environments where humidity accuracies must be maintained. Humidity sensors should be checked a minimum of once

per year for accuracy because the human body is not an accurate indicator of the humidity level. Any instrument used to check the accuracy of humidity sensors must be regularly checked against a known standard.

Humidity sensors may be capacitive sensors or bulk polymer sensors. A *capacitive sensor* is a humidity sensor that uses a thin film of hygroscopic element to alter the capacitance of a circuit. Capacitive sensors are extremely sensitive to a change in the surrounding humidity. A *bulk polymer sensor* is a humidity sensor that consists of a polymer saturated with a salt compound. Bulk polymer humidity sensors are stable over time and are often designated as 2% rh, 3% rh, and 5% rh accuracy sensors.

Humidity sensors are often combined in a single package with temperature sensors to simplify humidity and temperature sensing. Humidity sensors use the same 4 mA to 20 mA and 0 VDC to 10 VDC standard signal values as temperature sensors.

Pressure Sensors. A *pressure sensor* is a device which measures the pressure in a duct, pipe, or room and sends a signal to a controller. Pressure sensors are used in HVAC applications to provide an indication of the actual pressure inside a duct or pipe to a building automation system controller. See Figure 17-7. Piezoelectric elements (pressure-sensitive crystals) are used in some pressure-sensing applications. Bellows elements are used in many piping applications. Standard pressure sensor output signal values include 4 mA to 20 mA and 0 VDC to 10 VDC.

> *Avoid installing pressure sensors downstream of obstructions such as tees, elbows, and filters. Depending on the application, the turbulence created by these obstructions can affect the reading of the sensor.*

Care must be taken with the installation and location of pressure sensors. Pressure sensors located in an improper area sense an incorrect pressure. Tubing and accessories must be installed according to manufacturer recommendations. Also, ensure the sensor range is correct for the expected system pressures. For example, a 0″ wc to 2″ wc sensor should not be used if the duct static pressure is higher than this range on startup. Some pressure sensors require an additional power supply and may require three or four wires instead of two. Many of these pressure sensors are polarity-sensitive, requiring installation to manufacturer specifications.

PRESSURE SENSORS

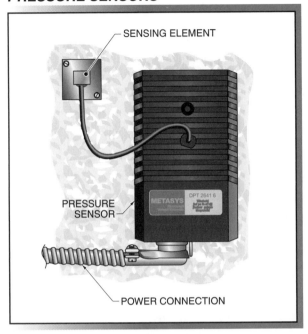

Figure 17-7. Pressure sensors sense the actual pressure inside a duct or pipe and send the reading to a building automation system controller.

Light Sensors. Outdoor and indoor lighting loads may use 20% or more of the power in a commercial building. An outdoor light sensor is used to efficiently control outdoor lighting loads. Indoor light sensors are used to control indoor lighting loads. Light sensors consist of a solid-state material that is exposed to light. See Figure 17-8. When light sensors are wired to a building automation system, they input the actual light level in lumens or footcandles to the building automation system. A *footcandle* is the amount of light produced by a lamp (lumens) divided by the area that is illuminated. Standard light sensor signals are 4 mA to 20 mA or 0 VDC to 10 VDC. The light level values are scaled in the building automation system software. For example, 20 mA represents the maximum footcandles measured by the sensor and 4 mA represents the minimum footcandles measured by the sensor.

Light sensors must be located properly. For example, most manufacturers recommend that outdoor light sensors be located on the north side of a building, out of direct sunlight. Some outdoor light sensors are deceived by high levels of background infrared radiation such as caused by hot asphalt. To avoid this, the light sensor should be installed high above the ground.

LIGHT SENSOR APPLICATIONS

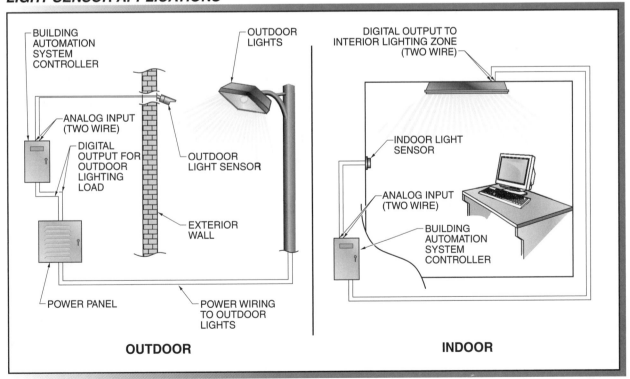

Figure 17-8. A building automation system controller turns outdoor and indoor lights ON and OFF based on the light level measured by light sensors.

Specialized Analog Inputs. Specialized analog inputs are available that can obtain almost any desired variable. Common specialized analog input devices include carbon dioxide, carbon monoxide, refrigerant, and gas sensors. Carbon dioxide sensors are used to sense indoor air quality (IAQ). Indoor air quality is related to the level of carbon dioxide due to human activity. See Figure 17-9. Carbon monoxide sensors are used to sense the amount of toxic exhaust fumes inside parking garages. Refrigerant sensors are used to sense the amount of refrigerant in the air due to leaks. Gas sensors are used to sense the amount of gas in the air.

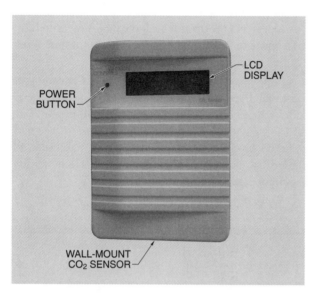

Figure 17-9. Carbon dioxide sensors are used to measure carbon dioxide and must have their calibration checked yearly or when a problem occurs.

Digital Inputs

A *digital (binary) input* is a device that produces an ON or OFF signal. Digital inputs differ from analog inputs in that analog inputs provide a readout of the actual value while digital inputs indicate if a value is above or below a certain setpoint. For example, an analog input device may provide a reading of 74.5°F. A digital input device may indicate that a temperature is above or below 70°F.

Digital inputs do not provide as much information as analog inputs. However, digital inputs are much less expensive than analog inputs and are often used as status points in a system. For example, digital input devices are used to indicate FAN ON/OFF, PUMP ON/OFF, FILTER OK/PLUGGED, or ROOM OCCUPIED/UNOCCUPIED status.

The input (analog or digital) selected for an application is often based on cost and functionality. For example, an analog input is used if a variable (72°F) is required. A digital input is used if only status (ON/OFF) is required. Digital inputs are commonly designated as dry contact closure. Dry contact closure inputs require no external power supply. Dry contact closure inputs only require wiring to the controller terminals to function. Digital input devices include thermostats, humidistats, differential pressure switches, flow switches, light level switches, accumulators, current-sensing relays, timed override initiators, and specialized digital inputs.

Thermostats. A *thermostat* is a temperature-actuated switch. A *limit thermostat* is a thermostat that maintains a temperature above or below an adjustable setpoint. Limit thermostats are normally used to indicate an improper temperature level and are not used as a primary temperature controller. Limit thermostats are normally inserted into a duct or strapped to a pipe. Limit thermostats have a setpoint that determines the temperature at which they switch. A building automation system reads only whether the thermostat is open or closed. The building automation system software indicates whether an open thermostat indicates a high or low temperature. Normally open and normally closed contacts are available for operations such as opening on a rise in temperature or closing on a rise in temperature. See Figure 17-10.

LIMIT THERMOSTATS

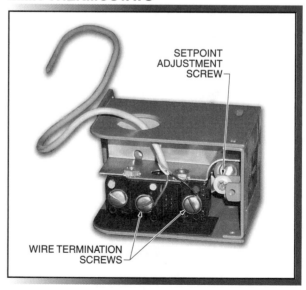

Figure 17-10. Limit thermostats maintain a temperature above or below an adjustable setpoint.

Humidistats. A *humidistat* is a device that senses the humidity level in the air. Humidistats are often used to indicate the humidity level in a duct or building space. Like digital thermostats, digital humidistats have a setpoint and normally open and normally closed contacts. Digital humidistats are normally less expensive than analog humidity sensors. Care must be taken in selecting the setpoint and installing the humidistat. See Figure 17-11.

DIGITAL HUMIDISTATS

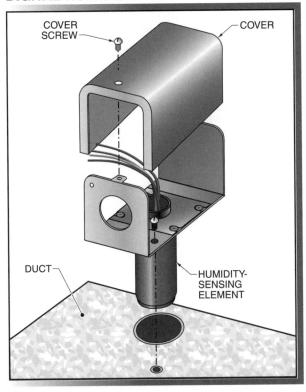

Figure 17-11. Humidistats are installed in ducts to prevent water condensation due to high levels of humidity.

Differential Pressure Switches. A *differential pressure switch* is a digital input device that switches open or closed because of the difference between two pressures. Differential pressure switches often use bellows or diaphragm elements. The setpoint of a differential pressure switch must be adjusted correctly and the correct contacts (normally open or normally closed) selected for a given application.

Differential pressure switches are often used to indicate a difference in pressure across a fan or pump or to indicate air handling unit filter condition. Differential pressure switches normally have a section of tubing located on the suction (inlet) side of a fan or pump and

another section of tubing located on the discharge side. See Figure 17-12. Differential pressure switch tubing is normally installed to avoid close bending, kinking, or long piping runs. Also, any air leaks must be located and corrected before differential pressure switches are put in service.

Differential pressure switches must not be located too close to a fan or pump where high pulsation and vibration occur. Also, avoid installing differential pressure switches too close to ductwork elbows, tees, or other obstructions that may affect the pressure sensed. Always follow manufacturer's installation recommendations.

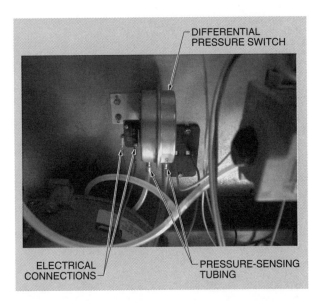

Figure 17-12. Differential pressure switches are often used to indicate a difference in pressure across a fan or pump or to indicate air handling unit filter condition.

Flow Switches. A *flow switch* is a switch that contains a paddle that moves when contacted by air or water flow. Flow switches can be used to indicate the flow status of a pump or fan. See Figure 17-13. Differential pressure switches are superior to flow switches because flow switches may be subject to problems due to fan or pump flow pulsation. The flow pulsation may lead to false indications at the building automation system controller. Alarms connected to a flow switch may provide false indication due to flow pulsation.

> *Differential pressure switches are used more often than flow switches in building automation system installations because of their reliability.*

FLOW SWITCHES

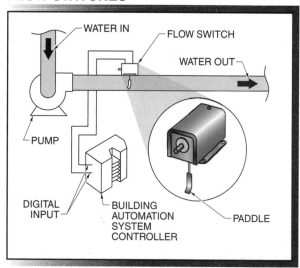

Figure 17-13. Flow switches use a paddle inserted in the air or water stream to indicate the flow status of a pump or fan.

Light Level Switches. A *light level switch* is a device that indicates if a light level is above or below a setpoint. Light level switches are less expensive than analog light sensors. Light level switch setpoint adjustment may consist of a shutter over the sensing element that is adjusted to trip at a higher or lower light level. A light level switch setpoint must be set at the proper level and the switch located properly for correct operation.

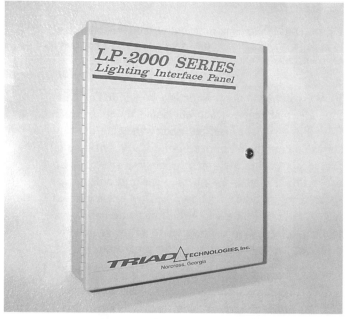

Light level switches may be wired to a building automation system controller or to a lighting control panel. Many lighting control panels are designed to communicate with the building automation system in commercial buildings.

Accumulators. An *accumulator (counter)* is a device that records the number of occurrences of a signal. Accumulators are commonly used to measure the electrical demand for a building or load. See Figure 17-14. Most modern utility electrical demand meters use pulses to indicate the amount of electricity being used. These pulses can also be sensed by an accumulator and wired as a digital input to a building automation system controller. Each ON/OFF pulse represents a specific amount of electricity used. For example, each pulse may indicate .5 kW of electricity. This pulse rate multiplier information, when sent to building automation system software, enables building electrical demand control and data trending. Accumulators can also be used with water and natural gas metering. The pulse rate multiplier must be obtained from the meter manufacturer.

ACCUMULATORS

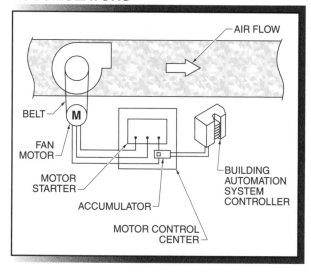

Figure 17-14. Accumulators record the number of occurrences of a signal and are commonly used to measure the electrical demand for a building or load.

> *Accumulators may be used to submeter areas in a building so each tenant can be charged for their utility use.*

Current-Sensing Relays. A *current-sensing relay* is a device which surrounds a wire and detects the electromagnetic field due to electricity passing through the wire. Current-sensing relays are digital inputs used to obtain the operational status of pumps, fans, compressors, and lighting loads. See Figure 17-15. Current-sensing relays are installed so the current-carrying wires of a motor pass through the center of the relay. When the motor starts, the relay trips, indicating to the building automation system controller that current is flowing. The current flow indicates that the motor is doing work. A fan motor that has a broken or missing belt drive draws a low amount of current and does not trip the relay. The current setpoint must be adjusted properly to avoid nuisance trips.

SPLIT-CORE CURRENT-SENSING RELAYS

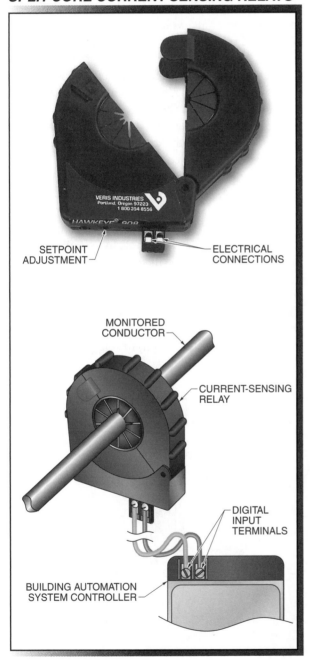

Figure 17-15. Current-sensing relays sense the current flowing in a conductor.

Timed Override Initiators. A *timed override initiator* is a device that, when closed, sends a signal to a controller which indicates that a timed override period is to begin. At night or when a building is unoccupied, a building automation system controls the temperature in building spaces to unoccupied setpoints such as 55°F for heating and 85°F for cooling. Building spaces are often occupied temporarily after normal occupied hours. A timed override initiator in the building space is wired to the building automation system controller. When the timed override initiator is made, it signals the building automation system to switch to the occupied setpoints for a specific time period. The timed override initiator is normally a momentary contact switch. See Figure 17-16. A momentary contact switch is a device that closes its contacts for a short period of time when depressed.

TIMED OVERRIDE INITIATORS

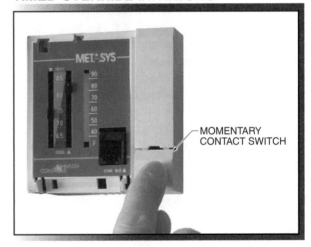

Figure 17-16. A momentary contact switch can be used as a timed override initiator.

Specialized Digital Inputs. Specialized digital inputs include card access controllers, 3ϕ monitors, and flame safeguard controllers. Card access controllers cause a digital input signal to a building automation system controller to indicate that an individual has used their access card. This turns ON their HVAC unit and lighting. Three-phase monitors are digital inputs used to monitor incoming electrical service. Three-phase monitors send a digital input signal to a building automation system controller if the electrical power is not at the correct level. The building automation system generates an alarm and turns all electrical loads OFF that may be damaged. A flame safeguard controller sends a digital input signal to a building automation system controller to indicate an alarm or fault condition.

BUILDING AUTOMATION SYSTEM OUTPUTS

A *building automation system output* is a device that changes the state of a controlled device in response to a command from a building automation system controller. Building automation system output devices may be either directly wired to the controlled device or wired to an interface such as a transducer or relay which changes the output signal. Building automation system outputs may be analog (positioning between 0% and 100%) or digital (ON/OFF).

Analog Outputs

An *analog output* is a device that produces a continuous signal between two values. Analog output devices produce signals in the ranges of 0 mA to 20 mA, 0 VDC to 5 VDC, or 0 VDC to 10 VDC. Analog output devices provide proportional control of damper and valve actuators and variable-frequency drives. Analog output devices lack standardization. For example, a manufacturer may produce output devices with a specific voltage range that may be incompatible with some actuators or other devices.

Analog outputs may be connected directly to controlled devices or may need their signal changed by a transducer. Electric motor drives produce a 0 mA to 20 mA or 0 VDC to 10 VDC analog output signals.

The controller analog output signal causes the electric motor drive to speed up or slow down the motor. Analog output signals are also connected directly to motor-driven valve and damper actuators to drive them open or closed.

Transducers. A *transducer* is a device that changes one type of proportional control signal into another. Most transducers are connected directly to controlled devices such as valve actuators, damper actuators, and variable-frequency drives. These controlled devices use common current and voltage ranges such as 4 mA to 20 mA and 0 VDC to 10 VDC. Transducers are available that can receive these input currents and voltages and output a voltage using a variable resistance such as 0 Ω to 135 Ω. These transducers allow the use of actuators that require the varying voltage created by a device with a variable-resistance output. Other transducers are available that use an analog input signal to trigger multistage heating and cooling units.

Pneumatic transducers normally have an analog input signal such as 4 mA to 20 mA from the building automation system and an output air pressure signal of 0 psig to 20 psig. See Figure 17-17. The transducer output pressure is often adjustable in the building automation system software to match the actual spring range of the actuator. Pneumatic transducers are used in applications that have existing pneumatic valve and damper actuators. The pneumatic transducers enable pneumatic actuators to be used in modern building automation systems, simplifying installation and reducing job costs.

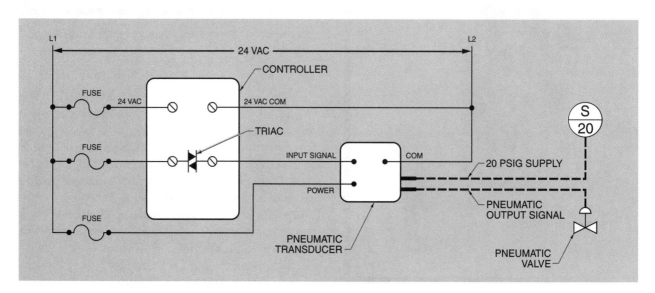

Figure 17-17. Pneumatic transducers normally have an analog input signal such as 4 mA to 20 mA from the building automation system and an output air pressure signal of 0 psig to 20 psig.

Digital Outputs

A *digital output* is a device that accepts an ON or OFF signal. The software of a building automation system controller evaluates the information received as inputs. Depending on its software, the controller sends a command causing a change of state at a controlled device. The change of state is performed by the controller output device. Various output devices are available for different controlled devices. Most digital output devices used in building automation systems are triacs, which require an external power supply. A *triac* is a solid-state switching device used to switch alternating current. Digital output devices include relays, incremental outputs, and pulse width modulated outputs.

Application-specific controllers are commonly used to control package rooftop units and heat pumps. Rooftop units and heat pumps normally contain low-voltage control systems rated at 24 VAC. Most building automation system controllers use triac outputs, which are rated as Class II (low-voltage) devices. Therefore, the application-specific controllers can directly control only the low-voltage control systems. The outputs of an application-specific controller are wired directly to the terminal board of the rooftop unit or heat pump. See Figure 17-18. In many cases, the existing rooftop unit or heat pump transformer can be used to provide power for the application-specific controller outputs. Care must be taken to avoid interference problems from spark ignition devices and to ensure the transformer is of the proper size. An HVAC technician can often use the standard color coding of the wires to make the proper connections.

When wiring building automation system controllers to packaged units, the wires from the building automation system are connected to the packaged unit terminal block. Most packaged units follow a standard terminal block color code.

WIRING APPLICATION-SPECIFIC CONTROLLER OUTPUTS

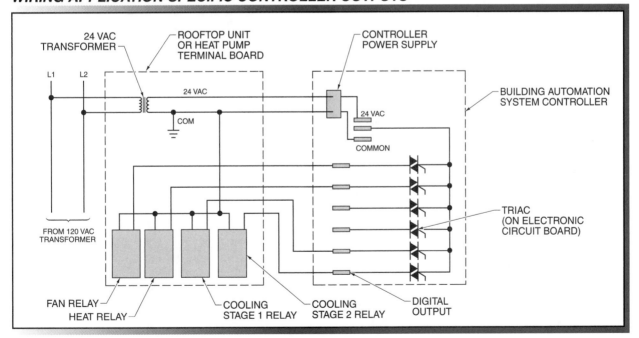

Figure 17-18. Building automation system controller output signals, normally rated at 24 VAC, can be wired directly to rooftop unit terminal boards.

Building automation systems use relays to control the flow of electric current between circuits.

Relays. The majority of controlled devices, such as motors, are line-voltage devices which operate at line voltage (120 VAC) and up. Building automation system controller outputs are generally rated at low voltage. Therefore, a relay must be used to allow the low-voltage-rated triacs in building automation system output devices to switch line-voltage devices. See Figure 17-19. Triac outputs are normally not powered, so an external power supply such as a 24 VAC transformer must be used. The relay can be supplied by the manufacturer or purchased separately.

Normally open and normally closed contacts are provided on a relay. Care must be taken to select the proper contacts to ensure correct output operation. The operation of the output can also be changed in the building automation system software. The software and wiring of the relay must be in agreement and provide for normal and fail-safe operation. Some high-quality relays provide manual hand/off/auto (HOA) switches that enable a technician to override the device ON/OFF condition. Digital inputs are commonly provided to allow the building automation system controller to monitor the status of the hand/off/auto switches.

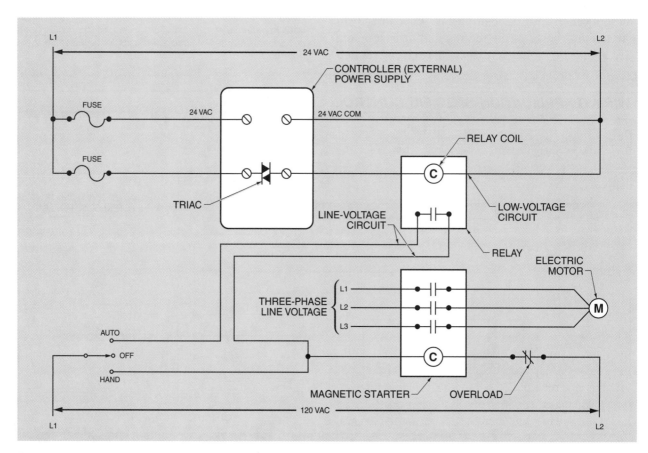

Figure 17-19. Relays are used to allow low-voltage-rated triacs in building automation system output devices to switch line-voltage devices.

Lighting control is a common application of building automation systems that is normally accomplished using digital outputs. Lighting loads are commonly controlled by latching relays and momentary outputs. A *latching relay* is a relay that requires a short pulse to energize the relay and turn ON the load. See Figure 17-20. Once the relay has been energized with a momentary pulse, the power flowing through the relay causes the relay to remain energized. Another short pulse causes the relay to de-energize. The momentary outputs can be only milliseconds long.

LATCHING RELAYS

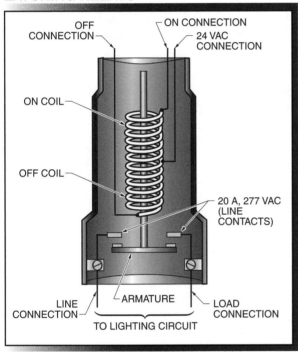

Figure 17-20. Latching relays are used to control lighting circuits and require a short pulse to energize the relay and turn ON the load.

Incremental Outputs. An *incremental output* is a digital output device used to position a bidirectional electric motor. Incremental outputs are used to position actuator motors which can be driven open and closed. Two outputs are required: one output drives the actuator open, and the other output drives the actuator closed. A disadvantage of using incremental outputs is that multiple outputs are used to drive one actuator. The position of the actuator is determined by the length of time the motor is driven in the open or closed direction.

Incremental outputs are commonly used in variable air volume damper applications. See Figure 17-21. In variable air volume damper applications, one output drives the damper open and the other output drives the damper closed. The stroke time of the actuator is the length of time that it takes the actuator to move from one extreme to the other. The stroke time should be recorded and entered in the building automation system controller software.

INCREMENTAL OUTPUTS

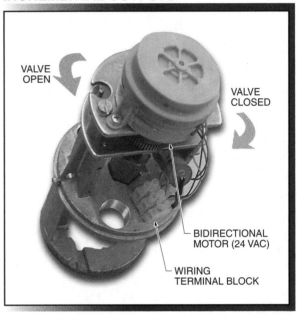

Figure 17-21. Incremental outputs are commonly used in variable air volume damper applications.

Pulse Width Modulated Outputs. *Pulse width modulation (PWM)* is a control technique in which a sequence of short pulses is used to position an actuator. Pulse width modulated outputs are used to position actuators using a digital output signal. A sequence of short pulses is sent from the building automation system controller to a transducer. The transducer receives the pulses and produces a corresponding variable output signal. See Figure 17-22. Pulse width modulation enables all building automation system outputs to be digital.

Different pulse width modulated device manufacturers may use slightly different pulse rates. For this reason, care must be taken when ordering or replacing pulse width modulated transducers.

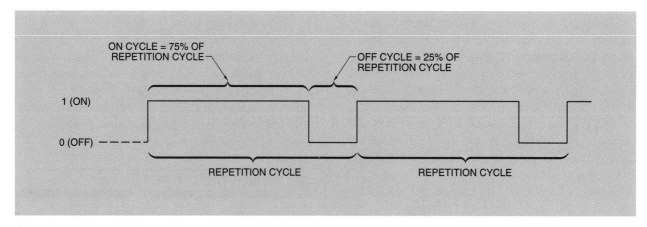

Figure 17-22. Pulse width modulated outputs use a sequence of pulses to position a device.

PSEUDOPOINTS

A *pseudopoint (data point)* is a point that exists only in software and is not a hard-wired point. Pseudopoints are used to hold information that is to be communicated on a network. Pseudopoints can be used to hold the lowest of three temperatures, the highest of three temperatures, or the status of an occupied/unoccupied point. Pseudopoints can be seen or measured with an electronic device such as a personal computer or keypad display because they only exist electronically.

An HVAC technician must distinguish between actual hard-wired points and pseudopoints because many input and output point lists contain both. Most building automation system software manufacturers provide a designation for pseudopoints to distinguish them from actual hard-wired points.

Controllers must be mounted and wired correctly for proper operation of a building automation system. Inputs and outputs are normally connected to a building automation system using twisted shielded wire. Most building automation system controllers use a 24 VAC power supply that has the proper capacity for the application. Building automation system controller mounting, wiring, and equipment testing should be performed according to manufacturer specifications.

BUILDING AUTOMATION SYSTEM INSTALLATION AND WIRING

Building automation system installation requires proper controller mounting, power supply selection, network configuration, and testing techniques. Correct wiring of building automation system input devices, output devices, and peripheral devices is required to effectively communicate information. All building automation system controllers should be wired in accordance with the National Electrical Code® (NEC®) and local regulations. Manufacturer recommendations should be followed regarding testing and diagnostic procedures.

Controller Mounting

Building automation system controllers require proper mounting to prevent controller damage. Building automation system controller installations can have ongoing problems due to improper controller mounting. Controller mounting methods and requirements vary with the manufacturer and application.

Controller Mounting Methods. Building automation system controllers must be mounted securely. In many applications, controllers can be mounted in existing pneumatic temperature control panels. Controllers

should be mounted to the surface of the control panel using the proper hardware. Many manufacturers use a thin DIN rail that mounts to the surface of the control panel. A *DIN rail* is a flat mounting rail that is attached to a control panel. The controllers are then attached to the DIN rail. See Figure 18-1.

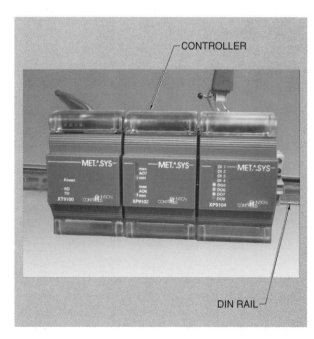

Figure 18-1. Building automation system controllers are often mounted using a DIN rail.

Controller Mounting Requirements. Building automation system controller mounting requirements vary due to the wide range of controller applications. Exterior-mounted building automation system controllers require an enclosure that prevents weather-related damage. For example, a controller mounted on a rooftop must be contained within a suitable waterproof enclosure. Exterior-mounted controller wiring and components must also be enclosed in waterproof material. Interior-mounted controllers must be located away from water lines and areas that are exposed to outside air to reduce exposure to harmful elements.

Many building automation systems are equipped with an operator interface that provides maintenance personnel with the ability to view controller functions and status. See Figure 18-2. Operator interfaces commonly require installation 60″ above the finished floor for viewing by maintenance personnel. Other operator interfaces are positioned in areas away from view, to prevent tampering.

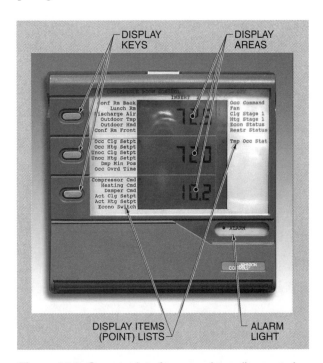

Figure 18-2. Operator interfaces can be wall-mounted so they are accessible to maintenance personnel.

A building automation system controller is mounted in an enclosure to provide adequate clearance on all sides of the controller to enable component accessibility. Overloading a control panel with the controller, components, and wiring can obstruct access by maintenance and repair personnel. If required, a larger panel can be used or location of the controller in the panel changed to provide access. A location should be chosen that minimizes the length of conductor runs to reduce wiring problems and cost. Also, manufacturer recommendations must be followed regarding minimum distances from high-voltage switching devices. Many manufacturers require a minimum distance to prevent electromagnetic interference at the controller. Common building automation system controller mounting requirements include the following:

- Separate building automation system controller conductors and power conductors by at least 1′.
- Wire building automation system controller connections using copper conductors of 18-gauge to 24-gauge.
- Run network field conductors and line-voltage conductors of 30 VAC or higher in separate conduits.

Controller Power Supplies

The most common building automation system controller power supply is a 120 VAC to 24 VAC step-down transformer. See Figure 18-3. The minimum size for the two conductors of the building automation system controller power supply is 18-gauge, stranded copper with ground. Building automation system controller power supplies use plug-in plastic connectors or screw and spade terminations.

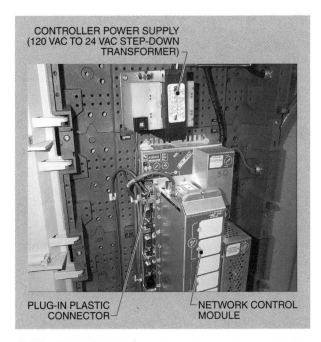

Figure 18-3. A building automation system controller power supply is normally a 120 VAC to 24 VAC step-down transformer.

Controller Power Supply Sizing. Building automation system controller manufacturers often recommend the use of a specific power supply. The size of a building automation system controller power supply can be determined by adding the power requirements of the components in the building automation system. See Figure 18-4. For example, a building automation system controller requires 10 VA of 24 VAC power. The controller has two humidity sensors and two electronic/pneumatic transducers, each requiring 3 VA of power. The total power requirement for the controller is 22 VA (10 VA + 4 × 3 VA = 22 VA). The next larger size transformer is selected as a factor of safety.

Controller Power Supply Requirements		
ACTUATOR	**TYPE**	**REQUIREMENTS FOR 24 VAC SUPPLY***
IDA2020	Incremental	3
IL-7150	Incremental	2.7
IL-7200	Incremental	5.5
IL-8020	Incremental	4
IL-8050	Incremental	6
K Series Electric Zone Valve	ON/OFF	7
A9100-A, M9200-C	Floating	6.5
A9100-G, M9200-F	Proportional (Voltage or Current)	6.5
IL-7152	Voltage (0 VDC to 10 VDC)	4.7
IL-7202	Voltage (0 VDC to 10 VDC)	7.5
IL-8022	Voltage (0 VDC to 10 VDC)	4
IL-8052	Voltage (0 VDC to 10 VDC)	6

* in VA

Figure 18-4. Charts are available for sizing building automation system power supplies.

A 100 VA transformer is the largest single transformer used in building automation systems. A 100 VA transformer can be used to power multiple controllers or devices. The controller manufacturer literature should be checked when multiple controllers are powered using the same transformer because equipment malfunction and damage may occur.

Controller Power Supply Isolation and Polarity. Isolation of a controller power supply may be required. Power supply isolation from ground is necessary to prevent ground loops, which cause improper controller operation. A *ground loop* is a circuit that has more than one point connected to earth ground, with a voltage potential difference between the two ground points high enough to produce a circulating current in the ground system.

The potential difference may be less than 1 VDC. This small potential difference can be significant because all building automation system components operate at less than 5 VDC and 1 VDC represents 20% of

the rated voltage. Many electronic devices such as diodes and transistors switch at less than 1 VDC. Ground loops are common in existing controllers.

Correct transformer polarity is essential within building automation systems because controller power supplies, sensors, and outputs are designed to operate properly when connected in the correct polarity. The two leads from a step-down transformer are labeled 24 VAC and COM (common). The lead that has a slightly higher potential to ground is the 24 VAC lead. The polarity of the power supply may be checked to determine the 24 VAC lead and common lead. See Figure 18-5. A 100 kΩ, ¼ W resistor is connected in parallel with a digital multimeter (DMM) to load the circuit.

Package Unit Power Supplies. Package units, such as rooftop units and heat pumps, have built-in power supplies. Package unit built-in power supplies are commonly used to power building automation system controllers. Package unit power supplies must be checked to determine if they are sufficient to power building automation system controllers. For example, the transformer must have the capacity rating to handle the building automation system power requirements. Isolation of the transformer must also be checked to determine if the transformer wiring is connected directly to ground. A ground-connected transformer, when connected to a building automation system controller or device, causes the controller or device to become grounded. This may damage the controller and personal computers connected to the controller.

Some air handling unit controllers have built-in 1:1 isolation transformers that reduce system problems caused by electrical interference.

TRANSFORMER POLARITY DETERMINATION

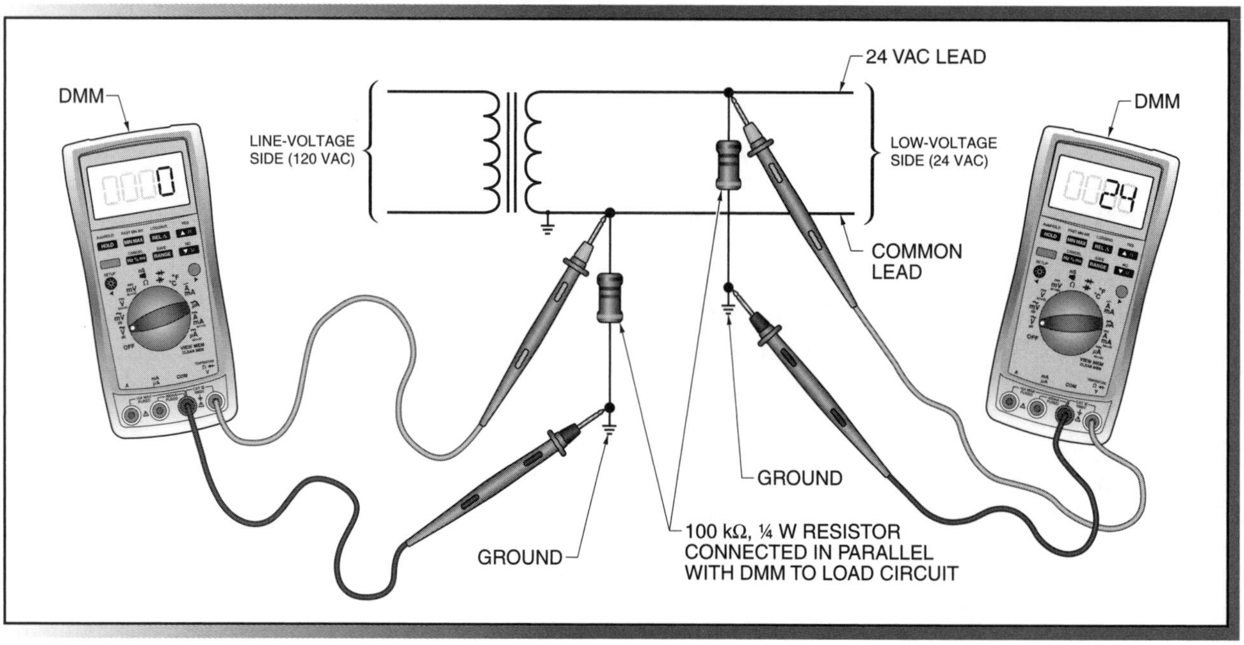

24 VAC LEAD
DMM
DMM
LINE-VOLTAGE SIDE (120 VAC)
LOW-VOLTAGE SIDE (24 VAC)
COMMON LEAD
GROUND
GROUND
100 kΩ, ¼ W RESISTOR CONNECTED IN PARALLEL WITH DMM TO LOAD CIRCUIT

Figure 18-5. The polarity of a controller power supply is checked with a DMM to determine the 24 VAC conductor and the common conductor.

Controller Power Supply Sharing. In many building automation systems, one large transformer can power multiple controllers. See Figure 18-6. The use of one large transformer eliminates the need for multiple small transformers. Using one large transformer is recommended when the building automation system controllers are located close together. Isolation of controllers is necessary to prevent communication and power problems.

Controller On-Board Batteries. Building automation system controllers have an on-board battery for retention of memory and clock functions. See Figure 18-7. The battery is self-charging when normal 24 VAC power is supplied to the controller. The battery is commonly disconnected from the controller during shipment to prevent power from draining and is enabled during the start-up procedure of the building automation system.

Multiple building automation system controllers commonly share the same power supply. As a general rule, the power supply should not be shared with devices such as actuators that also require 24 VAC. Manufacturer recommendations should be followed regarding power supply sharing.

CONTROLLER POWER SUPPLY SHARING

L1
L2
120 VAC
24 VAC
COM
POWER SUPPLY (120 VAC TO 24 VAC 100 VA STEP-DOWN TRANSFORMER)
CONTROLLER 1
CONTROLLER 3
CONTROLLER 2
CONTROLLER 4

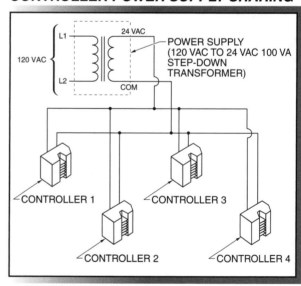

Figure 18-6. Multiple controllers are commonly powered from one power supply (transformer).

Communication Network Wiring

Communication network wiring enables building automation system controllers to communicate information. This information may include sharing of the outside air temperature value, time scheduling information, and alarm status. For example, the outside air temperature sensor located at the air handling unit controller may share information with the boiler controller. Building automation system controller wiring must be installed correctly for proper controller communication and proper control function. Many building automation systems use the RS-485 communication standard. *RS-485* is a multipoint communication standard that incorporates low-impedance drivers and receivers providing high tolerance to noise. The RS-485 communication standard has a maximum data transmission rate of 9600 bits per second (bps).

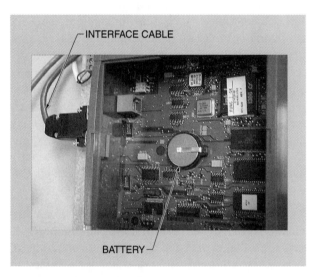

Figure 18-7. Building automation system controllers are equipped with a battery for the retention of memory and clock functions.

Conductor Type. The RS-485 communication standard requires a two-conductor twisted shielded cable or three-conductor twisted shielded cable. See Figure 18-8. The minimum conductor size used for building automation system controllers is 22-gauge wire. A 20-gauge or 18-gauge wire is recommended because 22-gauge wire can be easily kinked or damaged. Twisting the conductors cancels electromagnetic interference. Color-coding is used to ensure that the proper conductors are consistently connected to the same terminals at each controller. The RS-485 communication standard is polarity-sensitive and the connection terminals at each controller are normally labeled RS-485+, RS-485–, and COM.

Shielding and Grounding. Shielding reduces the effect of electromagnetic interference on communication network wiring. Electromagnetic interference is caused by the operation of large motors, compressors, high-voltage wiring, and lighting loads. Electromagnetic interference is reduced by keeping communication

network wiring away from large motors, compressors, high-voltage wiring, and lighting loads. Most manufacturers recommend that the shield conductor be connected to earth ground at the controller power supply or a centralized location.

BUILDING AUTOMATION SYSTEM CONDUCTORS

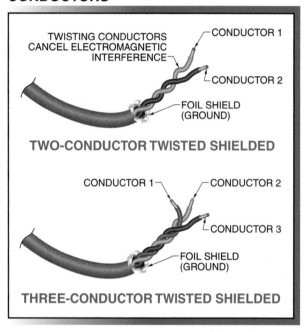

Figure 18-8. The RS-485 communication standard requires a two-conductor twisted shielded cable or three-conductor twisted shielded cable.

Optoisolation. Many building automation systems use optoisolated communication. *Optoisolation* is a communication method in which controllers use photonic (light-sensitive) components to prevent communication problems. In optoisolated communication, the information on the communication network is transmitted to the controller through hard-wired runs. The binary (0 or 1) information transmitted is converted to an optical signal that is processed by the controller. The building automation system controller is isolated from the communication network because there is no physical connection between them, reducing or eliminating communication problems.

Network Configurations. Building automation system controllers can be connected to a network in various configurations. The network is the physical connection that allows building automation system controllers to communicate. The network configuration used is based on the building layout, conductor or cable selection, and manufacturer recommendations. Network configurations include daisy chain, multidrop, and star configurations. See Figure 18-9.

BUILDING AUTOMATION SYSTEM NETWORK CONFIGURATIONS

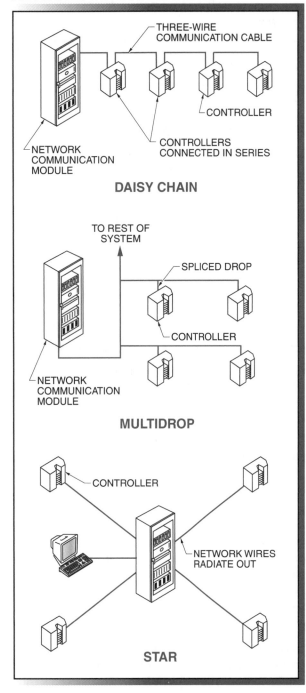

Figure 18-9. Building automation system network configurations include daisy chain, multidrop, and star network configurations.

A *daisy chain network configuration* is a configuration in which multiple controllers are connected in series. A daisy chain network configuration is a continuous chain of building automation system controllers. The daisy chain network configuration avoids using wiring splices that can fail and cause communication breakdown. Wiring terminations in daisy chain configurations are located at the controllers for easy accessibility.

A *multidrop network configuration* is a configuration in which controllers are connected to the network that runs throughout a building through spliced drops for each controller. The splicing in multidrop network configurations must be done correctly to reduce communication failure.

A *star network configuration* is a configuration in which communication network cables radiate out to remote controllers located within a building. Star network configurations are not as common as daisy chain configurations because of the long wire runs required. Star network configurations are normally used in buildings that have a centrally-located network communication module.

Network Addressing. To achieve accurate communication on a network, each building automation system controller must have a proper network address. A *network address* is a unique number assigned to each building automation system controller on an RS-485 communication network. RS-485 communication networks are limited to a maximum of 255 possible addresses. Each building automation system controller has a set of dual in-line package (DIP) switches. Each switch represents a value in the binary number system such as 1, 2, 4, 8, 16, 32, 64, and 128. Summing the switches that are in the ON position determines the network address of a building automation system controller. For example, if the switches that represent the numbers 1, 8, and 64 are in the ON position, the network address is 73. See Figure 18-10.

Input Wiring

Inputs are devices such as temperature sensors, humidity sensors, status switches, limit thermostats, and timed overrides. Input devices must be properly wired to a building automation system controller to enable the controller to process their information. Improperly wired input devices may produce inaccurate readings. Input wiring includes analog input wiring and digital input wiring.

CONTROLLER NETWORK ADDRESS

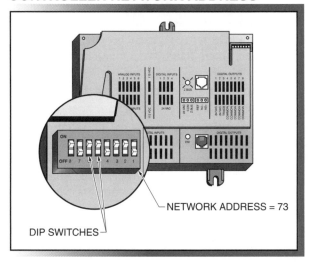

Figure 18-10. Building automation system controllers use DIP switches to determine controller network addresses.

Analog Input Wiring. Analog input devices are wired by running a pair of conductors to the analog input device and connecting them to the analog input terminals on the controller termination board. The analog input terminals are commonly labeled AI 1, AI 2, AI 3, etc., and COM. Analog input devices are polarity-sensitive, so the polarity must be checked and labeled. See Figure 18-11.

ANALOG INPUT WIRING

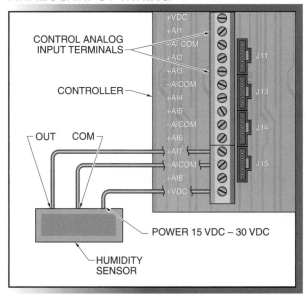

Figure 18-11. Analog input devices are connected to the analog input terminals (AI 1, AI 2, AI 3, etc. and COM) on a building automation system controller.

Some analog input devices require a separate power supply that many building automation systems provide at the terminals of the controller. The power supply may be +15 VDC or +30 VDC. A separate power supply requires the pulling of an additional pair of conductors which should be numbered or color-coded for identification. The conductors should be stranded and a minimum of 22-gauge, although 18-gauge or 20-gauge wire is recommended.

If long runs (hundreds of feet) of thin conductors are used, the increased resistance of the conductors reduces the accuracy of sensor readings.

Digital Input Wiring. Digital input devices are wired by running a pair of conductors to a digital input device (switch) and terminating them at the digital input device terminals on the controller. See Figure 18-12. The building automation system controller digital input terminals are commonly labeled DI 1, DI 2, DI 3, etc., and COM. Digital inputs are commonly configured as dry contacts. Dry contacts, when referring to digital inputs, are contacts that are not powered externally. An isolation relay is required to prevent electromagnetic interference for digital inputs requiring external power.

DIGITAL INPUT WIRING

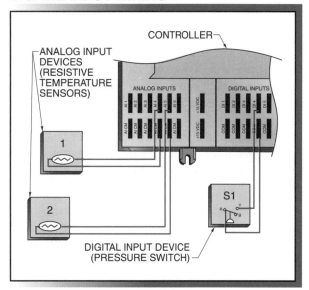

Figure 18-12. Digital input devices are connected to the digital input terminals (DI 1, DI 2, DI 3, etc. and COM) on a building automation system controller.

Proper care must be taken to ensure that the correct terminals are used at the switch to enable the building automation system software logic to agree with the switch contacts selected. For example, an airflow differential pressure switch is configured to make common (C) and normally open (NO) when the fan is running. The building automation system software logic must recognize closed contacts as fan ON and open contacts as fan OFF.

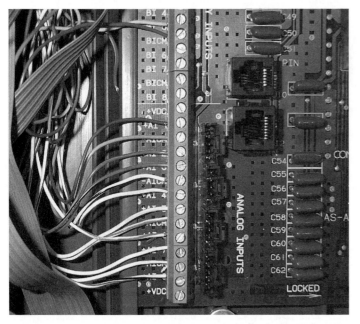

Inputs and outputs are normally terminated at a controller using screw terminals but may also be terminated by crimp-on spade lugs.

Output Wiring

Outputs are devices such as valve and damper actuators, variable-speed drives, lights, fans, heating and cooling stages, and pumps. Output devices must be properly wired to a building automation system controller so that the controller commands are properly executed by the devices. Output devices respond improperly if connected incorrectly. Output wiring includes analog output wiring and digital output wiring.

Analog Output Wiring. Analog output devices are wired by running two conductors from the building automation system controller to the outputs (actuators). Analog output devices are polarity-sensitive and must be wired accordingly. Controller analog output terminals are commonly labeled AO 1, AO 2, AO 3, etc., and COM.

Digital Output Wiring. Digital output devices may be wired using several different methods, depending on the application. Similar to digital inputs, digital output contacts supplied by some manufacturers are dry contacts. Dry contacts, when referring to digital outputs, require an external power supply. Dry contacts are Class II

rated (low-voltage) and the controller power supply is commonly a 24 VAC transformer. Package unit controls can be wired directly to the building automation system controller terminal board because they are both low-voltage. Often, standard color-coding is used, with red (R) representing power, white (W) representing heating, green (G) representing the fan, and yellow (Y) representing cooling.

An *incremental output* is a digital output device used to position bidirectional electric motors. Two outputs are required: one output drives the actuator open, and the other output drives the actuator closed. A common power terminal is shared by both outputs.

Care should be taken in selecting the correct contacts when digital output devices are wired to interface relays. If the wrong contacts are used, the device may turn OFF when commanded ON and turn ON when commanded OFF. Controller digital output terminals are commonly labeled DO 1, DO 2, DO 3, etc., and COM. Any digital output device problems or wiring errors are found during the checkout process of the start-up procedure.

> *Common output device wiring problems include an output device connected to the wrong terminals, an output device connected with the wrong polarity, the wrong voltage supply, loose or improper wiring terminations, electrical interference due to the proximity to high-voltage equipment, and incorrect controller programming for the specific output type used.*

Peripheral Device Wiring

A *peripheral device* is a device that is connected to a personal computer or building automation system controller to perform a specific function. Building automation system peripheral devices include modems, printers, and dumb terminals. Selecting the correct cable for a peripheral device requires knowledge of the application. Peripheral device cables may look similar and have the same number of connections, but have different internal functions.

Modem Wiring and Connections. Modems are commonly connected to a network communication module as a serial device. See Figure 18-13. A *serial device* is a device that transmits one bit of information (0 or 1) at a time. Serial devices normally include nine-pin male connectors. Modems are also connected to a phone line using a standard modular phone line (RJ11) cable. Modems are equipped with a power supply that must be plugged into a standard 120 VAC wall outlet.

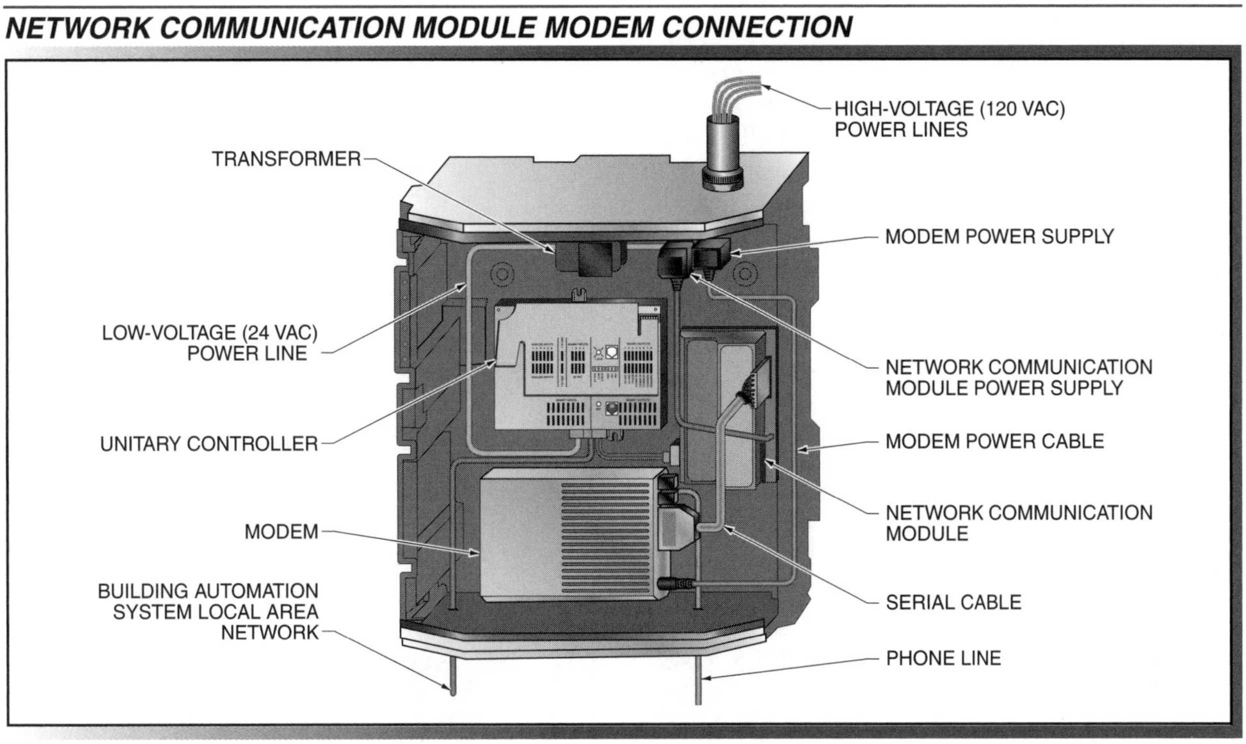

NETWORK COMMUNICATION MODULE MODEM CONNECTION

HIGH-VOLTAGE (120 VAC) POWER LINES

TRANSFORMER

MODEM POWER SUPPLY

LOW-VOLTAGE (24 VAC) POWER LINE

NETWORK COMMUNICATION MODULE POWER SUPPLY

UNITARY CONTROLLER

MODEM POWER CABLE

NETWORK COMMUNICATION MODULE

MODEM

SERIAL CABLE

BUILDING AUTOMATION SYSTEM LOCAL AREA NETWORK

PHONE LINE

Figure 18-13. A modem is commonly connected to a building automation system network communication module as a serial device.

Printer Wiring and Connections. Printers are connected to a network communication module as a parallel device. Printers may also be connected to personal computers (PCs). A *parallel device* is a device that transmits multiple bits of information (0s or 1s) simultaneously. Printers normally include 25-pin female connectors. See Figure 18-14.

Dumb Terminal Wiring and Connections. Many building automation systems use a personal computer as a dumb terminal. A *dumb terminal* is a display monitor and keyboard, with no processing capabilities. Dumb terminals are connected to network communication modules as serial devices.

BUILDING AUTOMATION SYSTEM TESTING AND DIAGNOSTIC PROCEDURES

Common procedures are used to test and troubleshoot building automation systems. Accurate test equipment is required to receive precise test results. Digital multimeters (DMMs) should be used to ensure accuracy. Common testing procedures used in building automation systems include transformer isolation to ground, RS-485 communication, communication network isolation, analog input resistance value, and digital input and output testing.

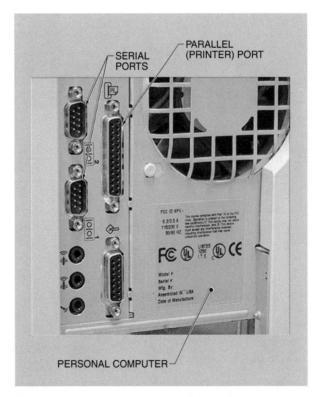

SERIAL PORTS

PARALLEL (PRINTER) PORT

PERSONAL COMPUTER

Figure 18-14. A printer is normally connected to a personal computer or building automation system network communication module as a parallel device.

Transformer Isolation to Ground Testing

Transformer isolation to ground testing is used to check the transformer power supply for proper isolation from ground. Building automation system controller communication may be disrupted if the power supply is not properly isolated from ground. A DMM is used to test a controller power supply to ground. The resistance between the secondary of the transformer (low-voltage side) and ground should be above the manufacturer-specified level.

RS-485 Communication Testing

RS-485 communication testing checks for proper levels of voltage within the building automation system RS-485 communication wiring. The three conductors tested are RS-485+, RS-485–, and common. RS-485 voltage levels are standardized. The voltage value of the RS-485+ and RS-485– conductors to COM should be between 1.5 VDC and 2.5 VDC. When the voltage values of the RS-485+ and RS-485– conductors are measured, the voltage should be between 3 VDC and 5 VDC. See Figure 18-15. A problem exists in the communication network wiring or controller if a DMM is used and these voltage levels are not present.

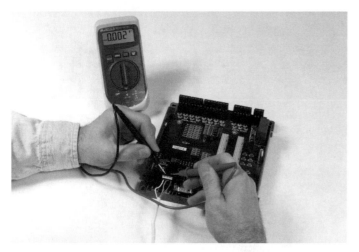

A DMM can be used for building automation system controller communication bus testing.

Communication Network Isolation to Ground Testing

Communication network isolation to ground testing is similar to transformer isolation to ground testing. In communication network isolation to ground testing, the RS-485 communication wiring is tested to ground

RS-485 COMMUNICATION TESTING

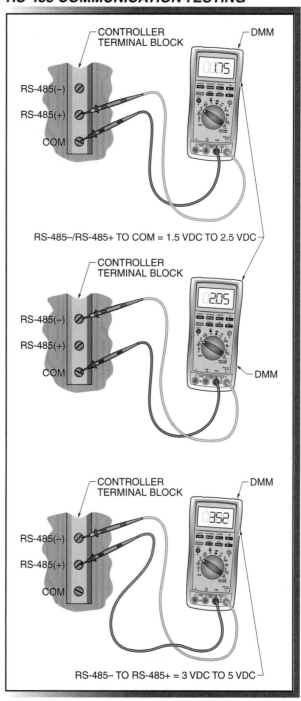

Figure 18-15. RS-485 communication wiring has predetermined voltage levels within a building automation system.

with a DMM. The communication network is not isolated properly if the resistance of this connection is below a specified level. A low resistance reading indicates incorrect grounding of an input(s), output(s), or power supply.

Analog Input Resistance Value Testing

Manufacturers supply charts that show the resistance and temperature characteristics of a temperature sensor (analog input). See Figure 18-16. A thermometer can be used to check the accuracy of a sensor. The temperature is measured and compared to the chart to determine if the sensor resistance is correct for the measured temperature. Thermistor sensors can be either positive temperature coefficient (PTC) or negative temperature coefficient (NTC) sensors. The resistance of a positive temperature coefficient sensor increases as the temperature increases. The resistance of a negative temperature coefficient sensor decreases as the temperature increases.

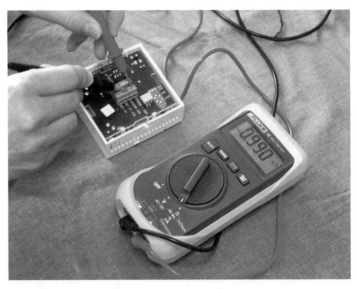

> **ⓘ** *If a nonlinear sensor or a sensor with a nonstandard range is used, the calculation programming block in the controller software can be used to enable the controller to read the correct values.*

The coefficient of a thermistor sensor is the amount of resistance change per degree Fahrenheit. For example, a thermistor sensor may have a coefficient of 2.2 Ω per 1°F change. This means that the sensor resistance changes 2.2 Ω for every 1°F temperature change. Thermistors also have a base (nominal) value. A common thermistor nominal value is 1000 Ω at 70°F. This means that the sensor has a base reading of 1000 Ω at 70°F. A coefficient of 2.2 Ω per 1°F for a positive temperature coefficient sensor results in a resistance increase of 2.2 Ω for every 1°F temperature change. This sensitivity makes the increased resistance due to long wire runs affect the reading of the sensor.

A DMM can be used to check temperature sensor resistance values. The readings should be at the correct value and within tolerance for a particular temperature.

Digital Input and Output Testing

Digital inputs and outputs are tested by checking the electrical continuity of the input and output conductors. Continuity testing determines if there is a complete circuit between the building automation system controller and digital input/output devices. The conductors between the controller and digital device can be disconnected and the resistance measured using a DMM. Normally open and normally closed contacts must be set for proper operation. For example, an energized output causes the device to turn either ON or OFF. Also, the conductors must be terminated at the proper points on the controller. Many digital inputs and outputs use a 24 VAC power supply. A DMM may be used to check for 24 VAC by connecting the meter probes to the desired controller terminals.

TEMPERATURE SENSOR RESISTANCE/TEMPERATURE CHARACTERISTICS

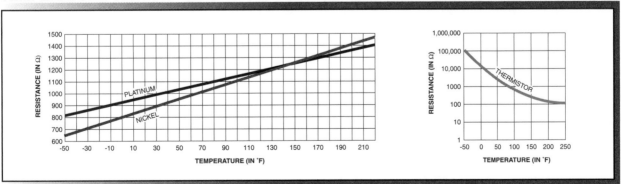

Figure 18-16. Temperature sensor manufacturers provide resistance/temperature characteristic charts.

Jumpers

A *jumper* is a conductor used to connect pins on a controller or device. The conductor is normally enclosed inside a piece of plastic. Jumpers are used on building automation system controllers to select the type of analog input or output signal received by the controller. For example, by covering different combinations of pins located on the rear of a controller, a particular type of sensor signal (resistance, voltage, or current) is selected for the controller. Analog inputs such as thermistors, resistance temperature detectors, current types, and voltage types are commonly selected on the controller using jumpers. Analog outputs and back-up batteries may also use jumpers for proper selection. Manufacturer literature is checked for proper jumper use. A jumper selection label is normally attached to the rear of a controller to aid in jumper placement. See Figure 18-17.

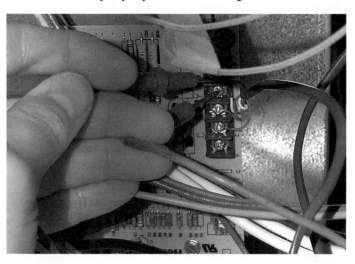

The operation of an output device can be checked by using a jumper wire across the proper terminals. If wired properly, the HVAC unit should turn ON and OFF.

CONTROLLER JUMPER USAGE

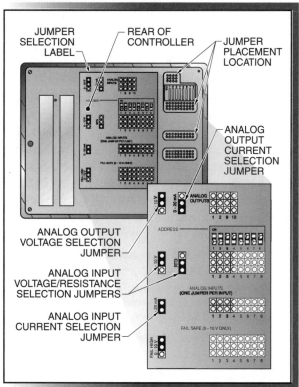

Figure 18-17. Jumpers are used on building automation system controllers to select the type of analog input or output signal received by the controller.

 Proper operation of an input or output device requires that the correct sensor signal (resistance, voltage, or current) be selected and the controller software programmed correctly for the specific type of input or output device.

HVAC Control Systems
Networking Fundamentals

Modern building automation systems communicate over networks. A network consists of hardware and software that allow reliable, high-speed data transmission in a building. Most buildings today use the Ethernet network topology. Hubs, switches, and routers are devices that enable all points on a network to communicate with each other.

NETWORK TOPOLOGY

A *network* is interconnected equipment used for sending and receiving information. A network may exist in a building or between equipment in one building and another remote location. A network consists of hardware (equipment), connecting media between the individual pieces of equipment, and software (programs) that permit communication between nodes. A *node* is a a a device, such as a computer or a printer, that has a unique address and is attached to a network. Each device requires specialized hardware and software that enables it to communicate on the network. Networks may be local area networks or wide area networks.

A *local area network (LAN)* is a communication network that spans a relatively small area. A local area network normally encompasses one business or building and does not require the linking of a large number of users in remote locations. A local area network is the most common network used in building automation systems.

A *wide area network (WAN)* is a communication network that spans a relatively large area. A wide area network is used to pass information across long distances involving multiple geographical locations and tens of thousands of users. Large organizations commonly have a wide area network to manage communication between many local area networks.

Network topology is the map (description) of the network configuration. Network topology includes hardware, software, and communication configurations. Each type of network has a different topology. Network topologies include Ethernet and Arcnet™ topologies.

Gateway Community College
Most new colleges are being designed and built to facilitate the installation and use of local area networks.

ETHERNET NETWORKS

Ethernet is a local area network architecture that can connect up to 1024 nodes and supports minimum data transfer rates of 10 megabits per second (Mbps). High-speed, 100 Mbps networks are becoming available and standards for 1000 Mbps networks are being developed. Ethernet is used in 90% of communication networks.

In an Ethernet network, data is divided into packets before being transmitted. To determine if the network is clear of data traffic before transmitting information, a node uses the standard carrier sense multiple access/collision detection (CSMA/CD) protocol. When the network is clear of traffic, the node transmits its packets of data and waits for a receipt notice from the receiving node. A *collision* is a data transmission overlap. A collision occurs when two nodes transmit simultaneously and the data is corrupted. The receiving node does not acknowledge receipt of corrupted data. The nodes that transmitted simultaneously each wait a randomly selected period of time before retransmitting.

An Ethernet network may be laid out in a star or common-bus configuration. A star configuration has each node connected to a central device. A common-bus configuration has each node connected to a common network wire that is connected to the central device. In a common-bus configuration, the nodes are not connected directly to each other. See Figure 19-1.

Building automation system components may be wired as a separate network independent of the main building network, or may be added to the same main building network as all of the other nodes of any type. When building automation system components are wired as a separate network, the building automation system network is easy to troubleshoot because all other nodes are excluded. This may also increase the building automation system network reliability by reducing network traffic. A separate building automation system network also eliminates any possible security issues on the main building network. Installation and maintenance costs may be increased because of the duplication of multiple network components and software. It is also possible that the specialized personnel needed for network maintenance would be unable to service dedicated building automation system nodes.

> *Common Ethernet components are sold at most computer hardware stores. These components can be purchased on short notice to get a network up and running.*

ETHERNET NETWORK CONFIGURATIONS

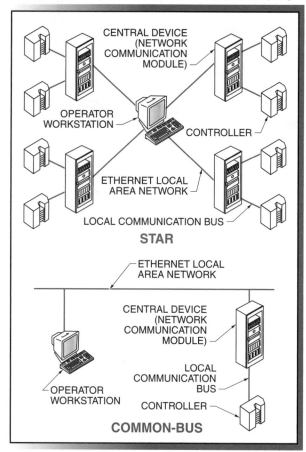

Figure 19-1. An Ethernet network may be configured in a star or common-bus configuration.

Building automation system components may also be added to the main building network. This requires that each building automation system controller, supervisory controller, and computer used to view system operation must contain the hardware, connecting media, and software to connect it properly to the network. The single network reduces installation costs but may complicate troubleshooting because of the number of nodes that are part of the network but not part of the building automation system.

Ethernet Hardware

Ethernet hardware components must be set up for the network to function correctly. Each node on the network must have an adapter card connected to the connecting media used for the network. An Ethernet network must have components (concentrators) that allow all of the nodes to communicate with each other.

Adapter Cards. An *adapter card (network interface card)* is a card (circuit board) installed in a network component to allow it to communicate with the network. An adapter card contains software that supports its connection to the network. A building automation system controller adapter card may be a proprietary device for each manufacturer. See Figure 19-2.

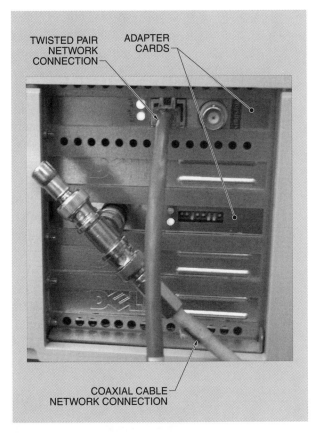

Figure 19-2. Adapter cards are installed in equipment to allow it to communicate on a network.

Concentrators. A *concentrator* is a network switchboard that allows a number of nodes to communicate with each other. A single connecting cable is all that is necessary to allow two nodes to communicate with each other. A concentrator is needed when more than two nodes need to communicate. Each node is connected to a concentrator that sends the messages from one node to another node. A concentrator also helps to manage network traffic and prevent network problems. Common building automation system concentrators include hubs, switches, and routers.

A *hub* is a concentrator that manages the communication between components on a local area network. Individual nodes on a network transmit information to a hub. The hub transmits the information to the appropriate node. Hubs are rated by the number of nodes that can be connected. For example, four to eight node capacities are the most common. Hubs can also be connected together to form larger networks. Hubs are plug-and-play components, meaning that no configuration is normally needed.

A *switch* is a concentrator that manages the communication between networks or parts of networks that operate at different data transmission speeds. Switches have the ability to automatically change the data transmission speeds between the different networks or parts of networks. If these speeds are not changed, the entire network transmits at the lowest speed. See Figure 19-3.

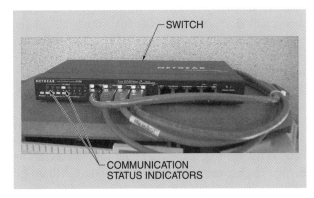

Figure 19-3. A switch manages communication between devices on a network that communicate at different speeds.

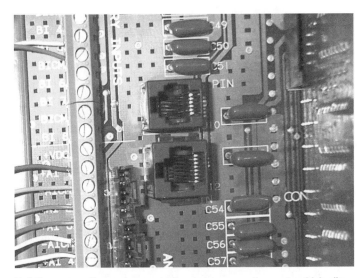

Building automation system controllers have connection ports which allow them to be connected to the building local area network.

A *router* is a concentrator that manages the communication between different networks. A router sends messages from the nodes of a local area network to the nodes of other networks. The messages are sent from a node to a hub and then to a router. The router sends the message to another router, which then sends it to the correct node of another network. A router has hardware and software that must be configured. Hubs, switches, and routers work together to connect different parts of a network. See Figure 19-4.

A router can also be used to separate parts of networks into logical subnetworks. A *subnetwork* is a local area network that is made up of components that communicate frequently. A subnetwork is created to minimize traffic on the main network.

Ethernet Connecting Media

Ethernet networks use a variety of connecting media to carry the network messages. Ethernet connecting media include unshielded twisted pair, thinnet, thicknet, and optical fiber. *Physical layer* is the cables and other network devices, such as hubs and switches, that make up a network.

Connecting media are rated by a category (CAT) rating number representing the quality of the media. The CAT rating number refers to the actual allowable cable length in hundreds of meters. Network wiring specifications list the required CAT rating for the connecting media. The following are connecting media CAT ratings:

- CAT 1—Older wire and cable used for telephone systems. Category 1 wire and cable is normally limited to voice transmission only.

- CAT 2—Indoor wire and cable systems used for voice applications and data transmission up to 1 MHz.

- CAT 3—Indoor wire and cable systems used for voice applications and data transmission up to 16 MHz. High-speed data networks use a minimum of CAT 3 wire or cable.

- CAT 4—Indoor wire and cable systems used for voice applications and data transmission up to 20 MHz.

- CAT 5—Indoor wire and cable systems used for voice and data transmission up to 100 MHz. Category 5 cable is one of the most commonly used cables in networks because it offers extended frequency transmission (above CAT 3 and CAT 4).

- CAT 5e—Indoor wire and cable systems used for voice and data transmission up to 100 MHz. Category 5e wire and cable allow long runs because wire is packaged tightly with greater electrical balance between wire pairs.

- CAT 6—Indoor wire and cable systems used for voice and data transmission up to 250 MHz or more.

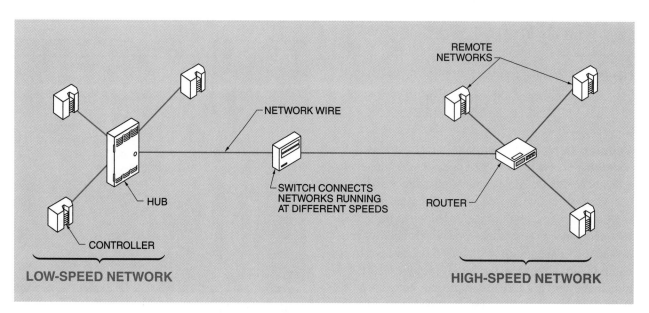

Figure 19-4. Hubs, switches, and routers work together to connect different parts of a network.

Ethernet communication cables are classified as XbaseY where X is the communication speed in megabits per second, base is baseband, and Y is an indicator of the type of media. *Baseband* is a data transmission method where data is sent without frequency modulation. This means that there is only one data channel on a cable. Y is an indicator of the type of media and may be the category or an abbreviation of the media.

Common values for X are 5 and 10. Common values for Y are 2, 5, and T. For example, 10base5 is a cable that supports baseband communications at 10 Mbps over a category 5 cable. A tester can be used to determine cable type and functionality.

When installing network connecting media, as few connections as possible are used because excessive connections lower system performance. In addition, always wire to the highest projected data-rated speed, allow at least 18″ of spare wire at all connection points, and never splice wires on communication cable runs. Finally, use nonmetallic staples when securing cable or wiring.

Unshielded Twisted Pair. *Unshielded twisted pair (UTP)* is a pair of wires that are twisted around each other with no electromagnetic shielding. The wires are molded into plastic connectors that resemble oversize phone jack connectors. Category 5 unshielded twisted pair consists of four twisted pairs of copper wire terminated by an RJ-45 connector. An *RJ-45 connector* is a connector that contains eight pins instead of four as in standard phone connectors. Most hubs use RJ-45 unshielded twisted pair connectors. Cables with RJ-45 connectors can be purchased commercially in different lengths or fabricated locally.

A common network communication cable is 10baseT. A 10baseT cable is a communication cable used for 10 Mbps baseband communication over an unshielded twisted pair. Commercially available 10baseT cables include patch cables and crossover cables. A *patch cable* is a cable used for connections from a node to a concentrator. Patch cables use four wires connected straight through from connector to connector. A *crossover cable* is a cable used for connections from node to node. Patch cables and crossover cables contain send and receive wires. In a crossover cable, the send wires from each connector pair are wired to the receive wires of the other connector pair. A crossover cable is used when a computer must be directly connected to a controller to monitor system performance or collect troubleshooting information. Patch cables and crossover cables are

color-coded or labeled for easy identification. The two cables are not interchangeable and cause a communication failure if misapplied. See Figure 19-5.

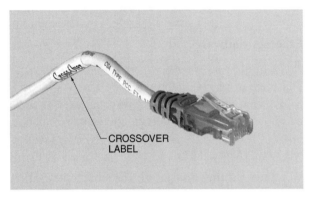

Figure 19-5. Crossover cables are used for connections from node to node and are labeled for easy identification.

Thinnet, Thicknet, and Optical Fiber. Thinnet, thicknet, and optical fiber are other types of connecting media used in networks. *Thinnet* is thin coaxial cable classified as 10base2. Thinnet was commonly used in early networks. *Thicknet* is thick coaxial cable classified as 10base5. Thicknet is also referred to as yellow cable because it is commonly available with a yellow plastic covering. *Optical fiber* is a glass fiber used to transmit information using light. Optical fiber has become popular for wide area networks and high-speed Ethernet networks. Optical fiber is rarely used in building automation systems and must be installed and serviced by professionals qualified in its use.

Many facilities are being wired with combination Ethernet and phone jack connections in each room.

Ethernet Software

Ethernet software is the program that enables nodes to transmit and receive information on a network. Each component on a network must be configured to communicate correctly. Software configuration tools include Windows® utilities and setup, Internet Protocol addressing, and submasking.

Windows Utilities and Setup. The most common personal computer operating system used today is Microsoft Windows. Windows has utilities that allow easy configuration of building automation system hardware and software components for Ethernet networks. Different versions of Windows are in use today. For example, Windows 95®, Windows 98®, Windows 98SE® (Second Edition), Windows NT® (various versions available), Windows 2000®, Windows ME® (Millennium Edition), and Windows XP® are all referred to as Windows. The operating system required for a building automation system component must be determined and matched to the operating system available on the network.

If not already present, the network adapter card is installed in the new building automation system component. Most Windows versions automatically detect the presence of a new adapter card and install the proper software drivers that enable the card to work properly. Some adapter cards require manual installation of the software from a floppy disk or CD-ROM provided by the card manufacturer. Manual software installation is done by using the Add/Remove Hardware function in the Windows Control Panel. In some cases, software that is loaded may need to be reconfigured. The settings are found using the Network and Dial-up Connections function in the Windows Control Panel. See Figure 19-6.

Internet Protocol Addressing. Internet Protocol (IP) addressing is a method of assigning addresses to nodes on a network. Each node must have an Internet Protocol address. An Internet Protocol address consists of four sets of three digits. Each address must be obtained from the network manager. A node address may be static or dynamic. A static address is a permanent address. A dynamic address is an address that can change

at any time. The network software keeps track of a dynamic address as it changes. Each building automation system manufacturer provides a software programming utility that allows a technician to define the Internet Protocol address of each node. See Figure 19-7.

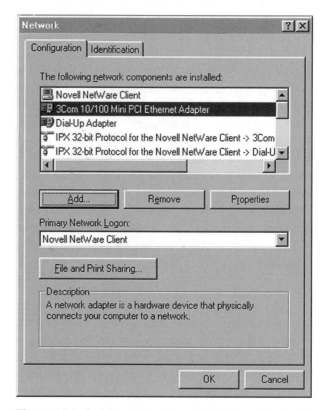

Figure 19-6. Building automation system component software configuration may be performed automatically or manually from within Windows.

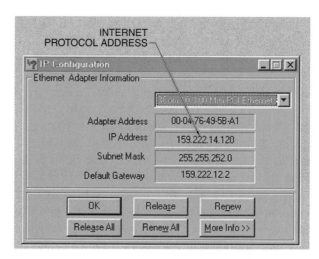

Figure 19-7. Internet protocol addressing assigns an address to each node on a network.

Submasking. Large networks are commonly split into subnetworks. Subnetworks help reduce overall network traffic by allowing components that communicate regularly to be placed on their own separate part of the network. *Submasking* is an addressing scheme that filters messages and determines if the message is to be passed to local nodes on a subnetwork or if it is to be sent on to the main network through the router. Submasking also consists of four sets of three digits, similar to Internet Protocol addressing, although the functions are different.

Ethernet Troubleshooting

Most modern building automation systems use an Ethernet network configuration. Problems with the network can cause improper building automation system operation. These problems may be due to software configuration errors or hardware failures.

Software Configuration Errors. Software configuration errors cause most of the problems with a network. All components must have software configurations set correctly to allow communication to take place. The configurations of all Windows components can be checked from the Windows Control Panel. Each manufacturer of network control modules provides a software setup utility. Windows comes with a winipcfg.exe utility that can be used to view Internet Protocol and submasking addressing. Winipcfg.exe may be activated from the Windows Run function.

A *ping* is an echo message and its reply sent by one network device to detect the presence of another device. This command is used in Windows by typing ping xxx.xxx.xxx.xxx at the MS-DOS prompt (command prompt), where the xs represent the Internet Protocol address of the node being interrogated. A reply means that the nodes can communicate. A request timed out means that the node did not respond and no communication is taking place. See Figure 19-8.

Hardware Failures. Hardware failures are also a possible cause of network problems. Unshielded twisted pair wires can be checked at each end with cable checkers. Network adapter cards normally contain LEDs that indicate network activity and/or presence of power. The ping command and Windows utilities can also be used to determine if the hardware is working properly. In addition, network diagnostic meters are available that can monitor cable integrity and polarity, test network adapter cards, and indicate amount of traffic. Network diagnostic meters can also print diagnostic reports for technician use.

Figure 19-8. The ping command can be used to test network connections.

ARCNET NETWORKS

Arcnet is an early network architecture that uses token ring data transmission. *Token ring* is a data transmission method in which a device obtains a token (ticket) that allows it to transmit on a network while the other nodes listen. When the transmission is finished, the node surrenders the token, which is taken by the next node that needs to communicate. This method of transmission is orderly and predictable; however, Arcnet is becoming obsolete.

Arcnet networks require network adapter cards similar to Ethernet networks. Arcnet networks also use concentrators similar to the hubs and switches found in Ethernet networks. Arcnet networks can use coaxial cable or unshielded twisted pair and are limited to a communication rate of 2.5 Mbps.

MIXED NETWORKS

A *mixed network* is a network in which new Ethernet network components are added to early Arcnet network components. Mixed networks require the technician to be familiar with both network architectures. Special components, such as Ethernet/Arcnet routers, are used to connect the Arcnet network components to the Ethernet network components. Each mixed network is a unique installation.

Many buildings are upgrading their networks from Arcnet to Ethernet. As time passes, Arcnet building automation system components become more difficult to find. In addition, new PCs may not have the correct computer hardware and/or software to handle Arcnet cards.

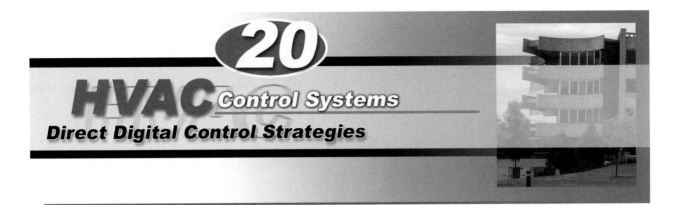

20

HVAC Control Systems
Direct Digital Control Strategies

Direct digital control strategies are used to automate the operation of HVAC equipment. Different direct digital control strategies are applied to control different types of HVAC equipment. Direct digital control algorithms used to control HVAC systems include proportional control, integral control, derivative control, and adaptive control algorithms.

DIRECT DIGITAL CONTROL STRATEGIES

A *control strategy* is a building automation system software method used to control the energy-using equipment in a building. A *direct digital control (DDC) strategy* is a control strategy in which a building automation system performs closed loop temperature, humidity, or pressure control. A *direct digital control system* is a control system in which the building automation system controller is wired directly to controlled devices and can turn them ON or OFF, or start a motion. See Figure 20-1.

Direct digital control systems contain sensors, controllers, and controlled devices. Sensors measure temperature, pressure, or humidity and provide the measured values as input to a controller. A controller compares the actual value to the programmed setpoint parameters. The controller also calculates the desired position of the controlled device and outputs a signal to it. The controlled device changes position and supplies heating, ventilation, air conditioning, lighting, or some process variable. Most building automation systems are direct digital control systems.

Direct digital control strategies are not designed to be final safety devices. For example, low-temperature limit controls, boiler high-temperature limit controls, and flame safeguard controls are required with direct digital control systems. In addition, the installation of a direct digital control system does not reduce the need for regular preventive maintenance.

DIRECT DIGITAL CONTROL SYSTEMS

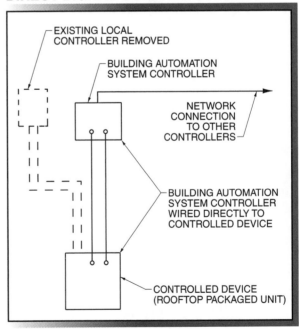

Figure 20-1. In a direct digital control system, a building automation system controller is wired directly to a controlled device.

267

Direct digital control systems use transducers in a building automation system. However, transducers only change signals from one type to another. For example, an electric/pneumatic transducer has a 4 mA to 20 mA input and a 3 psig to 15 psig output. See Figure 20-2. The signal change does not affect the building automation system, which still controls the pneumatic damper or valve actuator. Direct digital control systems include closed loop control and open loop control.

Early building automation systems were supervisory control systems that could only enable or disable the existing local controller.

Closed Loop Control

Closed loop control is control in which feedback occurs between the controller, sensor, and controlled device. *Feedback* is the measurement of the results of a controller action by a sensor or switch. For example, a building automation system controller measures temperature in a building space. The building automation system operates a controlled device to maintain a setpoint within the building space. The building automation system receives feedback on the control of the temperature in the building space by measuring the result of the action taken and changing the position or state of the controlled device.

Malfunction of one component in a closed loop control system results in other components within the system having the incorrect value or position. For example, a temperature sensor in a building space can provide feedback to a building automation system controller. The controller controls a hot water valve (controlled device). A malfunction in the temperature sensor, controller, or hot water valve can result in an incorrect building space temperature. See Figure 20-3.

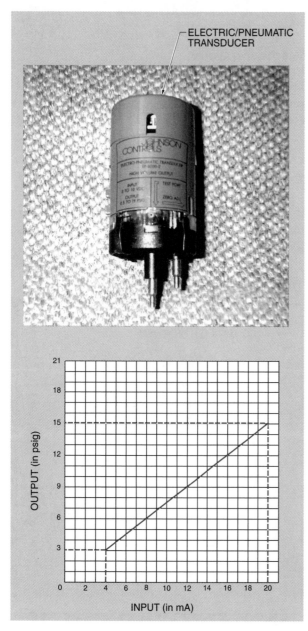

Figure 20-2. Electric/pneumatic transducers change an input voltage or current signal to a linear pneumatic output pressure signal.

CLOSED LOOP CONTROL

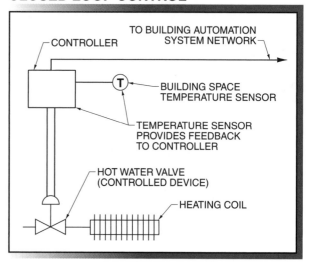

Figure 20-3. In closed loop control, temperature values or other feedback is provided to a building automation system controller.

Open Loop Control

Open loop control is control in which no feedback occurs between the controller, sensor, and controlled device. For example, a controller cycles a chilled water pump ON when the outside air temperature is above 65°F. In an open loop control system, the controller has no feedback regarding the status of the pump. See Figure 20-4. Without feedback, the controller does not receive a signal to verify that the pump is actually ON. For this reason, most essential control loops have feedback.

OPEN LOOP CONTROL

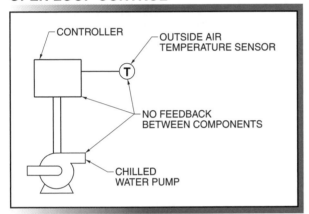

Figure 20-4. In open loop control, no feedback is provided to a building automation system controller.

Direct Digital Control Features

Direct digital control features determine exactly how a particular energy-using device in a building is controlled. Direct digital control features include setpoint control, reset control, low-limit control, high-limit control, economizer lockout control, lead/lag control, high/low signal select, and averaging control.

Setpoint Control. The most common direct digital control feature is the ability to maintain a setpoint in a building automation system. A *setpoint* is the desired value to be maintained by a system. The setpoint can be one of many controlled variables, such as temperature, humidity, pressure, light level, dewpoint, and enthalpy. The setpoint and the desired accuracy can be programmed into a building automation system controller. For example, a building automation system is required to maintain a temperature of 72°F in a commercial building. The 72°F temperature is the setpoint of the direct digital control system.

Customer demands regarding temperature setpoint and accuracy have increased in recent years. In the past, a range of ± 5°F may have been acceptable for some HVAC systems. With modern digital control system technology, a range of ±.2°F is acceptable today. Achieving this degree of accuracy depends on factors such as sensor quality, sensor location, and system calibration.

A *control point* is the actual value that a control system experiences at any given time in a system. For example, a temperature sensor in a building space measures a temperature of 74°F. The control point is 74°F. At any given time, a control point may differ from a setpoint. *Offset* is the difference between a control point and a setpoint. Depending on the accuracy of the controller, the actual value (control point) in a building space may be different from the value the controller is programmed to maintain (setpoint). See Figure 20-5.

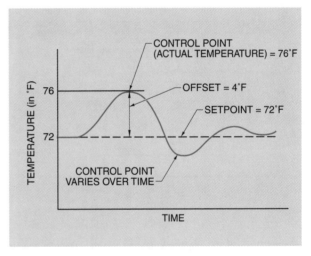

Figure 20-5. Offset is the difference between a control point and a setpoint.

Another common direct digital control feature is setup/setback setpoint control. Setup/setback setpoints are values that are active during the unoccupied mode of a building automation system. *Setup* is the unoccupied cooling setpoint. For example, a cooling setpoint is raised from 74°F during the day to 85°F at night. The setup setpoint is 85°F. *Setback* is the unoccupied heating setpoint. For example, a heating setpoint is lowered from 70°F during the day to 55°F at night. The setback setpoint is 55°F.

> *Setup/setback setpoint control requires time to bring the building temperature to the desired setpoint. Time-based schedules or optimum start/stop strategies allow an HVAC system to be turned ON prior to building occupancy to allow the temperature to reach the occupied comfort setpoint.*

Setup/setback setpoint control saves energy by preventing heating and/or cooling systems from operating when a commercial building is unoccupied. Setup/setback setpoint control also protects commercial buildings from abnormal temperatures by enabling the heating and cooling systems to operate if the building space temperature gets excessively hot or cold. Although the setup/setback setpoint control feature is commonly used with building space temperature setpoints, setpoints such as hot water, chilled water, and static pressure can also be controlled. See Figure 20-6.

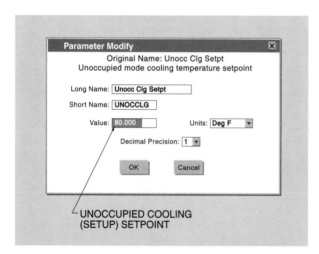

Figure 20-6. Setup/setback setpoints are values that are active during the unoccupied mode of a building automation system.

Reset Control. *Reset control* is a direct digital control feature in which a primary setpoint is reset automatically as another value (reset variable) changes. The most common example of reset control is the reset of a hot water heating setpoint as the outside air temperature changes. A *setpoint schedule* is a description of the amount a reset variable resets the primary setpoint. Another example of reset control is the reset of variable air volume system static pressure from the highest cooling demand from several zones.

Low-Limit Control. Low-limit control ensures that a controlled variable remains above a programmed low-limit value. The controlled device maintains a minimum value if the controlled variable drops below the low-limit value. Low-limit control is commonly used with mixed air damper controls. An excessively low outside air temperature causes a low mixed air temperature. For example, when the mixed air temperature of a commercial building drops below 45°F, the controller overrides the normal control logic and forces the outside air dampers closed. See Figure 20-7.

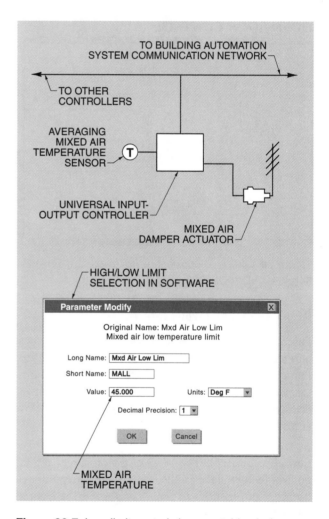

Figure 20-7. Low-limit control closes outside air dampers when mixed air temperatures are low due to excessively low outside air temperatures.

High-Limit Control. High-limit control provides capabilities similar to low-limit control. High-limit control is commonly used in applications that require a temperature, pressure, or humidity to not exceed a programmed value.

For example, a high-limit control can be used to prevent a hot water heating system water temperature from exceeding 210°F.

Economizer Lockout Control. An *economizer* is the use of outside air to provide cooling for a building space. Outside air has the ability to cool a building space if the outside air temperature is between 45°F and 65°F. The humidity of the air is also a factor. The mechanical cooling system is disabled when an economizer is used to provide cooling. *Economizer lockout control* is a direct digital control feature in which the economizer damper function is locked out or discontinued. Economizer lockout control methods include dry bulb and enthalpy economizer lockout control.

In the dry bulb economizer method, an outside air temperature sensor is used to control economizer operation. The economizer is used at 68°F and below and is locked out above 68°F. See Figure 20-8. In the enthalpy economizer method, the enthalpy of the outside air and return air is measured and the lowest value is used for cooling.

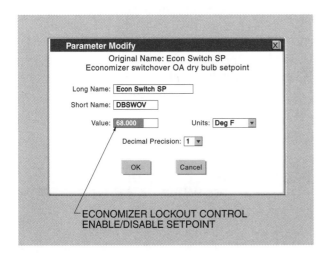

Figure 20-8. Economizer lockout control disables (locks out) the economizer when the outside air temperature rises above an enable/disable setpoint.

Lead/Lag Control. *Lead/lag control* is the alternation of operation between two or more similar pieces of equipment. The most common application of lead/lag control is the control of multiple hot water or chilled water pump installations. For example, a primary (lead) pump is energized by a building automation system when a pump is requested. A backup (lag) pump is energized if the primary pump fails to start. See Figure 20-9.

Often, lead/lag pump operation is rotated on a time schedule to have equal run time on the two pumps. In addition to pumps, refrigeration and air compressors are also common applications of lead/lag control.

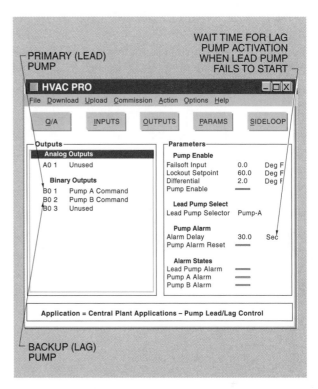

Figure 20-9. Lead/lag control is used with multiple pumps or compressors to equalize run time or turn ON a backup (lag) device if the primary device fails.

High/Low Signal Select. *High/low signal select* is a direct digital control feature in which a building automation system selects among the highest or lowest values from multiple inputs. The most common application of high/low signal select is the control of a building space temperature using multiple zone temperature sensors at different locations in a building. The highest signal represents the warmest building space and the lowest signal represents the coolest building space. The signal selection results are communicated to the building automation system reset control to determine the setpoint. The highest signal may be used to reset a cooling function to satisfy the warmest building space. The lowest signal may be used to reset a heating function to satisfy the coolest building space. Many technicians use high/low signal select instead of outside air temperature reset control.

Averaging Control. *Averaging control* is a direct digital control feature that calculates an average value from all selected inputs. Averaging control is used in applications where the highest and lowest values do not represent the actual conditions in a commercial building. For example, a temperature sensor located in a foyer experiences outside air conditions whenever a door is open. This sensor normally senses the coldest building space temperature and controls the heating reset setpoint most of the time. An average from several temperature sensors located throughout the building is more representative of the actual building conditions. Correct placement of the averaging control sensors are required for accurate results. See Figure 20-10.

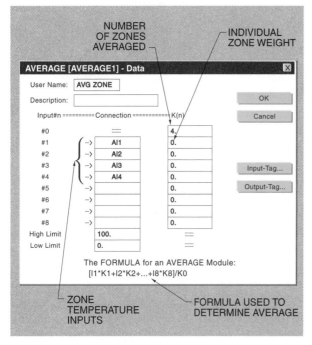

Figure 20-10. Averaging control calculates an average value from all selected inputs.

Direct Digital Control Algorithms

An *algorithm* is a mathematical equation used by a building automation system controller to determine a desired setpoint. Direct digital control algorithms enable a building automation system to achieve a high level of accuracy. Direct digital control algorithms are selected from a personal computer and sent to a building automation system controller CPU. The communication between a personal computer and building automation system controller CPU is referred to as downloading the controller. To achieve proper control, the correct algorithm must be selected and accurate setpoints input into the personal computer. Common algorithms used in direct digital control systems include proportional, integral, derivative, and adaptive control algorithms.

When averaging a number of sensors, ensure that the sensors are in the correct location and are not being affected by sunlight or drafts. A numerical limit may be placed on the averaging calculation if a sensor is damaged and giving a false reading. Zone values may need to be weighted if a problem zone affects the average more than another zone.

Proportional Control Algorithms. A *proportional (P) control algorithm* is a control algorithm that positions the controlled device in direct response to the amount of offset in a building automation system. Proportional control algorithms are the most common algorithms used in direct digital control systems. See Figure 20-11. The setpoint and the desired accuracy (throttling range) are entered into the controller. The controller is programmed to have a specific output when the controller reaches the setpoint. This is often done by using the mid-stroke position of the controlled device.

PROPORTIONAL CONTROL

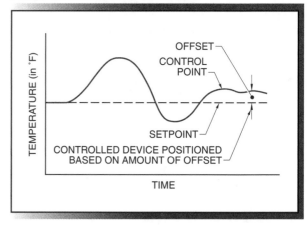

Figure 20-11. Proportional control algorithms position a controlled device in response to the amount of offset experienced in a building automation system.

Proportional control algorithms do not provide precise control. Proportional control algorithms assume that a controlled device must be at a particular position, such as mid-stroke, when at setpoint. However, a load may require that a controlled device be at a different position when at setpoint. This results in increased offset and energy expenditures. Proportional control algorithms are commonly used in basic control systems that do not require precise control.

Integral Control Algorithms. An *integral control algorithm* is a control algorithm that eliminates any offset after a certain length of time. Integration algorithms are used in calculus to determine the area under a curve. Integral control algorithms are commonly used in conjunction with proportional control algorithms. Proportional/integral (PI) control is the combination of proportional and integral control algorithms. When an offset remains after a specific length of time, the integral control algorithm repositions the controlled device to eliminate the offset. The integral control algorithm changes the position of the controlled device to meet the actual requirements of the load at a given time.

Derivative Control Algorithms. A *derivative control algorithm* is a control algorithm that determines the instantaneous rate of change of a variable. See Figure 20-12. Derivative control algorithms provide real-time data to a building automation system controller. The derivative control algorithm acts against the integral control algorithm. The derivative control algorithm increases the speed at which a controlled device eliminates an offset. A *proportional/integral/derivative (PID) controller* is a direct digital control system that uses proportional, integral, and derivative algorithms. In the HVAC industry, only extremely sensitive control applications require PID control. PI control is normally sufficient to achieve a setpoint.

> ⓘ *When using PID control algorithms, verify HVAC system response before attempting algorithm application. A PID control algorithm mirrors the response of the mechanical equipment. Some HVAC systems, such as static pressure and steam heat, have quick changes in pressure and heat while others, such as hot water heat in a fan coil, change slowly.*

DERIVATIVE CONTROL

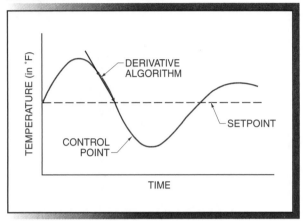

Figure 20-12. Derivative control algorithms determine the instantaneous rate of change of a variable.

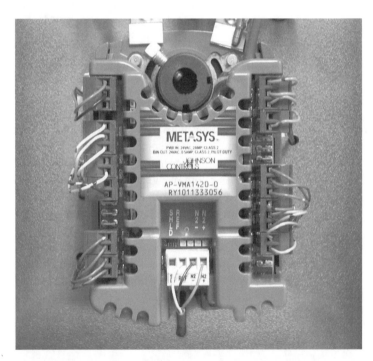

Some VAV terminal box controllers have built-in adaptive temperature and airflow algorithms that require little or no tuning, resulting in lower installation cost and higher customer satisfaction.

Adaptive Control Algorithms. The most sophisticated control algorithms used today are adaptive control algorithms. An *adaptive control algorithm* is a control algorithm that automatically adjusts its response time based on environmental conditions. This results in increased accuracy, stability, and simplicity. Adaptive control algorithms are a self-tuning form of PID control. See Figure 20-13.

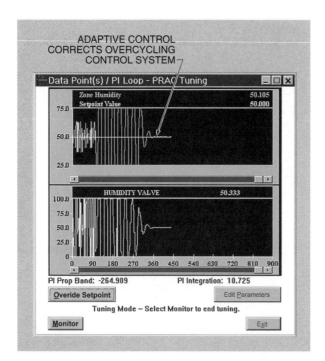

Figure 20-13. Adaptive control algorithms are self-tuning algorithms that are often used to correct an overcycling control system.

HVAC technicians must accurately tune PI and PID controls by adjusting the response times in the building automation system software to ensure accuracy and stability. Adaptive control algorithms have the ability to readjust to load changes or incorrectly programmed response times. Adaptive control algorithms reduce the amount of time a technician tunes a control loop. Also, the programmed response times do not have to be accurate because adaptive control algorithms have the ability to change response times to the proper value. Not all building automation systems provide adaptive control algorithms.

The Trane Company

Some absorption units use steam as an energy source for cooling. The steam valve PID control algorithm may require tuning to prevent undershoot and overshoot of the steam valve, which can cause crystallization.

Tuning PI and PID Control Loops

PI and PID control loops must be tuned (calibrated). *Tuning* is the downloading of the proper response times into the controller and checking the response of the control system. The control system must be adjusted and checked routinely if the response is incorrect. Each HVAC system responds differently and must be custom tuned. Tuning is the responsibility of a building automation system technician. Guidelines used when tuning PI and PID control loops include the following:

• Tune the control loop during a heavy load demand. During a heavy load demand, the setpoint is satisfied under the most adverse conditions. The integration time may be shortened if it is not possible to cause the control loop to experience a heavy load demand.

• Tune the control loop using proportional control only. A control loop cannot be tuned if it cannot be tuned using proportional control only. The proportional control algorithm sets the control system in the desired range of tuning. Before continuing the tuning process, turn OFF the integral and derivative algorithms in the software and attain good control.

• Double the throttling range setting in the control system software. Once the proportional control algorithm is stable, double the setting. The integral algorithm performs the final control.

• Begin using long integration times. Long integration times provide stable conditions. Short integration times result in quick responses but a high chance of cycling.

• Shorten the integration time slowly, checking the system response. As the times are shortened, the system becomes unstable. Lengthen the time until the system retains stability.

• Use the proportional control algorithm and integral control algorithm first when the derivative control algorithm is used. When the integral algorithm has been tuned in a stable manner, the derivative control algorithm can be used to ensure a quick response in the event of a rapid load change. Small derivative numbers make the system stable but react slowly. Large derivative numbers make the system unstable but react quickly. Increase the integral control algorithm time above that needed for stability, and then increase the derivative factor to ensure a timely response. When properly tuned, P, PI, or PID control loops provide long-term stable control of building automation systems.

Direct Digital Control System Applications

Direct digital control strategies must be applied correctly to a particular piece of equipment. Misapplication of direct digital control strategies causes occupant discomfort and dissatisfaction with the control system. In addition, direct digital control strategy misapplication can cause damage to essential HVAC units and/or the building structure. Common direct digital control system applications include unitary control of rooftop packaged units, air handling unit control of variable air volume systems, and universal input-output control of hot water heating systems.

New direct digital control strategy applications include dimming lights in response to an increase in ambient light coming through windows, rotating products at an angle to ensure proper drying, running new, high-efficiency units longer instead of old, low-efficiency units, operating a system at a high enough level to maintain a calculated load, and selecting devices to operate at a given time of day based on the energy cost at that time.

Unitary Control of Rooftop Packaged Units. In a rooftop packaged unit direct digital control system, a unitary controller is used in place of a standard thermostat control. The unitary controller maintains a specific heating and cooling setpoint during the day. The unitary controller uses inputs from a temperature sensor and air flow switch to control heating elements and an air conditioning compressor. See Figure 20-14. The direct digital control system provides setup and setback setpoints for the system unoccupied mode.

Possible additional control features include economizer lockout control for the outside air dampers, low-limit control to prevent damage caused by freezing if the mixed air temperature becomes too low, and reset control to increase the cooling setpoint as the outside air temperature increases. If the rooftop packaged unit has proportional heating and cooling, PID control can help the system achieve close setpoint control.

Air Handling Unit Control of Variable Air Volume Systems. In a variable air volume direct digital control system, an air handling unit controller uses input from a static pressure sensor to control supply and return fans through variable-speed drives. The air handling

unit controller provides a 4 mA to 20 mA output signal to two variable-speed drives. See Figure 20-15. The programming of the air handling unit controller software requires accuracy to provide stable and responsive control. The PID control variables may require tuning to achieve optimum setpoint control of the variable-speed drives. The common duct static pressure setpoint for variable air volume systems is 1″ wc. Some variable air volume air handling units also measure and control the volume of air in the supply and return duct. The return air flow volume follows (tracks) the supply air flow volume. The difference in volume is directly related to the internal pressure in the building.

UNITARY CONTROL OF ROOFTOP PACKAGED UNIT

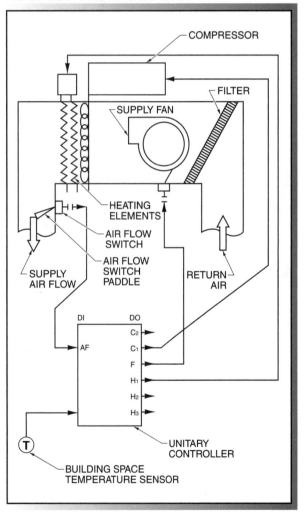

Figure 20-14. Direct digital control of a rooftop packaged unit enables a unitary controller to provide increased control capabilities compared to a standard thermostat control.

AIR HANDLING UNIT CONTROL OF VARIABLE AIR VOLUME SYSTEM

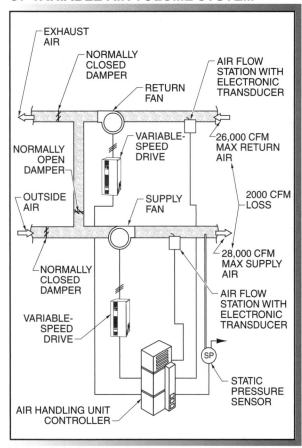

Figure 20-15. Direct digital control of a variable air volume system enables an air handling unit controller to perform static pressure control of supply and return fans.

Building automation systems are often used in critical applications such as fume hoods and pharmaceutical manufacturing. These applications have stringent code requirements regarding job documentation and software changes. Technicians must always follow the governing body code requirements or risk costly fines.

Universal Input-Output Control of Hot Water Heating Systems. In a hot water heating direct digital control system, a universal input-output controller provides an output signal to an electric/pneumatic transducer. The electric/pneumatic transducer outputs a proportional air pressure signal to a three-way hot water valve actuator. See Figure 20-16. The universal input-output controller uses a hot water supply temperature sensor to maintain a hot water temperature setpoint. The controller has an outside air temperature sensor that enables the controller to adjust the hot water setpoint as the outside air temperature changes.

UNIVERSAL INPUT-OUTPUT CONTROL OF HOT WATER HEATING SYSTEM

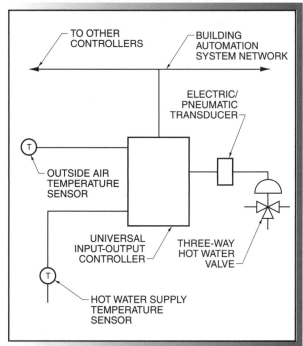

Figure 20-16. Direct digital control of a hot water heating system enables a universal input-output controller to modulate the hot water supply temperature based on the outside air temperature.

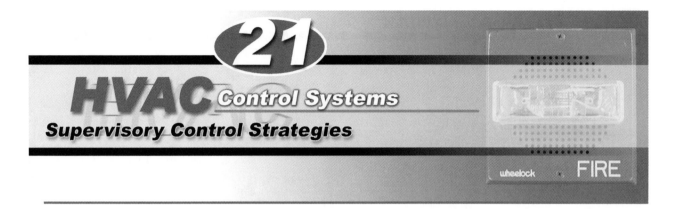

Supervisory control strategies use programmable software to control the energy-consuming functions of a building. Supervisory control strategies differ from direct digital control strategies in that they are commonly performed at the network communication module. Supervisory control strategies include life safety, time-based, optimum start/stop, duty cycling, and electrical demand control strategies.

SUPERVISORY CONTROL STRATEGIES

A *supervisory control strategy* is a programmable software method used to control the energy-consuming functions of a commercial building. Supervisory control strategies were the original control strategies developed for building automation systems.

Supervisory control strategies do not always require the replacement of the existing local controller. A building automation system controls the system operation by enabling or disabling the existing local controller. The network communication module commonly performs the supervisory control strategies. Supervisory control strategies include life safety, time-based, optimum start/stop, duty cycling, and electrical demand control strategies.

Life Safety Supervisory Control

Life safety supervisory control is a control strategy for life safety issues such as fire prevention, detection, and suppression. See Figure 21-1. Life safety supervisory control strategies have the highest priority of all control strategies. An HVAC technician must be familiar with life safety system wiring, software, and codes.

Time-Based Supervisory Control

Time-based supervisory control is a control strategy in which the time of day is used to determine the desired operation of a load. Time-based supervisory control strategies were the first supervisory control strategies developed for building automation systems. Time-based supervisory control strategies involve turning a load ON and OFF at a specific time. Time-based supervisory control strategies are the most widely used supervisory control strategies in building automation systems.

When troubleshooting a building automation system, the time schedule should be checked to determine if the HVAC load is active or inactive (occupied or unoccupied). The controller status must be determined because many operations are based on time scheduling.

Before the development of supervisory control strategies for building automation systems, timed-based control functions were performed by electromechanical time clocks. However, electromechanical time clocks do not allow for flexibility of a building automation system. In addition, electromechanical time clocks are unpredictable and are less accurate compared to supervisory control strategies.

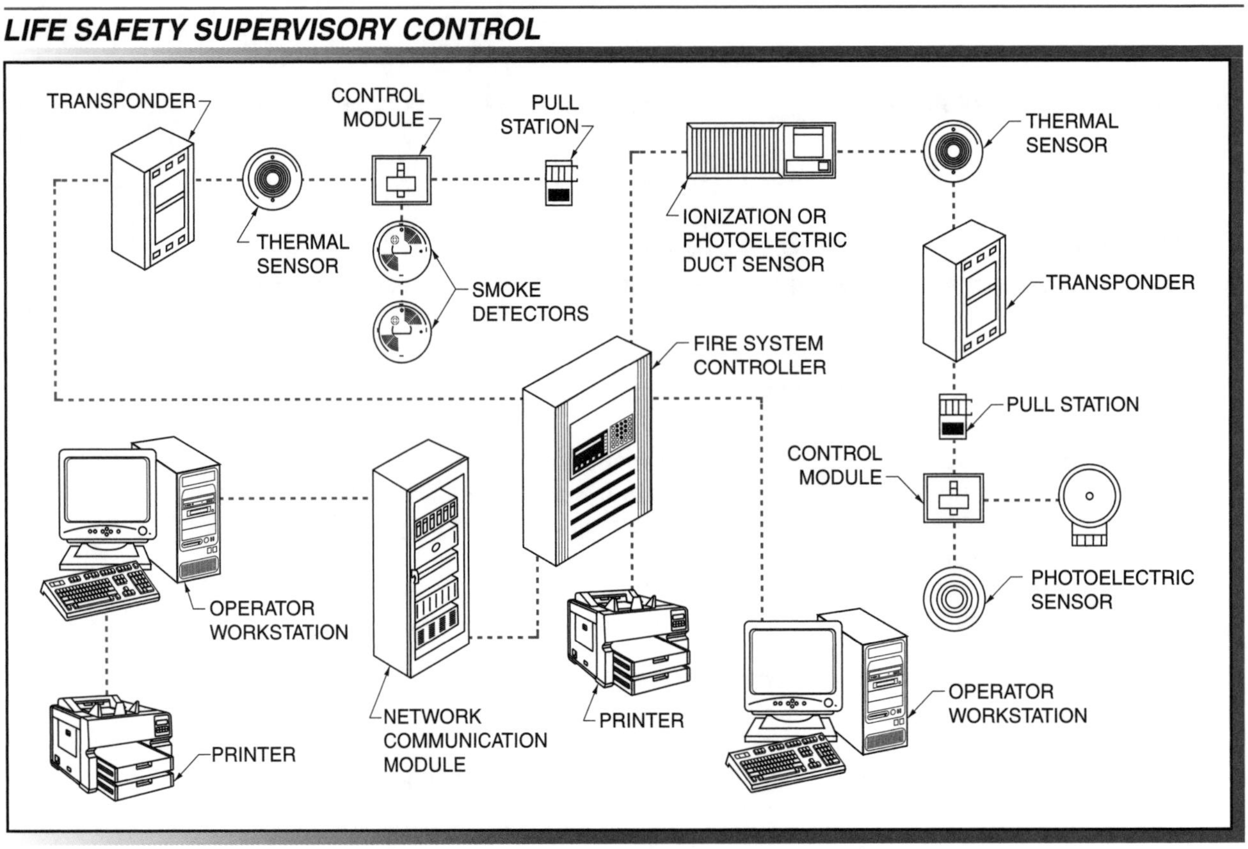

LIFE SAFETY SUPERVISORY CONTROL

Figure 21-1. Fire and smoke detection are the most common life safety supervisory control strategies.

The majority of a building automation system workload consists of adjusting time schedules for various areas of a building. Time-based supervisory control strategies were created to reduce the workload of HVAC technicians. Time-based supervisory control strategies include seven-day programming, daily multiple time period scheduling, holiday and vacation scheduling, timed overrides, temporary scheduling, daylight savings time changeover scheduling, time zone scheduling, schedule linking, and alternate scheduling.

Seven-day Programming. Seven-day programming allows an HVAC technician to individually program building automation system ON and OFF time functions for each day of the week. Seven-day programming is common today, but some early building automation systems use a 5+2 time-based supervisory control strategy. A 5+2 time-based supervisory control strategy recognizes Monday through Friday as normal workdays and Saturday and Sunday separately. Building automation systems should provide the capability to program all seven days independently to allow flexibility of use for commercial building tenants.

Daily Multiple Time Period Scheduling. Daily multiple time period scheduling allows building automation system software to accommodate unusual building occupancy. See Figure 21-2. The HVAC, lighting, and other devices (events) can be scheduled to operate during multiple independent time periods. For example, the scheduled operating hours of a rooftop unit in a commercial building are 8 AM to 5 PM. The building is also used at night for a continuing education class from 8 PM to 10 PM. These daily multiple time periods can be programmed into the building automation system software. Most building automation system software provides two or three separate programmable time periods per day.

Holiday and Vacation Scheduling. Holiday and vacation time-based scheduling enables the building automation system software to turn loads OFF during holidays and vacations. An HVAC technician must obtain a comprehensive yearly operation calendar to program holidays and vacations. A *permanent holiday* is a holiday that remains on the same date each year.

A *transient holiday* is a holiday that changes its date each year. For example, New Year's Day, which falls on January 1st each year, is a permanent holiday. Memorial Day, which falls on the last Monday in May each year, is a transient holiday.

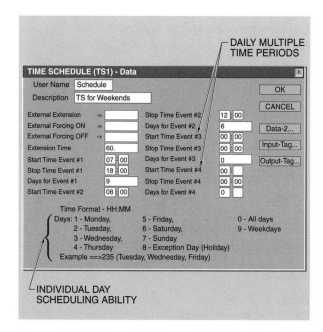

Figure 21-2. Daily multiple time period scheduling allows building automation system software to accommodate unusual building occupancy.

Timed Overrides. A *timed override* is a time-based supervisory control strategy in which the occupants can change a zone from an unoccupied to occupied mode for temporary occupancy. A quick response to building needs is required to accommodate building occupancy changes. A timed override can be activated by a pushbutton switch that is configured as a digital input and wired to the building automation system. See Figure 21-3. A timed override can also be activated by a personal computer. During the unoccupied mode, if the switch is pressed, the controller uses the occupied setpoints for a specific period of time.

Some building automation systems cancel the timed override period if the switch is depressed a second time. Other building automation systems wait until the time period has elapsed before switching back to unoccupied mode. Some building automation systems provide a data trend that records the amount of time spent in the override mode each month. Timed overrides should be used only when little advance notice is given for changes in building occupancy.

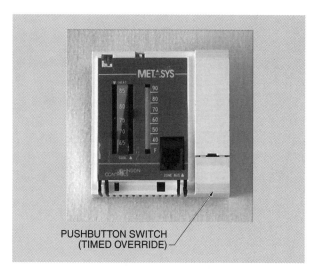

PUSHBUTTON SWITCH (TIMED OVERRIDE)

Figure 21-3. Timed overrides are used to respond to sudden changes in the occupancy of a commercial building.

Temporary Scheduling. Building automation systems provide the ability to program a one-time temporary schedule. Temporary schedules are commonly associated with a specific calendar date. See Figure 21-4. Temporary scheduling compensates for a specific event in a building without using a timed override. At the end of a temporary schedule on a specific date, the temporary schedule is erased and the normal time schedules resume. Temporary scheduling is commonly used for regularly scheduled weekly or monthly events. Temporary schedules take priority over standard time schedules.

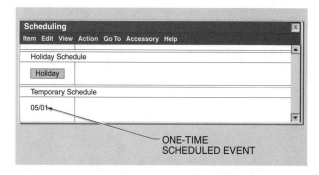

ONE-TIME SCHEDULED EVENT

Figure 21-4. Building automation systems provide the ability to program temporary schedules.

Daylight Saving Time Changeover Scheduling. Building automation systems provide automatic daylight saving time changeover. See Figure 21-5. Building automation systems are programmed with default values that correspond to changeover dates. Daylight saving time is changed as required, depending on the geographical location of a building.

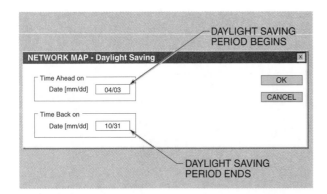

Figure 21-5. Building automation system software provides a conversion method between daylight saving time and standard time.

Time Zone Scheduling. Building automation system software can operate remote buildings located in different time zones. This feature is required for contractors who maintain buildings in different time zones. Buildings in different time zones are assigned an offset time value in positive or negative hours.

Schedule Linking. Schedule linking provides the ability for building automation systems to join loads that are used during the same time. For example, when a rooftop unit is energized for a particular zone, the lighting load must also be energized. Schedule linking enables both loads to be energized simultaneously.

Electromechanical time clocks, while still used in systems today, are being replaced by building automation system time-based supervisory control strategies.

Alternate Scheduling. Alternate scheduling enables building loads to be scheduled for use for more than one unique time schedule per year. See Figure 21-6. Alternate scheduling is commonly used during the Christmas shopping season. For example, one time schedule extends from January through November. Beginning on the day after Thanksgiving, the alternate time schedule takes effect, overriding the yearly schedule until the end of the Christmas shopping season.

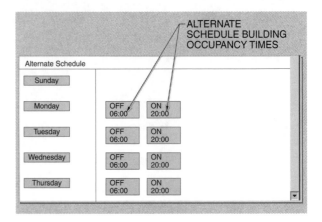

Figure 21-6. Alternate schedules can be used for events such as inventory and seasonal production schedules.

Optimum Start/Stop Supervisory Control

Optimum start is a supervisory control strategy in which the HVAC load is turned ON as late as possible to obtain the proper building space temperature at the beginning of building occupancy. *Optimum stop* is a supervisory control strategy in which the HVAC load is turned OFF as early as possible to maintain the proper building space temperature at the end of building occupancy.

Optimum Start Supervisory Control Strategy. Time-based start time does not represent the actual start time of a unit. Time-based start time represents the beginning of occupancy. The optimum start supervisory control strategy determines the actual start time of a unit. Actual start time is the time at which a unit must be turned on before occupancy to bring building space temperature to the desired level at the beginning of occupancy. The optimum start supervisory control strategy allows maximum energy savings and maintains comfort levels within a commercial building. Start times change daily depending on the building and programmed conditions. Two methods used in building automation systems to determine the actual start time are adaptive start control and estimation control.

The most common optimum start method used in building automation systems is adaptive start control. *Adaptive start control* is a control method that adjusts (learns) its control settings based on the condition of a building. Adaptive start control identifies the proper time to heat or cool a building before occupancy. The adaptive start control method changes an HVAC unit actual start time each day until the optimum start time is achieved.

Estimation control is a control method that uses the latest building temperature data to estimate the actual start time to heat or cool a building before occupancy. A disadvantage of estimation control is that the estimate used by a technician can be calculated or input incorrectly. In addition, the estimate may also cause building systems to start late, resulting in the building space temperature being at the incorrect level when occupancy starts.

> *Implementation of an optimum start strategy requires the proper setup time and experience. Until an optimum start strategy can be implemented, HVAC units can be started at a fixed time before occupancy to ensure occupant comfort when the building space is occupied.*

Optimum Start Time Factors. Optimum start algorithms use a variety of factors to calculate unit actual start time. The current outside air temperature influences the heating and cooling load. The current indoor air temperature and setpoint are evaluated to determine the temperature change required within a building space.

The scheduled occupancy time is a target to determine when the desired setpoint temperature must be reached. A seven-day, 10-day, or 14-day history is commonly used to determine the success rate of previous start times and as a readjustment point for current calculations. In adaptive start control, the start times get progressively closer to the ideal start time every day because of continual readjustment.

Thermal Recovery Coefficients. A *thermal recovery coefficient* is the ratio of an indoor temperature change and the length of time it takes to obtain that temperature change. Thermal recovery coefficient is expressed in °F/min. Thermal recovery coefficients are used to calculate the actual start time of HVAC

systems in commercial buildings. For example, a rooftop unit may start operation at 7 AM for 9 AM building occupancy. During this time, the indoor temperature increases from 60°F to 72°F. Therefore, it takes 120 min to increase the temperature 12°F. The thermal recovery coefficient is 12°F/120 min (.1°F/1 min).

Thermal recovery coefficient values can also be used as indicators of HVAC equipment efficiency and/or mechanical problems. For example, an HVAC system that has a thermal recovery coefficient of .2°F/1 min in one month and .1°F/1 min the next month may require filter replacement or preventive maintenance.

Optimum Stop Supervisory Control Strategy. The optimum stop supervisory control strategy turns an HVAC unit OFF before the end of building occupancy. For example, in winter, the optimum stop supervisory control strategy allows the building temperature to gradually decline until the end of occupancy. See Figure 21-7. The HVAC unit turns ON if the temperature fluctuates excessively. Optimum stop is commonly limited to a specific length of time, such as 15 min or 30 min. The optimum stop supervisory control strategy may be independent of the optimum start supervisory control strategy. Some software programs allow the optimum start supervisory control strategy to be used without using the optimum stop supervisory control strategy.

OPTIMUM STOP SUPERVISORY CONTROL STRATEGY

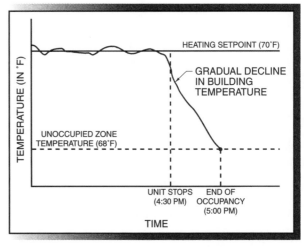

Figure 21-7. The optimum stop supervisory control strategy shuts an HVAC unit OFF and lets the building temperature decline gradually to the end of building occupancy.

Heat pump units must be duty cycled with great care. Loss of comfort and customer dissatisfaction result if heat pump duty cycling is not set up properly.

Duty Cycling Supervisory Control

Duty cycle is the percentage of time a load or circuit is ON compared to the time a load or circuit is ON and OFF (total cycle time). Duty cycle is normally given as a percentage of total cycle time. For example, an HVAC unit runs for 30 min out of a time interval of 60 min. The duty cycle is 50% (30 ÷ 60) × 100 = 50%.

The duty cycling supervisory control strategy is designed to reduce electrical demand charges in a commercial building. Duty cycling is used when HVAC units are oversized and can be shut OFF temporarily without adversely affecting building space temperature. HVAC loads such as cooling fans and compressors have their cycle times staggered so only one load is operating at a time. For example, two HVAC units have an electrical demand of 15 kW each. The electrical demand is 30 kW if both units operate simultaneously. By implementing the duty cycling supervisory control strategy, the operation of each unit is alternated so the electrical demand is 15 kW. See Figure 21-8.

The duty cycling supervisory control strategy is effective for reducing electrical demand and expenses. However, duty cycling increases motor wear due to an increased number of motor starts. Duty cycling also results in the loss of temperature control and loss of ventilation during the duty cycle OFF time. In addition, an increase in noise is caused by HVAC equipment cycling through its stages.

DUTY CYCLING

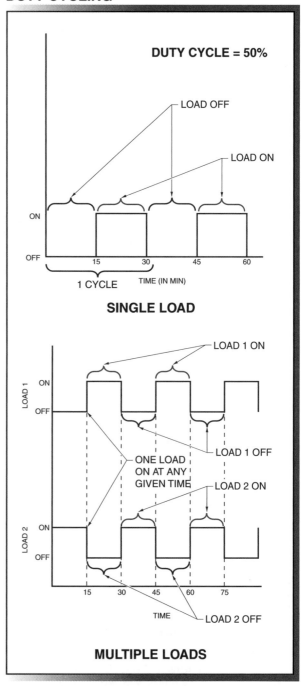

Figure 21-8. Duty cycling reduces the electrical demand in a commercial building.

Temperature-Compensated Duty Cycling. The loss of temperature control during the OFF cycle of an HVAC unit creates problems for occupants of a commercial building. Temperature-compensated duty cycling was developed to make the duty cycle ON time more responsive to the building space temperature. In temperature-compensated duty cycling, the length of the duty cycle ON time increases as the building space temperature deviates from setpoint. For example, a temperature of 74°F is desired in a building space and a limit of 78°F is established. The ON time of the duty cycle is increased by a specific length of time if the building space temperature increases above 78°F by 1°F.

> ⓘ *Duty cycling has long been a control strategy in building automation systems. Commercial buildings with a large number of electric baseboard heaters or small exhaust fans often use duty cycling as a control strategy. Because a duty cycle alternates the areas it heats, loss of comfort may occur when the load is OFF.*

Duty cycling is used in commercial buildings that have a large number of electric baseboard heaters or small exhaust fans. See Figure 21-9. The duty cycling order for the baseboard heaters can be defined during programming so the duty cycles are alternated around the building. For example, a building with multiple electric baseboard heaters may be set up so that one heater in an office is duty cycled for a short period of time and then another heater on the opposite side of the building is duty cycled.

ELECTRIC BASEBOARD HEATER

Figure 21-9. Duty cycling is used in commercial buildings that have multiple electric baseboard heaters.

Electrical Demand Supervisory Control

Electrical demand is the highest amount of electricity used during a specific period of time. The most common time period used by electric utility companies is 15 min or 30 min. The electrical demand charge is the highest 15 min or 30 min electrical demand period per month. *Electrical demand supervisory control* is a control strategy designed to reduce the electrical demand component of an electric bill. Electrical demand supervisory control strategies were developed to reduce building electrical demand by turning OFF electric loads in an orderly fashion.

Cooling units may be shed if care is taken to protect equipment and ensure comfort in a facility.

A *shed load* is an electric load that has been turned OFF by an electrical demand supervisory control strategy. A *restored load* is a shed load that has been turned ON by an electrical demand supervisory control strategy. Loads are shed (turned OFF) as the building electrical demand increases to a specific limit (target). The electrical demand decreases below the target when loads are shed.

A building automation system must have a method of measuring the electrical demand at the meter. To measure the electrical demand at the meter, an existing utility meter can be wired directly to the building automation system. ON/OFF pulses from the utility meter are configured as digital inputs to the building automation system. The pulses are commonly counted using an accumulator or counter. Each utility meter has a specific kilowatt value per ON/OFF pulse. Separate metering equipment may also be purchased and installed to measure electrical demand. Current transformers and watt transducers can be wired directly to a building automation system to measure electrical demand.

Shed Tables. A *shed table* is a table that prioritizes the order in which electrical loads are turned OFF. The electrical loads of a building are input into building automation system software shed tables. Building automation system manufacturers commonly provide low-priority and high-priority shed tables. A *low-priority load* is a load that is shed first when a high electrical demand period occurs. A *high-priority load* is a load that is important to the operation of a building and is shed last when demand goes up. High-priority loads are re-energized first when the electrical demand drops. See Figure 21-10.

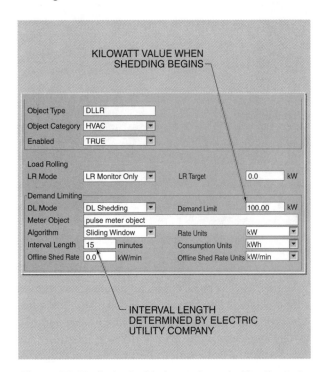

Figure 21-10. A shed table is used to prioritize the order in which loads are turned OFF.

In some applications, building automation system manufacturers supply a third shed table that fits between low- and high-priority load shed tables. A thorough analysis of the electrical loads of a building must be performed to document the priority of each load. All building loads do not have to be included in an electrical demand supervisory control strategy. Some loads are essential to the efficient operation of a commercial building and cannot be shed.

When using an electrical demand supervisory control strategy, allow time when tuning and monitoring system startup because these strategies may require adjustment before final implementation. Some electrical demand control strategies may not be desirable if building occupant comfort is affected. A compromise may be required regarding the particular loads to be shed and frequency of shedding.

When building electrical demand is above the target, the loads in the low-priority shed table are shed in order. If the building electrical demand stabilizes or drops, the loads are restored in reverse order. However, if the electric demand does not stabilize but increases, all loads in the low-priority shed table are shed and the electrical demand supervisory control strategy begins to shed loads in the high-priority shed table. The low-priority shed table is often referred to as first OFF/last ON. The high-priority shed table is often referred to as last OFF/first ON.

Rotating Priority Load Shedding. *Rotating priority load shedding* is an electrical demand supervisory control strategy in which the order of loads to be shed is changed with each high electrical demand condition. Load shedding reduces electrical demand in a commercial building, but can also create problems. For example, the first load in a low-priority shed table is electric hot water heat. This load is always the first load shed according to the building automation system software. If a commercial building has a constant high electrical demand, this load is constantly shed first and not available for use. The solution is to rotate the loads within the shed table. In rotating priority load shedding, load 1 is the first load shed the first time a high electrical demand condition occurs and load 2 is the first load shed the next time a high electrical demand condition occurs. See Figure 21-11. Some manufacturers rotate low- and high-priority shed tables while other manufacturers rotate one shed table and leave the other shed table fixed.

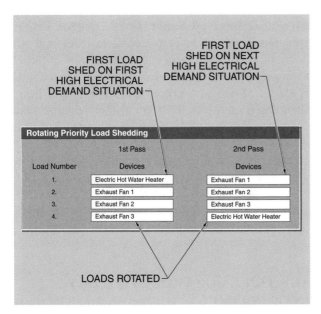

Figure 21-11. In rotating priority load shedding, the order of the loads to be shed is changed with each high electrical demand condition.

Electrical Demand Supervisory Control Strategies. Electrical demand supervisory control strategies reduce building electrical demand charges. However, electrical demand supervisory control strategies lack the flexibility to compensate for zones that are too hot or too cold before load shedding occurs, equipment that may be damaged by short cycling, and equipment that is essential to building operation.

> ℹ️ *Most load shedding turns loads OFF but can be configured to turn loads ON. For example, a diesel generator may be turned ON to supply electricity when a specific demand value is reached. In this case, running the generator to supply electricity is less expensive than paying the electric utility demand cost.*

Electrical demand supervisory control strategy techniques are used to allow specific events to occur. For example, maximum and minimum shed time software timers allow efficient control of shed loads. See Figure 21-12. The maximum shed time timer causes a load to be restored after it has been shed for a certain length of time. This load is restored regardless of the electrical demand in a commercial building at the

time. The maximum shed time timer is commonly used with loads that are essential to the building operation. These loads would not be involved in the electrical demand supervisory control strategy if the maximum shed time timer did not exist.

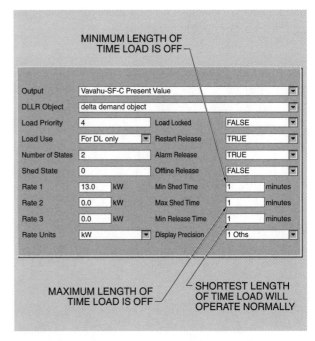

Figure 21-12. Maximum and minimum shed time software timers allow efficient control of shed loads.

The minimum shed time timer causes loads to be transferred to the electrical demand supervisory control strategy for a short period of time. The minimum shed time timer ensures that a shed load cannot be restored until a specific time period has elapsed. The minimum shed time timer reduces the possibility that a load is cycled ON and OFF repeatedly by the electrical demand supervisory control strategy. Electrical demand supervisory control strategies also have the ability to shed the cooling compressor of a package unit during supply fan operation. This allows air to be circulated, providing some relief while the cooling compressor is in the OFF position. Demand charges are reduced because the compressor uses more electrical power than the supply fan.

Electrical demand supervisory control strategies also have the ability to temporarily change the setpoint when building electrical demand is high. Instead of shedding a load when building electrical demand is high, the normal setpoints are changed by a specific number of degrees. For example, a rooftop unit is programmed for a normal cooling setpoint of 72°F. However, 4°F is added to the cooling setpoint when the building is in a high electrical demand period. The new, temporary setpoint is 76°F. This setpoint change causes the unit compressor to shut OFF. If the temperature in the building space changes outside of the new, temporary setpoint, the compressor turns ON regardless of the building electrical demand. This feature can also be programmed for heating applications.

Digital electric meters can be wired into a building automation system to indicate electrical demand and consumption.

Electrical Demand Supervisory Control Target Development. An effective electrical demand supervisory control strategy requires accurate monthly electrical demand targets. Load shedding is reduced if electrical demand targets are set too high. Load shedding is increased if electrical demand targets are set too low. The development of accurate electrical demand targets requires experience in evaluating prior electric bills. After the targets are developed, the targets may require adjustment by evaluating actual building electrical demand. The electrical demand targets can be lowered incrementally over a period of time until the optimum level is reached.

MULTIPLE CONTROL STRATEGY INTEGRATION

Supervisory control strategies and direct digital control strategies are commonly used for the same energy devices during the same period. Supervisory control strategies and direct digital control strategies can be integrated in an overall approach to effectively control HVAC units. Strategy priority is required when multiple strategies are combined for an HVAC unit. Manufacturers provide software indicators that appear on personal computers to advise an HVAC technician of the strategy that currently has priority. Tables that list priority requirements of each strategy are commonly provided in manufacturer literature.

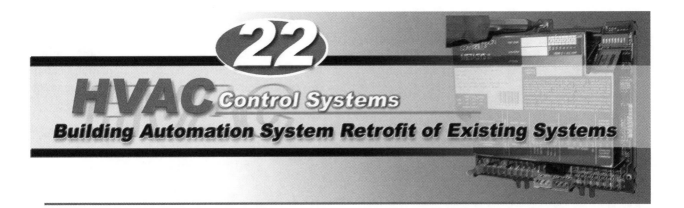

A retrofit replaces an old control system with a new one. Many possible retrofit applications exist for HVAC systems. A common retrofit application is the replacement of a pneumatic control system with a building automation system. A retrofit can be broken down into distinct stages to help facilitate efficient project completion.

BUILDING AUTOMATION RETROFIT OF EXISTING SYSTEMS

A *retrofit* is the process of upgrading a building automation system by replacing obsolete or worn parts, components, and controls with new or modern ones. A retrofit may be desired to reduce operating costs through improved energy efficiency, to reduce maintenance costs through elimination of obsolete or non-standard parts, or to enable more sophisticated control strategies with improved monitoring, alarming, communication, and remote control of setpoints. In addition, a retrofit may be desired because new building automation system mechanical equipment or an Ethernet network is being installed in a building and the new human-computer interface is incompatible with the old system.

A retrofit can be complex because of mechanical equipment location and operation, installation and wiring issues, problems with the existing control system, and required occupancy during retrofit. Subcontractors are used for many retrofit tasks, so efficient project management, planning, and coordination are required for a successful installation.

Retrofits are often managed by personnel having extensive mechanical and control systems knowledge.

Retrofit Stages

A retrofit can be broken down into distinct stages. The stages are the building survey, documentation, and planning; existing control removal; new building automation system installation; programming and commissioning; human-computer interface setup; and ongoing support.

Building Survey, Documentation, and Planning.
A *building survey* is an inventory of the energy-consuming equipment in a commercial building. A building survey describes the general building layout, HVAC mechanical equipment, energy issues, and problems with the existing control system. The surveyor records the location of all mechanical equipment, location of control panels and areas served, equipment manufacturers, model numbers, voltages, and current ratings. The surveyor records leaks, dirt, loose or missing access doors, and other circumstances that indicate the condition of the mechanical equipment. This survey information is used in the documentation and planning process. The condition of the existing HVAC mechanical equipment must be determined. See Figure 22-1. Any equipment that is not working properly must be repaired or upgraded during the retrofit.

Cleaver-Brooks

Figure 22-1. The condition of existing HVAC equipment is determined during a building survey.

The building automation system manufacturer or representative determines the wire run design and location, cable routing, and craft timing. As the new building automation system is being installed, the information from the manufacturer and the building survey is developed into accurate, as-built drawings of the new building automation system. Programming documentation indicates the sequences of operation, lists of control points and locations, setpoints, strategies, and interlocks for the building automation system.

A materials list is developed during this retrofit stage. A materials list includes the specific controllers, sensors, power supplies, cabling, and other hardware required for the new building automation system installation. A materials list also includes wiring, pneumatic, and riser diagrams and other drawings required to complete the job.

Existing Control Removal. After the building survey, documentation, and planning stage is complete, the existing control system must be removed. See Figure 22-2. Care should be taken when attempting to reuse any components. Communication wires, air lines, sensors, and transducers are inspected before reuse. Problems such as inaccurate sensors, incompatible sensor ranges, and poor condition adversely affect the new building automation system. Components containing hazardous materials such as mercury must be disposed of properly, following all environmental regulations.

The building owner requirements for occupancy during installation must also be considered. Some building owners require total usability of the building during the installation of the new building automation system. This means that the work may need to take place in the evening and on weekends.

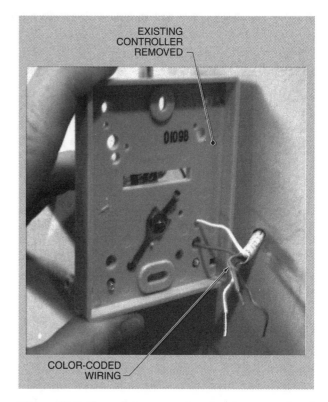

EXISTING CONTROLLER REMOVED

COLOR-CODED WIRING

Figure 22-2. The existing control system is removed and the wires are labeled before beginning installation of a new building automation system.

New Building Automation System Installation. The building automation system manufacturer provides detailed installation instructions that must be followed. In many cases, the new equipment can be installed in the same control panels used by the old system. Existing pneumatic piping and wiring may be directly connected to the new building automation system control equipment. In many cases today, the installation of the components is performed by qualified subcontractors. These subcontractors may include electrical, pipefitters, etc.

Programming and Commissioning. *Programming* is the creation of software. Building automation system programming is the creation of software for the individual controllers and operator interface devices. The programming for a new building automation system must be written and downloaded to the controller. *Downloading* is the process of sending a controller database from a personal computer to a controller. A *database* is the completed programming information of a controller. A *controller* is a device that receives a signal from a sensor, compares it to a setpoint value, and sends an appropriate output signal to a controlled device.

A database may be downloaded to a controller before starting the building automation system installation or it may be downloaded in the field during system installation. A database may be downloaded to a controller by using an operator workstation linked to a network controller. A database may also be downloaded to a controller by using a download interface. See Figure 22-3. Some building automation system manufacturers have tools that enable multiple databases to be downloaded simultaneously over a common network. Normally, the database has been developed before installation.

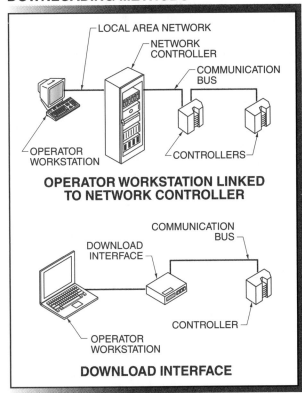

DOWNLOADING METHODS

LOCAL AREA NETWORK

NETWORK CONTROLLER

COMMUNICATION BUS

OPERATOR WORKSTATION

CONTROLLERS

OPERATOR WORKSTATION LINKED TO NETWORK CONTROLLER

COMMUNICATION BUS

DOWNLOAD INTERFACE

OPERATOR WORKSTATION

CONTROLLER

DOWNLOAD INTERFACE

Figure 22-3. A database may be downloaded to a controller by using an operator workstation linked to a network controller or by using a download interface.

Commissioning is the checkout procedure used to start up a new building automation system. During the commissioning procedure, inputs and outputs are temporarily placed in override to ensure that valves and dampers open properly. *Warning:* The technician must understand and follow all mechanical equipment operation and safety rules and procedures. Failure to do so may result in dmage to property or loss of life.Setpoints and real values are changed to check heating and cooling response. Portable operator terminals and modem and telephone tie-ins are set up and tested. The database may be modified when field problems are discovered. For example, an adjustment for an actuator spring range may need to be modified to obtain proper control.

Human-Computer Interface Setup. A *human-computer interface (HCI)* is the central computer that enables the building staff to view the operation of the building automation system. The human-computer interface is installed as part of the building automation system. This computer is normally a desktop PC with a large, high-resolution monitor. There may be multiple HCI computers installed in a large building.

Point mapping is the process of adding the individual input and output points to the database of the human-computer interface. Point mapping may be done by setting input and output points as a group or by setting each point individually. Software tools are available that enable many points to be added at the same time, saving time when point mapping.

Ongoing Support. After the building automation system installation is complete, final as-built drawings and other documentation are provided to the customer and training is scheduled. Warranty dates and parameters are set and long-term maintenance agreements may be signed and implemented. Some building automation system manufacturers provide a subscription service that provides automatic software updates.

Rooftop Packaged Unit Retrofit Application

A rooftop packaged unit combines heating and cooling operations in one piece of equipment. Most commercial buildings use rooftop packaged units for heating and cooling applications. Rooftop packaged units are easy to install and versatile, and require little floor space in the building. The existing controller is normally an electromechanical thermostat. A rooftop packaged unit may have multiple stages of heating and cooling. The control system of a common rooftop packaged unit, such as a 10-ton, two-stage cooling, single-stage gas heat unit may be retrofitted to a building automation system. See Figure 22-4.

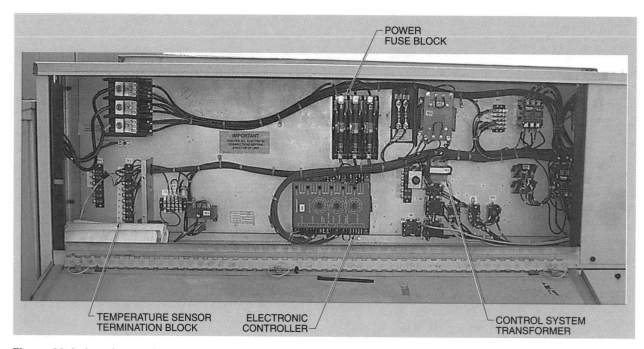

Figure 22-4. A rooftop packaged unit combines heating and cooling operations in one piece of equipment.

Materials List. In a rooftop packaged unit retrofit application, the specific controllers and sensors needed for a rooftop packaged unit retrofit are determined at the documentation and planning stage. The equipment selected depends on the required sophistication and number of input/output points needed. The materials list for this retrofit includes an application-specific controller; a zone temperature sensor package with override; a power supply package; the rooftop unit wiring diagram; a riser diagram for controller addressing; cabling for the room sensor, power, and communication bus; and required hand tools.

The controller electronics are installed after the base plate is wired and checked out. This reduces the chance of the electronics being damaged by improper power levels.

Existing Control Removal and New Building Automation System Installation. All manufacturer installation recommendations must be followed when retrofitting a rooftop packaged unit. Lock out and tag out the rooftop unit and any other hazards before beginning. Remove the existing thermostat from the wall and disconnect and label all wiring terminals. The installation of the new building automation system equipment is performed by applying the following procedure:

1. Install the new controller at the location of the old thermostat or inside the control section of the rooftop packaged unit. Install the zone temperature sensor package at the location of the previous thermostat if the controller is mounted inside the rooftop unit. Connect the zone temperature sensor package to the controller with a prefabricated cable.

2. Install the wiring to the new controller baseplate. See Figure 22-5.

3. Ensure the power supply voltage to the controller is at the recommended value.

4. Connect the power supply to the controller.

5. Terminate the wires for the network communication bus.

6. Address the controller by setting the dual in-line package (DIP) switches.

7. Double-check all wiring.

8. Apply power to the controller and rooftop unit.

Figure 22-5. In a rooftop packaged unit retrofit, the controller digital (binary) output is wired to the rooftop unit heating stage.

Programming, Commissioning, and Human-Computer Interface Setup. The new building automation system controller is programmed, the system commissioned, and the human-computer interface set up by applying the following procedure:

1. Determine that the controller is functioning properly by connecting the laptop or other service tool to the controller, downloading the database (if not done previously), opening the commissioning tool on the laptop or service tool, entering into the commissioning mode, and overriding all outputs to ensure that the fan, cooling stage(s), and heating stage(s) are functional.

2. Override all inputs to ensure that the heating or cooling equipment energizes at the proper temperature levels. If necessary, edit functions such as interstage delays and sensitivity to ensure smooth equipment operation. See Figure 22-6.

3. Add new controller information to the database at the human-computer interface.

4. Determine which input and output points are needed by the operator.

5. Set up alarms, trends, and graphics as needed. If the customer desires, set up a portable terminal for local use.

Variable Air Volume Air Handling Unit Retrofit Application

A *variable air volume air handling unit* is an air handling unit that moves a variable volume of air. Variable air volume air handling units were developed to provide comfort while saving energy. Many variable air volume air handling units originally contained pneumatic control systems. A common variable air volume air handling unit retrofit application is retrofitting an existing pneumatic control system to a building automation system. See Figure 22-7.

Materials List. In a variable air volume air handling unit retrofit application, the specific controllers, transducers, and sensors needed for the retrofit are determined at the documentation and planning stage. The equipment selected depends on the required sophistication and number of input/output points required. The materials list includes a universal input-output controller; power supply package; four electronic/pneumatic transducers—one each for the heating valve, cooling valve, outside air dampers, and fan volume control; a static pressure sensor; two 24 VAC to 120 VAC interface relays—one each for the supply fan and the return fan; a unit wiring diagram; a unit pneumatic diagram; a riser diagram for controller addressing; cabling for room sensor, power, and communication bus; and required hand tools.

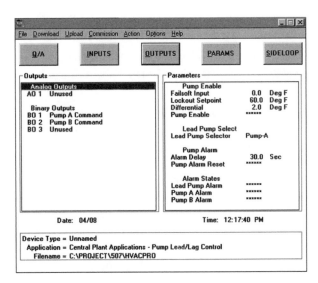

Figure 22-6. Commissioning uses software to override inputs, outputs, and setpoints to ensure proper operation.

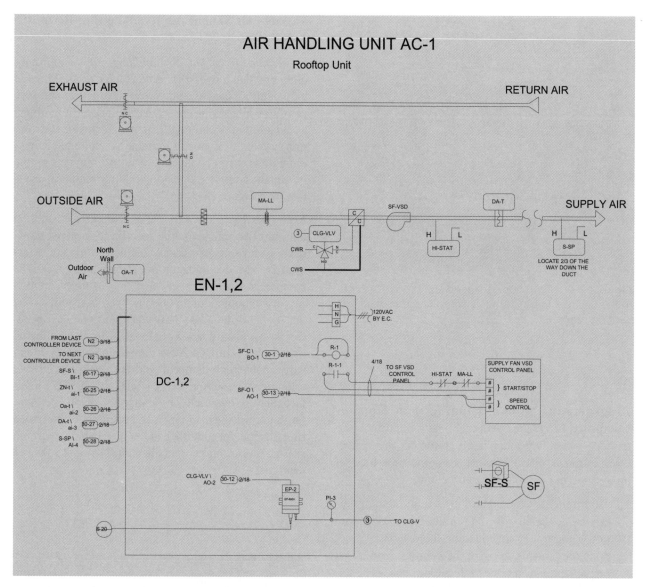

Figure 22-7. Variable air volume air handling unit diagrams are used when retrofitting an existing pneumatic control system to a building automation system.

Existing Control Removal and New Building Automation System Installation. All manufacturer installation recommendations must be followed. Lock out and tag out the air handling unit and any other hazards before beginning. Remove the pneumatic controls, such as receiver controllers, switching relays, signal selection relays, and pneumatic sensors. Carefully label main air and all branch line destinations. The installation of the new building automation system equipment is performed by applying the following procedure:

1. Install the universal input-output controller inside the old pneumatic panel. Install a new panel if required.

2. Ensure proper pneumatic pressure, then install and pipe the electronic/pneumatic transducers. See Figure 22-8.

3. Install and wire the sensors.

4. Install the wiring to the new controller baseplate.

5. Ensure the power supply voltage to the controller is at the recommended value.

6. Connect power for the controller.

7. Terminate the wires for the communication bus.

8. Address the controller by setting the DIP switches.

9. Double-check all wiring.

10. Apply power to the controller and air handling unit.

Figure 22-8. An electronic/pneumatic transducer is used to connect an electronic control to a pneumatic actuator.

Programming, Commissioning, and Human-Computer Interface Setup. The new building automation system controller is programmed, the system commissioned, and the human-computer interface set up by applying the following procedure:

1. Determine that the controller is functioning properly by connecting the laptop or other service tool to the controller, downloading the database (if not done previously), opening the commissioning tool on the laptop or service tool, entering into commissioning mode, and overriding all outputs to ensure that the fan, damper, cooling valve, heating valve, and fan volume control are functional.

2. Override all inputs to ensure that heating or cooling energize at proper temperature levels. Be careful to override the fan volume control slowly as damage to the unit may result from sudden changes. If necessary, edit functions such as sensitivity to ensure smooth equipment operation. Adjust transducers to obtain proper pneumatic spring range for the actuators. See Figure 22-9.

3. Tune the control loops for precise control.
4. Add new controller information into the database at the human-computer interface.
5. Determine which input and output points are needed by the operator.
6. Set up alarms, trends, and graphics as needed. If the customer desires, set up a portable terminal for local use.

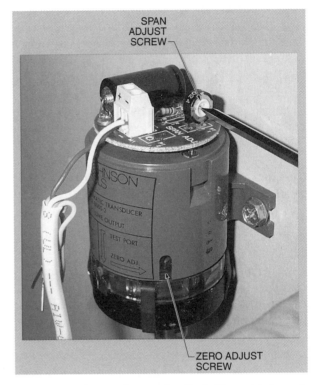

Figure 22-9. A potentiometer is adjusted to match the output of the electronic/pneumatic transducer to the spring range of the pneumatic actuator.

Alternate Retrofit Strategies. Alternate retrofit strategies are available for variable air volume air handling units. Common alternatives include using a variable frequency drive (VFD) in place of vane volume control, resetting the fan static setpoint based on zone demand, and controlling the fan volume in addition to static pressure.

A variable frequency drive is more commonly used today in place of inlet vanes. Variable frequency drives provide greater energy savings and more motor protection, and automatically ramp up the fan over time to reduce starting stress on the motor. The variable frequency drive may be either wired as an analog output from the building automation system or commanded directly from the building automation system if it is compatible with the system manufacturer.

Instead of operating the air handling unit based on a static pressure that does not change, a reset schedule may be used to change (reset) the setpoint as the zones demand more or less cooling. As the zones get warmer, they demand more air, which may be provided by increasing the duct static pressure. This is typically done in software.

Instead of controlling the fan based on static pressure, a velocity pressure sensor and flow sensor grid may be used to indicate air flow volume. The conversion from velocity pressure to volume is done in software. The advantages are that air volume is a much more realistic measurement of outside air for ventilation and total air flow. If a flow sensor grid is also mounted in the return duct, the difference between supply air volume and return air volume may be controlled, and the difference is then reflected as the building pressure.

Variable Air Volume Terminal Box Controller Retrofit Application

A *variable air volume terminal box controller* is a controller that modulates the damper inside a VAV terminal box to maintain a specific building space temperature. Each building space has a separate VAV terminal box controller. Normally, the existing controller in each zone is a pneumatic room thermostat. In addition, there may be a reset volume controller which controls the volume of air introduced into the building space. Pneumatic reheat valves may be controlled with a reversing relay. A common retrofit application is retrofitting the controls in each zone to a building automation system with the controller mounted on the VAV terminal box. See Figure 22-10.

Figure 22-10. A VAV terminal box controller modulates the damper inside a terminal box to maintain a specific building space temperature.

Materials List. In a VAV terminal box controller retrofit application, the specific controllers, actuators, and sensors needed for this retrofit are determined at the documentation and planning stage. The equipment selected depends on the required sophistication and number of input/output points needed. The materials list for this retrofit includes an application-specific controller; a zone temperature sensor package with override; a power supply package; an electromechanical actuator for the VAV terminal box damper (if not included with the controller package); a velocity pressure transmitter (if not included with the controller package); an electronic/pneumatic transducer for the reheat valve; a unit pneumatic diagram; a riser diagram for controller addressing; cabling for room sensor and communication bus; and required hand tools.

Existing Control Removal and New Building Automation System Installation. Always follow all manufacturer installation recommendations. Lock out and tag out any hazards before beginning. Remove the existing pneumatic room thermostat and the volume reset

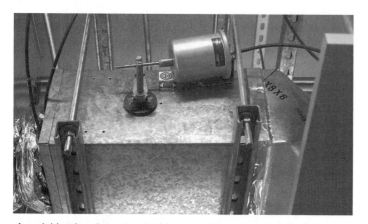

A variable air volume terminal box may have a pneumatic damper actuator and will need a transducer to allow the DDC controller to operate the actuator.

controller. Carefully mark main air and all branch line destinations. The installation of the new building automation system equipment is performed by applying the following procedure:

1. Mount the VAV terminal box controller directly to the terminal box sheet metal.
2. Install the sensor package at the location of the previous thermostat and cable it to the controller using a prefabricated cable.
3. Attach the electric actuator to the variable air volume damper shaft.
4. Attach the differential pressure transmitter to the crossflow tubing at the VAV terminal box. See Figure 22-11.
5. Install and pipe the electronic/pneumatic transducer to the reheat valve.
6. Install the wiring to the new controller baseplate.
7. Ensure the power supply voltage to the controller is at the recommended value.
8. Install and wire the controller power supply.
9. Terminate the wires for the communication bus.
10. Address the controller by setting the DIP switches.
11. Double-check all wiring.
12. Apply power to the controller.

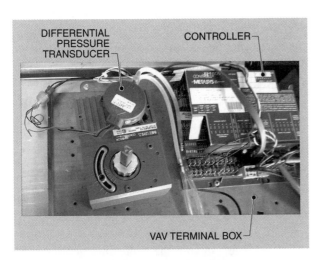

Figure 22-11. In a VAV terminal box retrofit application, a differential pressure transmitter is piped to a flow sensor in the air stream and wired back to the controller.

Programming, Commissioning, and Human-Computer Interface Setup. The new building automation system controller is programmed, the system commissioned, and the human-computer interface set up by applying the following procedure:

1. Determine that the controller is functioning properly by connecting the laptop or other service tool to the controller, downloading the database (if not done previously), opening the commissioning tool on the laptop or service tool, entering into commissioning mode, and overriding all outputs to ensure that the damper and heating valve are functional.
2. Override all inputs to ensure that heating equipment or the damper energize at the proper temperature levels. If necessary, edit functions such as sensitivity to ensure smooth equipment operation. Adjust the transducer to obtain the proper pneumatic spring range for the reheat valve actuator.
3. Air balance the VAV terminal box by connecting a flow hood to the air outlet and use the service tool to override the terminal box to minimum and maximum airflow values. The settings of the VAV terminal box must be adjusted if the box flow does not match the mechanical schedules.
4. Add new controller information into the database at the human-computer interface.
5. Determine which points are needed by the operator.
6. Set up alarms, trends, and graphics as needed. If the customer desires, set up a portable terminal for their local use upon completing the job.

Alternate Retrofit Strategies. Alternate retrofit strategies are available for VAV terminal boxes. Common retrofit alternatives include keeping the pneumatic actuator and installing a flow sensor to convert the terminal box to pressure-independent configuration. Instead of using an electric actuator, the pneumatic actuator can be kept in place. A transducer is needed for the interface.

A cross tube flow sensor may be obtained from the terminal box manufacturer that can be installed at the supply duct inlet. When the crossflow sensor tube is installed, the controller can then provide the proper volume of air regardless of the inlet supply pressure, allowing true volume control.

Variable air volume system retrofits require precise planning, timing, and coordination by the project manager. Precise planning, timing, and coordination is required because of the number of different VAV terminal box types available, possible previous retrofits, lack of accurate drawings and prints, and the effects a retrofit may have on the comfort of individuals in occupied building areas.

Boiler Control Retrofit Application

Many commercial buildings use steam or hot water boilers to heat the building. These boilers may also be used to provide heat for manufacturing or other processes in the building. A flame safeguard system is burner control equipment that monitors a burner start-up sequence and the main flame during normal operation, and provides an air purge to rid the combustion chamber of unburned fuel during a shutdown. A programmer is a controller that manages the firing cycle. See Figure 22-12. Electromechanical systems are commonly used to activate the flame safeguard system.

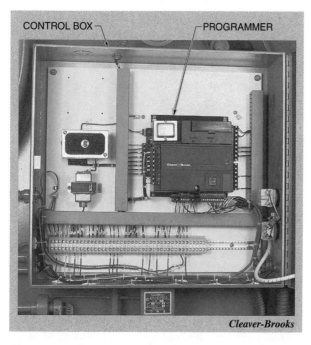

CONTROL BOX ⌐   ⌐ PROGRAMMER

Cleaver-Brooks

Figure 22-12. A flame safeguard system for a boiler includes a programmer that may be controlled by a building automation system network controller.

A boiler control retrofit application normally consists of retrofitting an existing electromechanical boiler control system to a building automation system. The building automation system gives an ON/OFF signal to the boiler programmer when heat is needed. Many hot water boiler systems have a three-way valve which controls the temperature of the water pumped through the building heating coils. The water temperature setpoint may be reset by the lowest or average zone temperatures.

Materials List. In a boiler control retrofit application, the specific controllers, relays, transducers, and sensors needed for the retrofit are determined at the documentation and planning stage. The equipment selected depends on the required sophistication and number of input/output points needed. The materials list for this retrofit includes a universal input-output controller; two temperature sensors—one for hot water supply and one for outside air; a power supply package; a 24 VAC to 120 VAC interface relay for the boiler start circuit; an electronic/pneumatic transducer for the hot water three-way valve; a wiring diagram for the flame safeguard controller; a pneumatic diagram for the three-way hot water valve; a riser diagram for controller addressing; cabling for temperature sensors, power, and communication bus; and required hand tools.

Existing Control Removal and New Building Automation System Installation. Always follow all manufacturer installation recommendations. Lock out and tag out the boiler and any other hazards before beginning. Remove the existing electromechanical controllers from the panel. If the three-way valve is pneumatic, remove the controller and sensor. The installation of the new building automation system equipment is performed by applying the following procedure:

1. Install the new controller inside the old electromechanical panel. Install a new panel if necessary.
2. Install the temperature sensors at the locations of the previous sensors and cable them to the controller.
3. Install and pipe the electronic/pneumatic transducer.
4. Install the wiring to the new controller baseplate.
5. Install and wire the 24 VAC to 120 VAC interface relay.
6. Wire the controller to the appropriate terminals on the flame safeguard controller.
7. Ensure the power supply voltage to the controller is at the recommended value.
8. Run wires for the controller power.
9. Terminate wires for the communication bus.
10. Address the controller by setting the DIP switches.
11. Double-check all wiring.
12. Apply power to the controller and boiler.

Programming, Commissioning, and Human-Computer Interface Setup. The new building automation system controller is programmed, the system commissioned, and the human-computer interface set up by applying the following procedure:

1. Determine that the controller is functioning properly by connecting the laptop or other service tool to the controller, downloading the database (if not done previously), opening the commissioning tool on the laptop or service tool, entering into commissioning mode, and overriding all outputs to ensure that the boiler and heating valve are functional.

2. Override all inputs to ensure that heating equipment energizes at the proper temperature. If necessary, edit functions such as sensitivity to ensure smooth equipment operation. Adjust the transducer to obtain the proper pneumatic spring range for the three-way hot water valve.

3. Add the new controller information into the database at the human-computer interface.

4. Determine which points are needed by the operator.

5. Set up alarms, trends, and graphics as needed. If the customer desires, set up a portable terminal for their local use upon completing the job.

Alternate Retrofit Strategies. An alternate retrofit strategy for a boiler control application is to install a flame safeguard controller that is compatible with the building automation system communication bus. Commands can then be issued directly to the flame safeguard controller.

Chiller Control Retrofit Application

Commercial buildings commonly have large water chillers as part of a rooftop unit. The chillers are used to cool the building. The chiller may be reciprocating, screw, scroll, or centrifugal. A common chiller control retrofit application is the retrofitting of the pneumatic control system of an old centrifugal chiller with a building automation system.

Materials List. In a chiller control retrofit application, the specific controllers, transducers, sensors, and relays needed are determined at the documentation and planning stage. The equipment selected depends on the required sophistication and number of input/output points needed. The materials list for this retrofit includes a universal input-output controller; three temperature sensors—one each for chilled water supply, chilled water return, and outside air temperature; a power supply package (if no power supply exists); a 24 VAC to 120 VAC interface relay for the chiller; an electronic/pneumatic transducer for the chiller inlet vanes; a chiller wiring diagram; a unit pneumatic diagram; a riser diagram for controller addressing; cabling for sensors, power, and interface relay; and required hand tools.

Existing Control Removal and New Building Automation System Installation. Always follow all manufacturer installation recommendations. Lock out and tag out the chiller and any other hazards before beginning. Remove the existing pneumatic or electromechanical controls. Installation of the new building automation system equipment is performed by applying the following procedure:

1. Install the new controller inside the old pneumatic panel. Install a new panel if necessary.

2. Install the sensors in wells inside the water lines. Install and pipe the electronic/pneumatic transducer for the chiller inlet vanes.

3. Install the wiring to the new controller baseplate.

4. Ensure the power supply voltage to the controller is at the recommended value.

5. Run wires for the controller power.

6. Install and wire the 24 VAC to 120 VAC interface relay.

7. Terminate the wires for the network communication bus. See Figure 22-13.

8. Address the controller by setting the DIP switches.

9. Double-check all wiring.

10. Apply power to the controller and chiller.

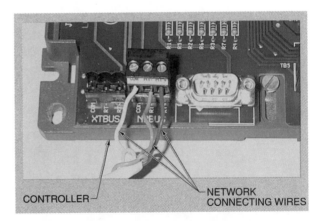

CONTROLLER — NETWORK CONNECTING WIRES

Figure 22-13. In a chiller control retrofit, the controller is wired to the network communication bus.

Programming, Commissioning, and Human-Computer Interface Setup. The new building automation system controller is programmed, the system commissioned, and the human-computer interface set up by applying the following procedure:

1. Determine that the controller is functioning properly by connecting the laptop or other service tool to the controller, downloading the database (if not done previously), opening the commissioning tool on the laptop or service tool, entering into commissioning mode, and overriding all outputs to ensure that the chiller starts and the inlet vanes operate properly.

2. Override all inputs to ensure that the chiller starts at the proper temperature. If necessary, edit functions such as sensitivity to ensure smooth equipment operation. Adjust transducers to obtain proper pneumatic spring range for the actuators.

3. Add the new controller information into the database at the human-computer interface.

4. Determine which points are needed by the operator.

5. Set up alarms, trends, and graphics as needed. If the customer desires, set up a portable terminal for their local use upon completing the job.

Alternate Retrofit Strategies. Most new chillers have a built-in microprocessor control panel. This panel controls the setpoints, faults, and operation of the chiller. The building automation system normally issues a simple start/stop command to the chiller. A common alternate retrofit strategy for a chiller control system is to install a translator panel to allow the building automation system to view and command points inside the chiller microprocessor control panel.

Special Retrofit Situations

Many other possible retrofit applications exist for HVAC systems. Common retrofit applications include the retrofitting of an old building automation system to a new system, and retrofitting of new mechanical equipment with microprocessor control panels to a building automation system.

Retrofit of Old Building Automation Systems. During a retrofit, it may be desired to integrate a new building automation system with an older, existing system. It may not be cost-effective to remove and replace the entire old system. While each manufacturer may do this differently, it is common to use a computer or software bridge that allows the new building automation system to view and adjust the old building automation system. In some cases, this can be done with equipment from different manufacturers.

Retrofit of New Mechanical Equipment. Many building automation system retrofits involve retrofitting the mechanical equipment at the same time. New mechanical equipment normally has its own microprocessor control panel. See Figure 22-14. A translator panel or special software bridge may enable points inside the microprocessor control panel to be viewed, alarmed, or even controlled by the building automation system. This may be customized for each application.

Carrier Corporation

Figure 22-14. Many building automation system retrofits involve retrofitting mechanical equipment which has its own microprocessor control panel.

Building system management functions provide information for efficient management of building HVAC systems. Alarms alert maintenance personnel to improper equipment operating conditions. Data trending is used to record information at specific times and intervals. Additional software packages can be used with building automation systems to provide other functions such as job costing and inventory control.

BUILDING SYSTEM MANAGEMENT

Building automation systems provide control, monitoring, and evaluation of building mechanical systems. These systems allow maintenance technicians to manage building systems using information provided by the building automation system.

Building system management functions provide information to maintenance technicians who maintain or manage a commercial building. Building system management functions provide a centralized location for controlling mechanical systems. Alarm monitoring, equipment repair, complaint response, inventory control, and proper preventive maintenance procedures can be efficiently performed when building system management functions are used properly. Building system management functions include alarm monitoring, data trending, generating preventive maintenance work orders, building system documentation, and building system management software.

> ⓘ *Building system management programs can be implemented during the installation of a building automation system because the system provides equipment performance and maintenance condition data.*

Alarm Monitoring

Alarm monitoring is the most common building system management function used in commercial buildings. Alarms are used to alert the maintenance technician of improper building conditions. See Figure 23-1. The most common alarms used in commercial buildings are associated with temperature sensors. Alarms can also indicate improper levels of humidity and pressure, or flow failure of a fan or pump. Alarms quickly alert maintenance personnel to failures and/or problems.

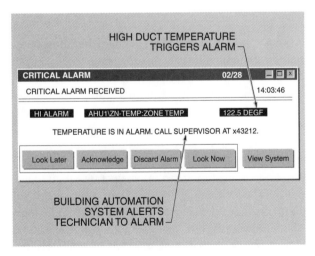

Figure 23-1. Alarms alert building maintenance personnel to improper HVAC system operation.

A common alarm point for equipment operation is a low temperature limit element installed in ductwork. A low temperature present at any point along the element causes the low-limit thermostat to stop the supply fan and send an alarm signal to the building automation system.

Alarms can be set up to monitor most inputs and outputs of a building automation system. However, a maintenance technician may be overloaded with information if numerous inputs and outputs are monitored. For example, a dangerous condition may result if a critical alarm is overlooked in the midst of a large number of noncritical alarms. Alarms are classified to ensure critical alarms are addressed promptly by maintenance personnel.

Alarm Classification. Most modern building automation systems allow the maintenance technician to classify alarms into different categories. The most common alarm classification categories are critical alarms and noncritical alarms. Critical alarms concern devices vital to proper operation of a commercial building. Critical alarms are reported immediately to multiple maintenance personnel for quick response. Many building automation systems can operate a digital output such as a rotating light or horn when a critical alarm is triggered. Noncritical alarms concern elements of building operation that are not vital. Noncritical alarms may or may not be reported immediately because quick response is not vital to the proper operation of the equipment or building.

Overalarming occurs when an excessive number of points are set up to alarm. Overalarming causes frequent alarms that are eventually ignored. In this event, critical situations may not be handled properly. Underalarming occurs when few points are set up to alarm, raising the possibility of critical equipment failure without an alarm being signaled.

Alarm values entered in the building automation system software represent the variable (temperature, humidity, pressure, flow) desired condition. Alarm differentials should be used to prevent an alarm from quickly changing state between alarm and no-alarm status. An *alarm differential* is the amount of change required in a variable for the alarm to return to normal after it has been in alarm status. See Figure 23-2. For example, a differential of 2°F to 4°F is normally used for temperature sensors. If a temperature sensor is set to alarm at 80°F, a 2°F differential will not allow the alarm to change to no-alarm (normal) status until the temperature value drops to 78°F or below. (80°F − 2°F = 78°F).

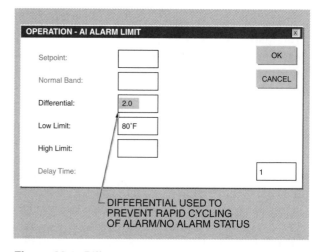

Figure 23-2. Differentials are set to prevent alarms from rapid cycling due to small changes in the variable monitored.

Most building automation system software also has an alarm time delay feature. The time delay gives an HVAC unit time to operate before a change in alarm status. In addition to alarm setpoints and time delays being set up properly, the correct location that an alarm should be reported to must also be set. Most building automation systems allow different categories of alarms to be reported through different operator interface methods. See Figure 23-3.

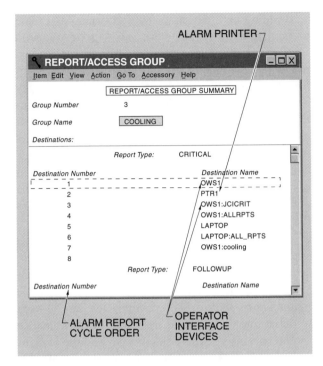

Figure 23-3. Alarms must be routed to the correct destination to ensure proper response.

Alarm Reporting. The most common operator interface device that receives alarm notification is a printer. Printers provide a permanent document of the alarm status. Critical alarms can be quickly transmitted to an on-site personal computer and dialed out to an off-site personal computer. Critical alarms are also commonly transmitted to maintenance personnel pagers.

Data Trending

Data trending is the use of past building equipment performance data to determine future system needs. Data trending is used to record temperature, humidity, pressure, and other equipment or building condition variables. In data trending, values and conditions of a commercial building are recorded during a specific time interval. The time interval used to record these values is programmable by a maintenance technician. A time interval of 20 min is commonly used for long-term data trending. Long-term data trending is used by a technician to view temperatures and other values that occur overnight or during weekends. Short time intervals such as 1 min or 2 min can indicate improper equipment operation and are used to troubleshoot equipment problems.

Multiple inputs and outputs can be recorded in the same data trend. This capability can be used to show relationships such as outside air temperature versus

heating/cooling equipment operation. In many cases the data trends can be used to fine-tune setpoints and equipment operation. See Figure 23-4.

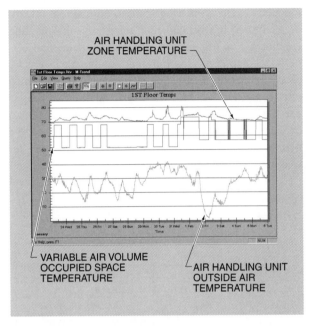

Figure 23-4. Multiple data trends can be plotted simultaneously to show relationships between temperature and equipment operation.

Data trending can also be started and stopped at specific times and dates. For example, complaints that a building space is excessively cold when occupants arrive are regularly received during the morning hours from a specific area of a commercial building. A data trend is created to record the building space temperature, outside air temperature, and equipment ON/OFF status beginning before occupancy and ending at midmorning. The data trend interval time is 5 min. The results of this data trend are used to change the actual start time of a unit or correct mechanical equipment problems. Data trends can also be used to make a decision regarding the purchase or replacement of HVAC units. Data trends can also be imported into a spreadsheet program and used to create graphs and charts.

> *Data trending provides a wealth of valuable data when set up and interpreted properly. Data trending information can be used to track equipment energy use, efficiency per unit of energy, and the need for equipment maintenance.*

Preventive Maintenance

Modern building automation systems have the ability to generate preventive maintenance work orders for maintenance technicians. *Preventive maintenance* is scheduled inspection and work (lubrication, adjustment, cleaning) required to maintain equipment in peak operating condition. See Figure 23-5. Preventive maintenance is performed routinely on building equipment to ensure proper operation. Corrective maintenance is performed after equipment has failed. Many commercial buildings use the corrective maintenance approach rather than a proper preventive maintenance program. Preventive maintenance is usually less expensive than the corrective maintenance approach.

EQUIPMENT LOCATION

PREVENTIVE MAINTENANCE WORK ORDER		
Work Order No: **46**	Requisition No:	
Issue Date: **6\21** Time: **15:49**	Skill: **Mechanical**	
Equipment Name: **Air Unit #3**	EQUIPMENT LOCATION	
Manufacturer: **Trane**	Building: **3E**	Floor: **3rd**
Description: **Air Handling Unit, 4 HP Century Motor**	System:	
Model No: **V4517** Serial No: **100-AHU01**	Date Work Performed:	

PREVENTIVE MAINTENANCE CHECKLIST		
MECHANICAL	COMMENTS	
Lockout/Tagout power supply		☐
Inspect motor rotation (bearings)		☐
Inspect fan rotation (bearings)		☐
Inspect fan blades		☐
Inspect V-belts		☐
Inspect motor and fan sheaves		☐
Inspect dampers		☐
Inspect damper actuators		☐
Replace unit filters		☐
Vacuum debris from unit		☐

LUBRICATION	COMMENTS	
Grease motor bearings		☐
Grease fan bearings		☐
Oil damper pivots		☐
Oil damper pneumatics		☐

WORK DESCRIPTION

Figure 23-5. Modern building automation systems are capable of generating preventive maintenance work orders after a specified number of equipment run hours.

Building automation system software can create preventive maintenance work orders associated with a specific input or output value. For example, most preventive maintenance activities include bearing lubrication and filter replacement. A work order may be produced to alert maintenance technicians to lubricate the motor bearings and replace the filters of a rooftop HVAC unit. The number of unit run hours is used to trigger the generation of the work order. Other values that can be used to trigger the generation of a preventive maintenance work order are the number of motor starts or the pressure drop across the filters.

Work orders may be sent to a specific technician and printed for future reference. In addition, the message can indicate the type and number of filters, lubricant, and tools needed to effectively complete the maintenance activity. This information can be used to determine the required inventory of maintenance products. This information can also be used to indicate the preventive maintenance workload of each maintenance technician.

> *Building automation systems have the ability to provide information to schedule preventive maintenance tasks. To effectively execute preventive maintenance tasks, adequate budgeting and funding is required. This includes adequate staff to perform the maintenance tasks and stocking sufficient quantities of consumables such as lubricants, filters, and spare parts.*

Building System Documentation

All building automation systems are capable of producing documentation of the information provided to maintenance technicians. The documentation can be used to indicate system operation, control, and reliability. For example, system operation documentation, if allowed by an existing customer, can be used to show a potential customer the efficient operation of an existing installation. This information can also be used in installations that require documentation for government agencies.

Manufacturers of products for critical industries such as jet engines, electronics, or pharmaceuticals normally require documentation to enable the tracing of the manufacturing conditions associated with each part produced.

The documentation function is also commonly used to record warranty conditions. The equipment serial number and other information such as equipment purchase, installation, and startup dates can be included in the documentation. This information is valuable for future warranty claims and preventive maintenance activities. See Figure 23-6.

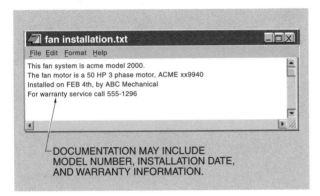

Figure 23-6. Building automation system preventive maintenance documentation may include model numbers, installation, and warranty information.

Building System Management Software

Many building automation systems use building system management software which visually communicates building and equipment conditions to maintenance technicians. Building or equipment values are provided using illustrations and icons instead of in a tabular or textual format. See Figure 23-7. Graphic representation of information is often easier to understand. Some systems use photographs of a particular piece of equipment. The actual temperature, humidity, pressure, or equipment status values are superimposed on the photograph.

Many building system management software programs use graphics created in standard packages such as Windows® Paint®. Many manufacturers provide a library of existing graphics. A maintenance technician can also draw his or her own graphics.

> ℹ *Some building system management software programs allow the importing of digital photos and engineering drawings. In addition, many HVAC manufacturers provide information and instructions using Adobe® Acrobat® files. This allows the same graphic to show both equipment operation and preventive maintenance procedures.*

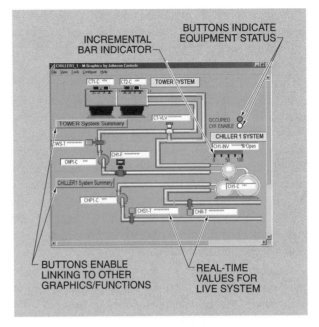

Figure 23-7. Building system management software programs are used to visually represent conditions of HVAC equipment in a commercial building.

Pump preventive maintenance requirements may be indicated on maintenance management software programs. Information provided may include required procedures, tools, and consumables.

A third-party computerized maintenance management system (CMMS) that integrates information from building automation systems is also available. Some CMMSs are used to manage inventory levels for preventive maintenance products such as filters and lubricant. For example, a CMMS can be used to maintain minimum levels of products and alert a maintenance technician when supplies must be reordered. The system can also provide a yearly budget for preventive maintenance products.

Some CMMSs allow preventive maintenance hours to be efficiently scheduled throughout a year. For example, extra preventive maintenance can be scheduled during low cooling or heating demand periods and less preventive maintenance can be scheduled during high cooling or heating demand periods. This allows additional time for maintenance technicians to handle increased heating and cooling concerns. Another feature available with CMMSs is the ability to add in hourly labor cost factors and determine the preventive maintenance dollar value performed by each technician.

Predictive maintenance software programs are also available. *Predictive maintenance* is the monitoring of wear conditions and equipment characteristics and comparing them to a predetermined tolerance to predict possible malfunctions or failures. Predictive maintenance attempts to detect equipment problems before failure occurs by use of vibration and other sensors.

Any software program should be researched thoroughly before purchase and implementation. The building automation system manufacturer may be consulted for recommendations.

Third-party computer maintenance management systems (CMMSs) should be checked to determine if they are compatible with existing operator workstation PCs, if they will require upgrading when the building automation system software is upgraded (typically once or twice a year), if there is a limit to the amount of data that can be entered into the software, and if data can be exchanged between the building automation system software and CMMS.

A commercial electric utility bill includes electrical consumption, electrical demand, and power factor/fuel recovery charges. Time of day rates and ratchet clauses alter the customer electricity cost. Utility bills may be analyzed to calculate payback for energy savings measures. Surveys are performed to list energy-using equipment and building information.

UTILITY RATES

Most building automation systems are implemented to reduce energy expenditures. The potential energy savings must be estimated accurately. An experienced energy engineer should be consulted when performing a detailed energy audit. Utility rate structures are used to estimate the potential cost savings of a building automation system. Most commercial electric utility bills are broken down into electrical consumption, electrical demand, and power factor/fuel recovery. See Figure 24-1.

Electrical Consumption

Electrical consumption (usage) is the total amount of electricity used during a billing period. Electrical consumption is measured in kilowatt-hours (kWh). A commercial building commonly uses more electricity during some parts of the month than others. At night, when most lighting and HVAC units are turned OFF, the amount of electricity used is lower than during the day. Electrical consumption measures the total amount of electricity used during the usual monthly billing period. See Figure 24-2. Residential electric bills also include electrical consumption. The difference between a residential electric bill and a commercial or industrial electric bill is that a residential bill does not normally include a charge for electrical demand.

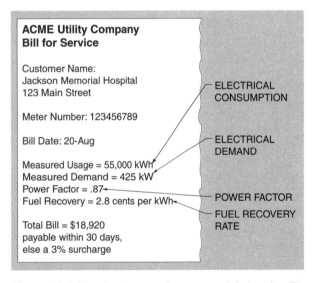

ACME Utility Company
Bill for Service

Customer Name:
Jackson Memorial Hospital
123 Main Street

Meter Number: 123456789

Bill Date: 20-Aug

Measured Usage = 55,000 kWh ── ELECTRICAL CONSUMPTION
Measured Demand = 425 kW ── ELECTRICAL DEMAND
Power Factor = .87 ── POWER FACTOR
Fuel Recovery = 2.8 cents per kWh ── FUEL RECOVERY RATE

Total Bill = $18,920
payable within 30 days,
else a 3% surcharge

Figure 24-1. The three parts of a commercial electric utility bill are electrical consumption, electrical demand, and power factor/fuel recovery rates.

When analyzing utility expenditures, enlist the help of utility company representatives. Utility representatives often can help diagnose bills, offer tips, and provide statistical information. Time should be taken to utilize these resources when undertaking a utility analysis.

ELECTRICAL CONSUMPTION

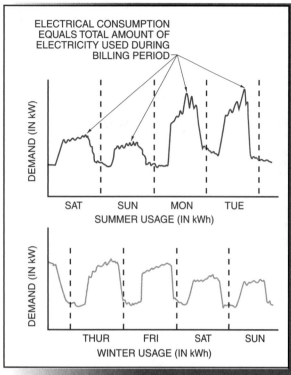

Figure 24-2. Electrical consumption (usage) is the amount of electricity used during a billing period.

Electrical Demand

Electrical demand is the highest amount of electricity used during a specific period of time. Electrical demand is measured in kilowatts (kW). The time interval used to determine electrical demand is commonly 15 min or 30 min. See Figure 24-3. *Instantaneous demand* is electrical demand calculated on a moment-by-moment basis. *Integrated demand* is electrical demand calculated as an average over the time interval.

An electric utility constantly meters the demand value and charges a facility for the highest measured time interval per monthly billing period. Methods used to determine the beginning and ending of the 15 min or 30 min interval include fixed interval and sliding window.

Fixed Interval. The fixed interval method uses a fixed time period to determine electrical demand. For example, electrical demand may be based on time periods from 10:00 AM to 10:15 AM or 10:15 AM to 10:30 AM. The fixed interval method uses definite start and stop times that do not change. Some utilities send a pulse across the incoming power service lines which signals the beginning of each new period.

Sliding Window. The sliding window method computes electrical demand on a minute-by-minute basis. Any 15 min or 30 min interval, regardless of when it occurs, can be a new peak demand period for the month. For example, a sliding window can be used from 10:01 AM to 10:16 AM or 10:02 AM to 10:17 AM. The value is reset each month when the meter is reset.

ELECTRICAL DEMAND

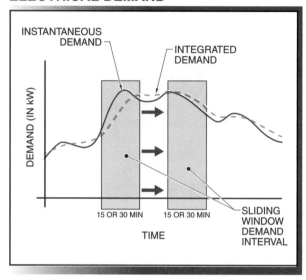

Figure 24-3. Electrical demand is the highest amount of electricity used during a specific period of time.

Power Factor and Fuel Recovery Rates

Power factor (PF) is the ratio of true power used in an AC circuit to apparent power delivered to the circuit. Power factor is a measure of electrical efficiency and is commonly expressed as a percentage. True power equals apparent power only when the power factor is 100%. When the power factor is less than 100%, a circuit is less than 100% efficient and has an increased operating cost because not all power is performing work. Some incoming power is used to perform work, while the rest is used to produce electromagnetic fields. The power used to produce electromagnetic fields is not directly measured or charged by a utility. Special meters are used to indirectly measure the power factor. Penalties may be charged based on poor power factors, which occur in large, energy-intensive buildings having numerous 1φ AC motors and fluorescent lamps. See Figure 24-4.

Fuel recovery rate is the amount of money a utility is permitted to charge to reflect the constantly changing cost of energy. Fuel recovery rate charges are often added to the utility bill for a commercial building. Fuel recovery rate charges are adjusted in some cases on a month-to-month basis. The fuel recovery rate is often billed by multiplying the fuel recovery charge by electrical consumption. In addition to these charges, state or local taxes may be added.

DETERMINING POWER FACTOR CHARGES

DEMAND CHARGE ADJUSTMENT–DEMAND CHARGE INCREASED OR DECREASED BY FACTORS BASED ON POWER FACTOR.

EXAMPLE:
1% INCREASE FOR EACH 1% BELOW 80% POWER FACTOR.

1% DECREASE FOR EACH 1% ABOVE 90%.

NO ADJUSTMENT IF POWER FACTOR BETWEEN 80% AND 90%.

POWER FACTOR	DEMAND MULTIPLIER
70%	1.15
75%	1.10
80%	1.00
90%	1.00
95%	.95
100%	.80

Figure 24-4. Penalties (demand multipliers) may be charged to large facilities that have poor power factors.

Tiers and Negotiated Rates

A *tier* is the classification of a customer depending on the type of service and the amount of electricity used. Tiers are based on the fact that the more electricity a customer uses, the lower the rate the customer pays per unit of electricity. For example, a small commercial customer may pay $.10 per kWh and a large industrial customer may pay $.015 per kWh. A variation on this is to charge graduated rates depending on the amount of electricity used. For example, the first 50 kW of demand costs $10 per kW and from 50 kW to 100 kW costs $8 per kW.

Tiers may also be established depending on customer and service classification. An automotive plant may be classified as an industrial customer based on their usage of 1,000,000 kWh per month. As an industrial customer, they pay a lower rate for every kilowatt-hour than the amount charged a customer in a different tier. In general, most electric utility companies categorize customers into small commercial, large (general) commercial, and industrial customers. Small commercial customers pay the most per unit power.

Large (general) commercial customers pay less per unit than small commercial customers. Industrial customers that use very large amounts of power pay the smallest amount per unit of electricity. See Figure 24-5.

ELECTRIC SERVICE

INDUSTRIAL SERVICE

APPLICABLE TO ANY CONSUMER HAVING A DEMAND OF LESS THAN 10,000 kW AND USING MORE THAN 500,000 kWh PER MONTH DURING THE CURRENT MONTH OR ANY OF THE PRECEDING 11 MONTHS. NO RESALE OR REDISTRIBUTION OF ELECTRICITY TO OTHER USERS WILL BE PERMITTED UNDER THIS SCHEDULE.

MONTHLY RATES

KILOWATT DEMAND CHARGE (IN DOLLARS PER kW)		
	SUMMER	WINTER
FOR FIRST 50 kW	$9.62	$8.62
FOR ALL EXCESS OVER 50 kW	$8.34	$7.43

KILOWATT HOUR CHARGE (IN CENTS PER kWh)		
	SUMMER	WINTER
FOR FIRST 40,000 kWh	4.58	4.15
FOR NEXT 60,000 kWh	3.27	2.95
FOR NEXT 200 kWh PER kWd BUT NOT LESS THAN 400,000 kWh	2.86	2.66
FOR NEXT 200 kWh PER kWd	2.182	2.043
FOR ALL EXCESS	.91	.82

Figure 24-5. Electrical customers may be classified in tiers based on the type of service and/or amount of electricity used.

Large industrial customers in an area may decide to negotiate directly with the utility company instead of paying the listed rate. Negotiated rates are often a fraction of the standard industrial customer rate. Negotiations are often tied to political and social factors such as economic development zones and tax negotiations. The decision to relocate a firm to a particular area may be tied to the results of these negotiations.

Ratchet Clauses

A *ratchet clause* is documentation that permits an electric utility to charge for demand based on the highest amount of electricity used in a 12-month period, not the amount actually measured. Normally, the demand charge on an electric bill reflects the interval for a single billing period. Some utilities use a ratchet clause to set the demand charge for up to a year to the highest monthly interval. See Figure 24-6. A ratchet clause allows an electric utility to charge a customer if they set a new demand peak. This amount is the amount paid each month for the next year.

RATCHET CLAUSE/ADJUSTMENT

PURPOSE OF RATCHET IS TO PENALIZE AND DISCOURAGE HIGH SUMMER PEAKS.

BILLED DEMAND IS BASED ON HIGHEST MONTHLY PEAK ON A SEASONAL OR YEARLY BASIS OR ACTUAL PEAK, WHICHEVER IS HIGHEST.

TIME PERIOD–SOME UTILITIES APPLY RATCHET ONLY TO SUMMER MONTHS, JUNE–SEPT, OTHERS USE ANY PREVIOUS 11 MONTHS.

DEMAND BILLED–USUALLY 50%, 60%, 75%, OR 80%, OF THE HIGHEST DEMAND IN PREVIOUS 11 MONTHS, OR PREVIOUS SUMMER. SOME UTILITIES MAY USE DIFFERENT PERCENTAGE PER SEASON, SUCH AS 100% OF HIGHEST SUMMER PEAK OR 50% OF HIGHEST WINTER PEAK, WHICHEVER IS HIGHER.

EXAMPLE: WITH 80%–11 MONTH RATCHET

MONTH	ACTUAL kW	BILLED kW
JUNE	313	313
JULY	342	342
AUGUST	343*	343
SEPTEMBER	312	312
OCTOBER	307	307
NOVEMBER	299	299
DECEMBER	258	258
JANUARY	263	**274**
FEBRUARY	264	**274**
MARCH	254	**274**
APRIL	251	**274**
MAY	266	**274**

*HIGHEST PEAK—343 kW (80%) (343) = 274 kW

Figure 24-6. Ratchet clauses discourage high demand charges by charging a minimum amount as a percentage of the highest monthly bill during the previous year.

Summer/Winter Rates

Summer/winter rates may be used because it takes different levels of electrical generating capacity to meet fluctuating seasonal needs. Typically, the amount charged for electricity during the summer is higher than the amount charged for electricity during the winter. To avoid confusion, summer and winter periods are clearly defined. For example, summer rates may be defined as taking effect for 5 months beginning with the May billing month and ending with the October billing month. The winter rates apply in all other months not specified during the summer rate period.

Time-of-Day Rates

In many cities, electrical generating capacity is stretched to the limit on hot, humid, summer days. A time-of-day rate structure may be used to increase or decrease the cost of power depending on the time of day in which it is used. See Figure 24-7. In a typical time-of-day rate structure, the peak (day) rate is defined as 7 AM to 5 PM. The power used in this period is the most expensive. From 5 PM to 11 PM, the cost per unit of power is less than the peak rate. During the period from 11 PM to 7 AM, the cost per unit of power is the lowest because of minimal use. In some areas, the peak rates are so expensive that they entice many customers into using alternate equipment strategies such as gas- or steam-fired chillers and thermal storage systems.

ACME Electric Utility
Time of Day Rates

Summer Demand	Winter Demand
Peak 7 AM to 5 PM $11.50 per kW	Peak 7 AM to 5 PM $10.00 per kW
Mid Peak 5 PM to 11 PM $10.50 per kW	Mid Peak 5 PM to 11 PM $8.00 per kW
Off Peak 11 PM to 7 AM and all weekends $6.50 per kW	Off Peak 11 PM to 7 AM and all weekends $5.00 per kW

Figure 24-7. Time-of-day rates are used to change the amount charged for power during various periods during the day.

Natural Gas Billing and Metering

Natural gas is an energy source commonly used in commercial buildings. *Natural gas* is a colorless, odorless fossil fuel. An odorant is added to natural gas in the distribution process to enable the detection of leaks. The billing for natural gas is calculated as the sum of the customer charge, distribution (delivery) charge, environmental recovery cost, and natural gas cost. See Figure 24-8. The customer charge is a fixed monthly fee charged by the gas company for establishing and maintaining service. The distribution (delivery) charge reflects the cost of transporting the gas through the pipeline. The environmental recovery cost is the cost for environmental remediation (cleanup) for past gas production. The natural gas cost is the cost of the gas itself (per therm).

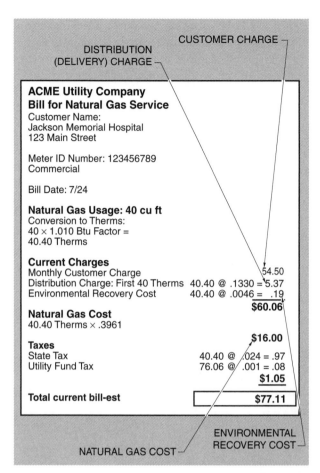

Figure 24-8. Natural gas charge is determined by multiplying the natural gas usage by the natural gas recovery charge.

The unit of measurement for natural gas is based on the cubic foot. Natural gas may be measured in hundred cubic feet (ccf), thousand cubic feet (Mcf), or million cubic feet (MMcf). Natural gas may also be measured in therms. A *therm* is the quantity of gas required to produce 100,000 Btu. A *British thermal unit (Btu)* is the amount of heat energy required to raise the temperature of 1 lb of water 1°F. Natural gas customer charge may be based on the customer tier. Large customers commonly negotiate long-term rates directly with utility companies, similarly to electrical demand. The distribution charge, environmental recovery cost, and natural gas cost are often based on the number of therms.

Energy Utilization Index and Energy Cost Index

Statistical averages may be used after a professional energy engineer analyzes the utility bills of a facility. The *energy utilization index (EUI)* is the amount of heat energy (in Btu) used in a commercial building divided by the number of square feet in the building. The *energy cost index (ECI)* is the amount spent on energy in a commercial building divided by the square feet in the building. The energy utilization index and energy cost index are useful when comparing buildings that have similar uses. For example, a school system may have an ECI of $75/sq ft/yr. A neighboring school system may have an ECI of $65/sq ft/yr. Research can be done to determine why the discrepancies exist for comparison purposes.

Statistical Use of Utility Bills

Utility bills disclose useful information when analyzed properly. Utility bills are also used to justify mechanical system retrofits and expenditures on new control systems. New energy systems can be compared with the cost of operating existing boilers and chillers. For example, a new chiller may have an efficiency of .9 kW per ton and an existing chiller may have an efficiency rating of 1.2 kW per ton. A savings of .3 kW per ton of chiller capacity is realized (1.2 kW per ton – .9 kW per ton = .3 kW per ton). If the existing chiller uses 600 kW of power and the new chiller uses 450 kW, a savings of 150 kW is realized (600 kW – 450 kW = 150 kW). At $10 per kW, $1500 (150 kW × $10 per kW = $1500) is saved per month in demand. The cost of the new chiller is used to determine the number of months needed to recoup the cost of a new chiller. The same basic calculations can be used for control retrofits and other energy saving expenditures. Utility bills can also be used to compare energy-efficient commercial buildings versus non-energy efficient commercial buildings. Regional comparisons may be used to determine the cost of operation in one area versus the cost of operation in another area.

Natural gas is metered at the service entrance to a building. The indicator dials are read from left to right to find natural gas consumption.

Utility Cost Calculation

Utility costs to a facility are normally billed monthly. In general, the utility cost is calculated by multiplying the amount of energy consumed by the cost charged per unit of energy. Electrical billing is usually more complicated than natural gas billing.

Cost Calculation — Electricity. A commercial building monthly total electrical bill is calculated from the electrical consumption (usage), power factor/fuel recovery charge, and electrical demand. See Figure 24-9. The monthly total electrical bill for a commercial building is found by applying the following procedure:

1. Determine total usage charge. Usage charge is determined by multiplying the usage (in kWh) by the electric energy rate for the time of day and/or season. The usage charge may vary based on the amount of electricity used. Total usage charge also includes the power factor/fuel recovery charge. The fuel recovery charge is the amount the utility is permitted to charge to reflect the constantly changing cost of energy. The fuel recovery charge is calculated by multiplying the usage (in kWh) by the fuel recovery rate (in $/kWh). These values are added to obtain the total usage charge.
2. Determine demand. Electrical demand is determined by multiplying the demand (in kW) by the demand rate(s) (in $/kW). Demand rates may vary based on time of year and/or the amount of electricity used.
3. Calculate total electrical power cost. Total electrical power cost is calculated by adding the total usage charge and total demand.

ACME Electric Utility

Large Commercial Rate: This rate applies to customers having a demand of 100 kW or greater during the previous 12 months.

Usage:	Summer	Winter
First 40,000 kWh	9 cents/kWh	8 cents/kWh
Next 60,000 kWh	7 cents/kWh	6 cents/kWh
All Excess	5 cents/kWh	4 cents/kWh

Power Facor/Fuel Recovery Charge .12 cents/kWh

Demand:	Summer	Winter
1st 50 kW	$9.50/kW	$8.50/kW
All Excess	$7.50/kW	$6.50/kW

Summer rates begin in May and end in October.

Figure 24-9. The total monthly electrical bill for a commercial building is based on the electrical consumption (usage), power factor/fuel recovery rate, and electrical demand.

For example, one year's worth of electric utility bills are obtained for a facility in the large commercial rate tier. The first bill is for the month of November. The consumption (usage) is 116,000 kWh. The demand is 395 kW. What is the total electrical power cost for the facility?

1. Determine total usage charge.

first 40,000 kWh: 40,000 × $.08 per kWh = $3200

next 60,000 kWh: 60,000 × $.06 per kWh = $3600

all excess 16,000 kWh: 16,000 × $.04 per kWh = $640

usage charge = $7440

fuel recovery charge = 116,000 kWh × $.0012 per kWh = $139.20

total usage charge = $7579.20

2. Determine demand.

first 50 kW: 50 × $8.50 = $425

all excess: 345 × $6.50 = $2242.50

total demand = $2667.50

3. Determine total electrical power cost.

total electrical power cost = total usage charge + total demand

total electrical power cost = $7579.20 + $2667.50

total electrical power cost = **$10,246.70**

Cost Calculation — Natural Gas. A commercial building total natural gas bill is the sum of the customer charge, distribution (delivery) charge, environmental recovery cost, and natural gas cost. State and local taxes may also be added to the total. The monthly total natural gas bill for a facility is found by applying the following procedure:

1. Convert usage (in cu ft) to therms. Natural gas usage (in cu ft) is converted to therms by multiplying the cubic foot quantity by a Btu factor.
2. Determine distribution (delivery) charge. Distribution charge is found by multiplying the number of therms by the distribution charge per therm.
3. Determine environmental recovery cost. Environmental recovery cost is found by multiplying the number of therms used by the environmental recovery cost per therm.
4. Determine natural gas cost. Natural gas cost is found by multiplying the number of therms used by the natural gas cost per therm.
5. Determine total natural gas bill. The total natural gas bill is found by adding the customer charge, distribution charge, environmental recovery cost, and natural gas cost.

For example, one year's worth of natural gas bills are obtained for a facility. The first bill is for the month of May. The customer charge is $65.45. The distribution charge is $.1330 per therm. The environmental recovery cost is $.0046 per therm. The natural gas usage is 400 cu ft. The natural gas cost is $.3961 per therm. What is the total natural gas bill for the facility?

1. Convert usage (in cu ft) to therms.

therms = cu ft × Btu factor

therms = 400 × 1.01

therms = 404

2. Determine distribution charge.

distribution charge = therms × distribution charge/therm

distribution charge = 404 × $.133

distribution charge = $53.73

3. Determine environmental recovery cost.

environmental recovery cost = therms × environmental recovery cost/therm

environmental recovery cost = 404 × $.0046

environmental recovery cost = $1.86

4. Determine natural gas cost.

natural gas cost = therms × natural gas cost/therm

natural gas cost = 404 × $.3961

natural gas cost = $160.02

5. Determine total natural gas bill.

total natural gas bill = customer charge + distribution charge + environmental recovery cost + natural gas cost

total natural gas bill = $65.45 + $53.73 + $1.86 + $160.02

*total natural gas bill = **$281.06***

BUILDING SURVEYS

A comprehensive building survey is required for the building automation system purchase and estimating process. A *building survey* is an inventory of the energy-consuming equipment in a commercial building. A building survey identifies the characteristics of each load and determines the appropriate control strategy for the load. Building automation system installation and programming may be adversely affected if a building survey is incomplete or flawed. A building survey is also used to begin the estimating process. Maintenance technicians require knowledge of the elements included and effective methods for performing a complete building survey.

Building Survey Elements

Most manufacturers provide a standard survey form or software to conduct effective building surveys. See Figure 24-10. The survey form includes space for the name and address of the building. Most important is a list of all major energy-consuming equipment in the building. Generally, this consists of the HVAC equipment and lighting loads. Depending on customer needs and control capabilities, other equipment can also be listed. The equipment should be identified with the nomenclature used by maintenance personnel. The location of each piece of equipment should also be documented for ease in installation.

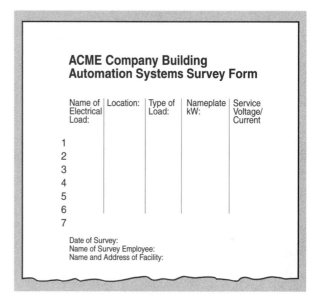

Figure 24-10. Building survey forms are often used to identify equipment included in energy saving strategies.

The nameplates of the HVAC units are used to obtain the voltage, current, phase, horsepower, and air conditioning tonnage. The heat energy (in Btu) input value of a load should be documented if gas heat is used. Electric heat values are listed in kilowatts. For lighting loads, the area served, type of lighting, wattage, voltage, and current of each service should be listed. The type of control currently used for each load should also be included. The maintenance technician should determine the importance of each load and the approximate times that each load must be on. This information is used to develop time-based scheduling, optimization factors, electric demand control, and/or duty cycling scheduling. The location, type, and information for each electric and natural gas meter should also be listed.

Modern electric meters can be read by technicians when checking utility consumption.

Performing Effective Building Surveys

The listing of equipment information during a building survey is normally straightforward. However, the individual performing the building survey can also obtain a large amount of information by asking questions, listening carefully, and performing careful observation. A building survey is used to provide information to the sales engineer. Questions commonly asked during a building survey include:

• What are some major problems?

• What parts of the building get excessively hot or cold?

• What is disliked regarding the present control system?

The condition of the current equipment should also be documented. Items such as missing doors, loose access panels, and empty refrigerant cylinders next to equipment may indicate poorly-functioning equipment. The areas that are identified as continually too hot or cold should be noted. These areas may have a mechanical or control problem that requires correction. Also, observe and ask whether the existing controllers such as thermostats are exposed to continual readjustment. Continual controller readjustment results in poor temperature control and excessive energy use. This condition is corrected by using a building automation system with no access to change the setpoints. The maintenance technician can be consulted as to whether anyone adjusts the thermostat setpoint when it is time to

go home. Maintaining the occupied setpoints during occupied times (day) and unoccupied times (night) wastes energy. This can be corrected by installing a building automation system. Also, notice whether any windows are open to the outside air. This can indicate poor temperature control and energy waste.

Observing the dress of individuals in a building can indicate current building condition comfort levels. Employees who are overdressed may indicate that a building space is excessively cold. The use of desk fans in the summer may indicate that a building space is excessively hot.

In many cases, when buildings are remodeled, their mechanical systems are divided and reallocated in an incorrect manner. This makes it difficult to provide excellent temperature control in various areas. Areas that are poorly remodeled may prevent the mechanical system from delivering the proper temperature control regardless of system operation. These areas must be identified prior to the installation of a building automation system.

BUILDING AUTOMATION SYSTEM PROPOSALS

Proposals are written to sell HVAC equipment and services such as furnaces, water treatment equipment, and preventive maintenance. Jobs may be obtained through a bid process or by negotiating privately. The proposals for each are written differently to reflect the different situations. Jobs that are put out for bid usually have rigid rules for writing the proposal. Privately-negotiated jobs allow a great amount of creativity when writing the proposal.

Estimating, control sequencing, and public relations are all involved in writing effective proposals. The customer concerns should always be documented in the proposal. The sales engineer normally develops building automation system proposals. However, a maintenance technician should have knowledge concerning the information contained in a proposal and a basic understanding of the proposal process.

The purpose of a written proposal is to sell the customer a building automation system. References should be used effectively. Past customers can be used to inform prospective customers about building automation system benefits. Existing customers must be contacted for their permission to be referenced. When possible, use references that are the same type of business as the prospective customer. Permission may be obtained from an existing customer to allow tours for a prospective customer. This allows the prospective customer to obtain first-hand information on a building automation system installation.

When writing a proposal, the use of technical words, phrases, and acronyms should be kept to a minimum. Many individuals that read proposals have a limited technical background and may be intimidated by technical terms. Potential customers may feel that they are being taken advantage of if excessive technical terminology is used. The proposal should be proofread and commented on by someone with a limited technical background. Components included in a proposal include the introduction, executive summary, proposal body, exclusions, signature sheet, appendices, and documentation.

Introduction

The introduction to a proposal, normally in the form of a cover letter, lists the length of time the inspector has been in business, the goals of the business, and high-profile local jobs that have been completed. Appreciation is given for the opportunity to provide the proposal to the customer, as well as assurances that the customer will be pleased with the results. The introduction serves as a positive beginning to the rest of the proposal.

Executive Summary

The *executive summary* is a one-page summary of the significant parts of the proposal. The executive summary is often the most important part of a proposal because management personnel often need only a short overview of the important parts of the proposal. The in-depth technical details are left for later inspection. The executive summary normally includes the cost of the job, the dates of installation and startup, and the location of the building requiring system installation. A critical part of the executive summary is the estimated payback time (in months). Along with the job cost, this is the most important part of the proposal. In many cases, a list of the installed controllers and equipment is also provided.

The executive summary of a building automation system proposal should always show an effect on profits. This is because many executives read only the executive summary and rely on the advice of their engineering personnel for information on the technical merits of the equipment.

Motor control circuits are often checked for power consumption when performing a building automation system survey.

Proposal Body

In the proposal body, a list is provided of the controlled equipment, number and location of the building automation system controllers, temperature and other sensors, and any interface devices. Listed along with the controlled equipment are the control strategies, which are used to provide control and/or energy savings. The installation times and completion dates are also included.

Exclusions

An *exclusion* is an item in a proposal that is the responsibility of the customer and not the contractor. Common exclusions include installation of the dedicated telephone line used by the building automation system for connection purposes and clarification of any customer installation work or aid provided. For example, a customer who desires to perform some of the installation tasks must have this spelled out precisely in the exclusions portion of a proposal. Additional exclusions include ensuring access to the building on nights and weekends for installation purposes or the use of existing control system components.

Signature Sheet

The signature sheet allows a customer to sign a proposal, thereby agreeing to its contents. The signature sheet contains space for the customer and contractor signatures. A clause is inserted into the signature sheet which guarantees the price given to the customer for a specific period of time. The guarantee clause induces the customer to sign the proposal before a deadline is reached. This clause protects the contractor from a customer that waits for an extended period of time and expects to get the same price previously quoted.

Appendices and Documentation

Appendices and documentation information includes description sheets of the building automation system equipment, graphs showing current energy consump-tion and demand, and estimates of energy savings. Appendices and documentation information may also include drawings of possible control configurations.

When developing a building automation system proposal, use spreadsheets and graphs whenever possible. A well developed graph makes it easy to understand sharp increases or decreases in energy costs that may not be apparent to building owners or managers.

25

HVAC Control Systems

Building Automation System Troubleshooting

An HVAC service technician may need to troubleshoot a building automation system. Although the majority of common problems can be solved using standard procedures, the building automation system manufacturer's specific procedures should be consulted when troubleshooting their equipment.

BUILDING AUTOMATION SYSTEM TROUBLESHOOTING

Like all control systems, building automation systems require troubleshooting when problems arise. Operating personnel may not be expected to perform in-depth troubleshooting. The majority of service technicians' day-to-day work in a facility is in correcting common problems, and the service technician must be able to fix a problem when called to the job site. Although the vast majority of problems can be corrected by service technicians, some complex problems can only be diagnosed and corrected by consulting with the system manufacturer's engineers.

The building automation system manufacturer should be consulted regarding specific procedures for troubleshooting their equipment. Although standard procedures apply to the majority of common problems, each building automation system manufacturer has different troubleshooting procedures.

To properly troubleshoot a building automation system, a service technician must have job prints and diagrams; factory cut sheets for each component (on paper or in an electronic version); knowledge of the general layout of the job site and of the location of devices, equipment, etc.; in-depth knowledge of the mechanical system operation; and a written sequence of operation for each HVAC system.

A service technician must also possess the following equipment and be proficient in using it: digital multimeter (for measuring resistance, voltage, and current levels in the system); a laptop computer loaded with appropriate software from the manufacturer; diagnostic equipment such as network (LAN) meters; and, if also diagnosing mechanical system problems, tools such as refrigeration gauges. The service technician also needs hands-on knowledge of personal computer setup and operation, including loading software, peripheral setup for devices such as printers and modems, and basic computer repair.

Common building automation system problems include an HVAC unit that is OFF when it should be ON, a building space temperature that is excessively hot or cold, improper or no sensor reading, a controlled device that cycles constantly, a controller that is off-line from the network, and too few or too many alarms.

HVAC Unit OFF When It Should Be ON

As a service technician performs preventive maintenance, the technician may notice, on the building automation system workstation computer in the maintenance office, that one of the HVAC units is OFF, when normally it is ON. Possible causes include overriding the unit OFF in the control software, and an unknown software control strategy that is keeping the unit OFF.

Overridden Unit in Control Software. An HVAC unit being OFF when it should be ON may occur when a technician overrides the unit OFF in the control software. An override may be performed to perform routine maintenance on the unit (proper lockout/tagout procedures must be followed), and when finished, the technician may not remove the override. This is a common, day-to-day occurrence.

If an overridden unit in the control software is suspected, the building automation system software is checked to identify the cause of the HVAC unit being OFF. Most manufacturers provide an indicator such as an O in the building automation system control software that can be accessed from the workstation computer that indicates an override is in effect. See Figure 25-1. A technician must release the override for the unit to operate. Follow-up communication should take place with the technician who initiated the override regarding releasing future overrides.

BUILDING AUTOMATION SYSTEM SOFTWARE INDICATORS

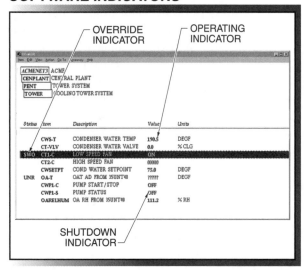

Figure 25-1. Building automation system software normally provides an override indicator on the workstation computer.

Unknown Software Control Strategy Keeping Unit OFF. An HVAC unit that is OFF when it should be ON may be the result of an unknown software control strategy keeping the unit OFF. For example, demand control or duty cycling control strategies may shut a unit OFF. A service technician may think that the unit should be ON but the control strategy is keeping the unit OFF.

If an unknown software control strategy is suspected, the technician should check the building automation system workstation computer. Many manufacturers provide indicators for the loads on each controller to show which control strategy (or strategies) is keeping the unit OFF or in a certain condition. See Figure 25-2. Changes can be made as required after checking the indicators at the controllers and the settings of the appropriate control strategy.

CONTROL STRATEGIES

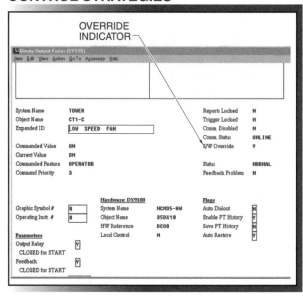

Figure 25-2. Checking an HVAC controlled device on the workstation computer can indicate which control strategy is keeping a device ON or OFF.

Building Space Temperature Excessively Hot or Cold

Complaints about the temperature being excessively hot or cold in a building space are the most common troubleshooting calls. A logical process must be taken to determine the cause because there are many possible causes. Possible causes include incorrect setpoints and malfunctioning mechanical equipment.

Incorrect Setpoints. The primary cause of excessively hot or cold temperature complaints is having the incorrect setpoints in the building automation system controller software. In the simplest case, an improper occupied cooling setpoint such as 80°F causes the building space temperature to be excessively hot. While basic heating and cooling setpoints are easily understood, others are more subtle. For example, heating and cooling equipment is often locked out (disabled) based on the outside air temperature. This lockout setpoint may prevent HVAC equipment from operating if set improperly by only a few degrees. This causes the building space temperature to be excessively high or low.

Another subtle setpoint problem is a low static pressure setpoint in a variable air volume system. The standard setpoint for a variable air volume system is 1″ wc. If the static pressure setpoint is too low, there is not enough air volume in the duct to supply all the building space VAV terminal boxes. In this case, one of the VAV terminal boxes may be starved (not supplied with enough air). This causes the building space temperature to be too hot, even though the VAV terminal box, controller, and software operate properly.

If incorrect setpoints are suspected, the controller software should be checked for any temperature setpoints that are excessively high or low. See Figure 25-3. In addition, many building automation system controllers provide a method for the building occupants to change the setpoints (biasing). The biasing may be mistakenly adjusted.

> *Advanced building automation system controllers contain built-in diagnostics that can provide a warning when a VAV terminal box is open 100% but cannot obtain the required air flow volume.*

Malfunctioning Mechanical Equipment. Excessively hot or cold temperatures may be caused by malfunctioning mechanical equipment that provides heating or cooling to the building spaces. While not directly a problem with the building automation system, a problem with the HVAC mechanical equipment may still lead to occupant complaints. In many cases, an occupant cannot differentiate between problems with the building automation system and problems with HVAC mechanical equipment. A fully-qualified technician should be able to solve problems that arise due to the building automation system or the mechanical equipment. This means that technicians should be familiar with the tools and procedures required for troubleshooting a building automation system and the tools and procedures used to diagnose mechanical equipment malfunctions.

If malfunctioning mechanical equipment is suspected, the HVAC equipment should be checked for heating or cooling devices that are ON or valves that are 100% open. A wide-open valve can indicate that the HVAC equipment is not capable of satisfying the heating or cooling demand for the building space. If the HVAC unit is a variable air volume system, the discharge air temperature should be checked at the air handling unit to ensure it is at 55°F. The air flow should also be checked at the VAV terminal box to ensure it is at the correct volume. The building automation system workstation computer can be used to check boiler or chiller steam, hot water, or chilled water supply to the building space. See Figure 25-4. If the system has a cooling tower, the tower should be checked for proper operation. Any system not operating correctly can cause the temperature to be excessively high or low in the building space.

CONTROLLER SETPOINTS

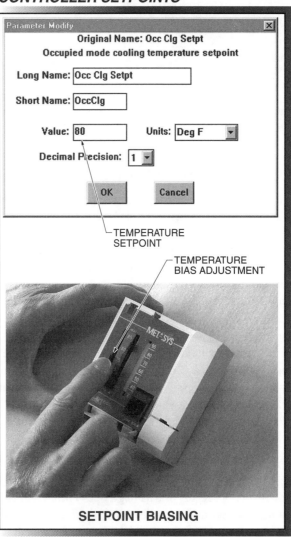

SETPOINT BIASING

Figure 25-3. When controller setpoints are excessively high or low, building occupant comfort is affected, leading to occupants making unwanted biasing changes.

Figure 25-4. Mechanical equipment condition may be checked using the building automation system workstation computer before checking the system in person.

Improper or No Sensor Reading

Building automation systems use temperature, pressure, and humidity sensors to provide HVAC control and activate an alarm when improper conditions occur in a building. Other specialty sensors such as carbon dioxide and oxygen sensors are also available. The proper selection, installation, and operation of these sensors is required for proper building space temperature control.

Any inappropriate sensor readings may cause problems with building space temperature control. A technician checking a building using the workstation computer may notice that a sensor is reading an inappropriate value—either excessively high or low—or the sensor may not be giving any reading. Improper or no sensor readings may be caused by physical damage to the sensor or incorrect sensor information.

Physical Damage to Sensor. A sensor that reads properly, but in a short period of time reads improperly, may have experienced physical damage. See Figure 25-5. The damage can occur from collisions with vehicles, or from the sensor being painted or being cleaned with steam.

If physical damage to the sensor is suspected, a DMM is normally used to check the operation of the sensor. For example, a resistive temperature sensor is designed to have a resistance of 1000 Ω at 70°F. The same sensor also has a positive temperature coefficient of 2.2 Ω per °F. At 75°F, the sensor should have a resistance reading of 1011 Ω ($1000 + 2.2 \times 5 = 1011$ Ω). To check the sensor resistance, a technician can use a DMM set to measure resistance. The sensor resistance should be close to 1011 Ω. The sensor should be replaced if it is outside the tolerance permitted by the manufacturer.

Problems caused by physical damage to a sensor can be avoided by relocating the sensor or using a protective enclosure to prevent damage. For example, a sensor located in a building space that has been damaged by the occupants may be replaced with a new sensor installed and located in the return air duct. This prevents easy access by the occupants. If a sensor must be protected, many manufacturers sell lockable protective enclosures that protect the sensor from damage.

Incorrect Sensor Information. Improper or no sensor readings may also be caused by incorrect sensor information in the control software. For example, a humidity sensor can be set up in the control software as a 0% to 100% relative humidity sensor. However, the sensor may actually be a 25% to 75% relative humidity sensor. In this case, a humidity reading of 25% rh at the sensor is read at the controller as 0% rh, resulting in improper humidity control.

MALFUNCTIONING SENSORS

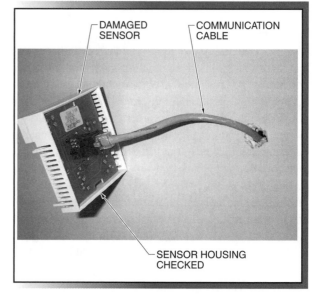

Figure 25-5. A sensor damaged by impact can lead to lack of comfort and erratic sensor readings.

If incorrect sensor information is suspected, the sensor itself should be checked. It is possible that the wrong sensor was installed. Obtain the correct sensor if needed. In addition, the building automation system sensor software should be checked to ensure that it is set for the actual sensor installed. Change any settings that are incorrect. See Figure 25-6. Finally, the engineering drawings should be checked to ensure the sensor listed matches the actual sensor installed. Update the drawings to ensure their accuracy.

BUILDING AUTOMATION SYSTEM SENSOR SOFTWARE

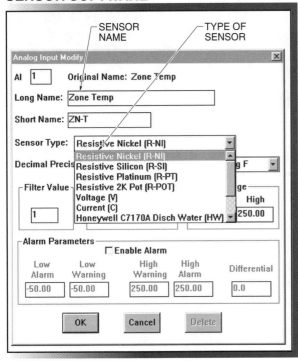

Figure 25-6. A sensor must be defined properly in the building automation system software to achieve proper control.

Controlled Device Cycling Constantly

A technician may discover that a heating or cooling valve is constantly cycling from fully open to fully closed. A variation of this problem is that the staged heating or cooling equipment is being turned ON and OFF too rapidly. Possible causes include incorrectly set control loop parameters and a mechanical system problem.

Incorrectly Set Control Loop Parameters. Constant cycling of controlled devices is commonly caused by building automation system software that has incorrectly

set control loop parameters. Each control loop has parameters such as a setpoint, loop action (whether direct or reverse), a setting for sensitivity, and a setting for integration if using PI control that must be set correctly.

If incorrectly set control loop parameters are suspected, each setting should be checked using a personal computer with the appropriate software. See Figure 25-7. Many manufacturers provide software that is able to tune control loops to provide the proper control and eliminate constant cycling of controlled devices. Some software programs enable the parameters to be changed manually while others provide automatic tuning.

CONTROL LOOP PARAMETER SOFTWARE

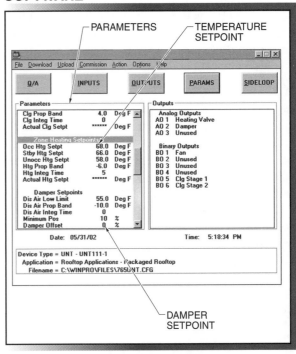

Figure 25-7. Incorrect control loop settings in controller software can cause controlled devices such as valves, dampers, variable speed drives, and compressors to cycle improperly.

If the controlled device is digital, such as gas heat or cooling compressors, the software should be checked for differentials and time delays. For example, a temperature differential of 1°F or 2°F commonly exists between individual stages of heating or cooling. If this parameter is adjusted to a higher value, the heating or cooling stages do not energize at a temperature that provides comfort. In addition, a time delay commonly occurs between the heating or cooling stages to prevent equipment from operating when not desired. If this parameter is set too long, the equipment may not energize in time to satisfy the comfort in the building space.

Mechanical System Problem. Constant cycling of controlled devices may also be caused by a problem with the mechanical system. The rapid cycling between opening and closing can be caused by a malfunctioning pump or by the mechanical system providing water at the wrong temperature. When inlet water temperature is fluctuating rapidly, the control valves cycle rapidly in response.

If a mechanical system problem is suspected, the mechanical systems should be checked using the workstation computer or portable operator terminal to determine any problems with the mechanical systems. When necessary, the mechanical systems must be checked on-site to ensure proper operation. See Figure 25-8.

Controller Off-line from Network

Building automation system controllers are connected to a communication network (bus) that runs through a building. The communication network enables controllers, workstation computers, printers, modems, and other devices to exchange information. The communication network must function properly. If the communication network is down for any reason, problems occur such as control being lost or critical alarms not being received.

Alnor Instrument Company

VAV terminal boxes and diffusers are often balanced by using a flow hood. Flow hood measurement information may be fed to building automation system controllers to electronically balance the system.

BUILDING AUTOMATION SYSTEM MECHANICAL SYSTEMS

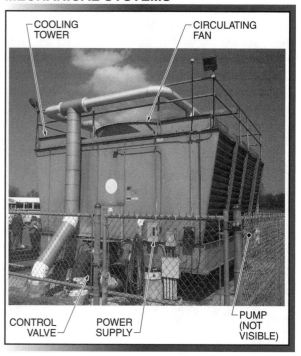

Figure 25-8. HVAC mechanical system observation may indicate the reasons for system failure or reasons for improper system conditions.

Before beginning any communication network troubleshooting, a technician must be familiar with the characteristics of the specific network used by the manufacturer. Proper voltage levels, communication software operation, distances, and network configurations must be known and understood.

A controller being off-line from the communication network is normally discovered when an operator observing the building systems on the workstation computer notices that the building automation system cannot communicate with one or more controllers. There is normally an indicator in the building automation system software that indicates that the controller or controllers are off-line (cannot communicate). Possible causes of a controller being off-line from the communication network include a damaged controller, controller power loss, and a damaged communication bus.

Damaged Controller. If a building automation system controller is damaged, the damage may cause it to be off-line from the network. A damaged controller is indicated by an individual controller being off-line, instead of large parts of the network. The controller that is off-line should be checked.

A controller may be damaged by a lightning strike, water, impact, or improper installation. Lightning strike damage is normally immediately apparent . Water damage is caused by drains or piping located too close to the controller. Impact damage may be caused by individuals performing other maintenance work and physically striking the controller. Improper installation may cause controller damage if the controller is mounted without regard to manufacturer recommendations.

Controller Power Loss. A building automation system controller that loses power cannot communicate and will be off-line from the network. Controller power loss may be caused by the power supply being destroyed or the power supply being overloaded.

A destroyed power supply may be the result of a lightning strike or component failure. If a destroyed power supply is suspected, the power supply should be checked for 24 VAC (typical) output using a DMM. If the DMM reads the rated power supply voltage, the controller should work correctly when power is restored. If the DMM does not read the rated power supply voltage, replace the power supply. See Figure 25-9.

CONTROLLER POWER SUPPLY

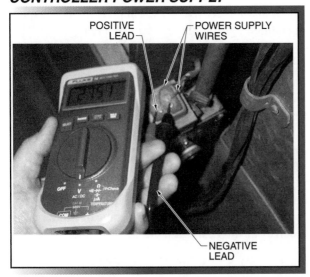

Figure 25-9. The power supply of a controller that is not communicating may be checked using a DMM.

An overloaded power supply indicates that the power supply was sized improperly. In this situation, if one controller is removed from the circuit, the others should work correctly. Additional or larger capacity power supplies may be used to correct an overloaded power supply problem.

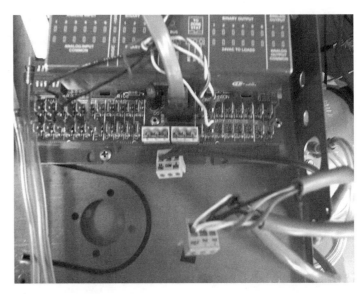

The network communication wires to an off-line controller should be checked because the connector may have been removed or may be loose.

Damaged Communication Bus. A controller being off-line from the network may also be caused by a damaged communication bus. Damage to the communication bus may cause more than one controller to be off-line from the network. It is possible that the communication bus to a particular controller is damaged even though the controller is powered and appears to be working properly. A damaged communication bus may be caused by improper installation/polarity, maintenance damage, or using the wrong connecting media.

Improper installation/polarity may be caused when additions are made to a building automation system. The network may have been installed improperly or the polarity may be wrong. This can cause the entire network to be down. In this case, the network is checked for bad connections and splices. Maintenance damage may be caused by individuals working on a controller and causing damage to the network wiring. In addition to damage, the network may have been wired using the wrong connecting media or the wrong size connecting media. This can happen if individuals try to save money by using the wrong connecting media. Always follow manufacturer recommendations regarding connecting media.

If a damaged communication bus is suspected, the communication bus is checked for proper voltage levels, wiring, and terminations. Depending on how the bus is run, technicians may be able to locate a problem spot by checking in the middle of the bus, and then progressively checking the middle of the remaining sections until the problem spot is found. See Figure 25-10.

COMMUNICATION BUS

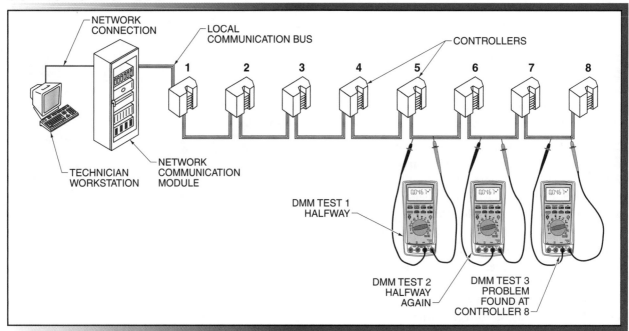

Figure 25-10. A communication bus problem may be found by checking the system in segments until the problem is located.

Too Few or Too Many Alarms

Many buildings or facilities with building automation system controls have problems in setting up and managing alarms. Some applications are set up having few or no alarms, causing important failure conditions to be unreported. Conversely, some applications are set up having too many alarms. An inappropriate number of alarms, either too many or too few, can be caused by failure to prioritize alarms properly and excessively close alarm parameters.

Failure to Prioritize Alarms Properly. Too many or too few alarms may be caused by a failure to prioritize alarms properly when setting up the system. Improper alarm prioritization is caused by too many or too few alarms being set up as critical.

When too many alarms are set up as critical, almost every sensor or condition triggers a critical alarm. As the multiple alarms are constantly reported, human nature is to regard them all as unimportant. For example, a storeroom for office supplies is set up so that if the temperature goes into alarm, a critical alarm is reported. The temperature in an office supply room is rarely critical.

When too few alarms are set up as critical, important conditions that need to be alarmed are not set up. When a potentially critical failure occurs, it is not reported or corrected. For example, an air handling unit that provides HVAC for a hospital operating room fails but no alarm is received. The temperature in a hospital operating room should be a critical alarm.

If improper alarm prioritization is suspected, a detailed functional analysis is required of all equipment and systems in the building. The equipment and systems are prioritized by deciding whether an alarm for that equipment or system is critical, nice to have, or unimportant. See Figure 25-11. Sensors and equipment that are not vital to the operation of a building must be set to generate minor alarms. Alarm values are changed in the control software to match their priority. For example, an office supply room alarm priority may be reduced or eliminated, while a hospital operating room receives the highest alarm priority possible.

Excessively Close Alarm Parameters. Few or excessive alarms may also be caused by alarm parameters being set up with such close tolerances that the sensors cycle in and out of alarm constantly, even in normal operation. For example, common tolerances for a zone temperature may be a low alarm setpoint of 68°F and a high alarm setpoint of 78°F. If the zone temperature alarm setpoints were instead set at 72°F and 75°F, the room sensor may alarm excessively. The excessive alarm may lead to an unnecessary response by maintenance personnel.

ALARM PRIORITY

ONLY VITAL EQUIPMENT AND SYSTEMS MUST GENERATE CRITICAL ALARMS

Figure 25-11. Identifying critical equipment or systems for alarm priority is required to prevent either too many or too few alarms.

If excessively close alarm parameters are suspected, the alarm parameter folder should be opened in the control software for the device and the alarm setpoints, differentials, and parameters changed as required. See Figure 25-12. The parameters should be matched to the actual operation of the mechanical equipment and systems. For example, most variable air volume air handling units produce 55°F air at a duct static pressure of 1″ wc. The alarm parameters for the discharge air temperature should be approximately 50°F for low alarm and 60°F for high alarm. Supply duct static pressure should be approximately .8″ wc low alarm and 1.2″ wc high alarm.

Many alarms also have differentials and time delays. A differential is often provided so that the sensor does not cycle into and out of alarm rapidly. For example, if the discharge air temperature of a variable air volume

air handling unit goes into alarm above 60°F, the temperature may have to decrease by 2°F before the alarm stops. A time delay may also be used which prevents the issuing of an alarm to the appropriate destination before a certain length of time has elapsed. For example, a variable air volume static pressure alarm may have a 2 min alarm delay. The alarm is not generated unless the alarm condition exists for a minimum of 2 min. This feature is also used to reduce nuisance alarms and to permit normal equipment startup without automatically issuing an alarm.

ALARM PARAMETER FOLDERS

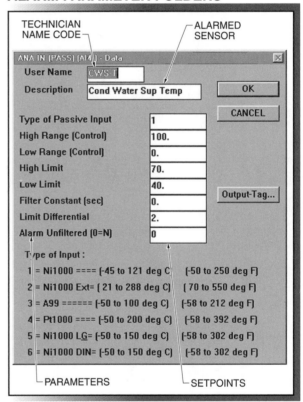

Figure 25-12. The alarm parameter folder can be checked for improper alarm setpoints and differentials that can lead to poor alarm reporting by the system.

After being generated, alarms must be routed to the proper destinations. Alarm destinations may include on-site or off-site workstation computers, printers, or pagers. Alarms may be routed to multiple destinations in a particular order or may be filtered to allow alarms with different priorities to be sent to different destinations. When troubleshooting alarm delivery problems, always check the alarm routing software.

HVAC Control Systems

Advanced HVAC Control Technologies

The development of new technologies and communication capabilities has changed the building automation system controls industry. Many of the advances have coincided with computer industry developments such as the use of the Internet and personal digital assistants (PDAs). Other developments such as BACnet™ have been driven directly by the building automation system industry.

ADVANCED HVAC CONTROL TECHNOLOGIES

Advanced HVAC control technologies are being developed to solve existing control system deficiencies. For example, building automation system control loops require extensive fine-tuning. A technician can spend many hours tuning particularly difficult control loops in order to properly control HVAC equipment. Proportional/integral/derivative (PID) control loops are common closed loop control loops. See Figure 26-1. Hours of tuning and adjusting control loops results in higher building automation system controller installation costs but better system control. Poor system control is the result of insufficient time spent fine-tuning a system.

Another control loop tuning problem is that the tuning parameters are often valid for only one particular condition. For example, an air handling unit cooling loop tuned at an outside air temperature of 80°F does not react the same when the outside air temperature is 95°F. Mechanical equipment wear and lack of maintenance can also cause control loops to react improperly. Improper system operation can also be the result of incorrect initial controller programming, causing the controller to be unable to properly adjust the system.

Advanced HVAC control technologies help reduce control system installation costs and improve control system accuracy. Advanced HVAC control technologies include fuzzy logic, artificial intelligence, Intranet and web communication, BACnet™, and Echelon™.

Applications such as electric economizer dampers achieve accurate flow control when using advanced HVAC control technologies.

CONTROL LOOPS

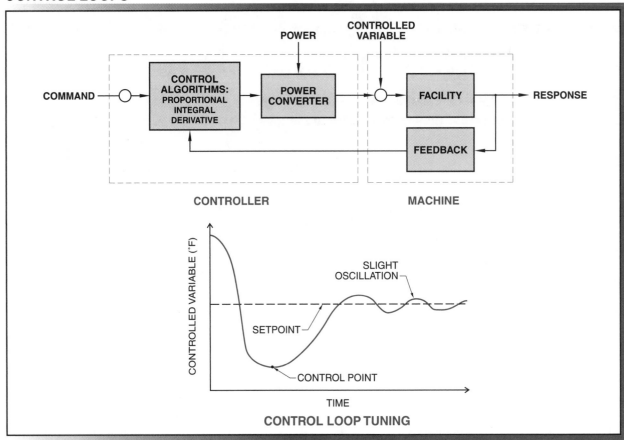

Figure 26-1. Control loops must be fine-tuned to provide optimal control of an HVAC system.

Fuzzy Logic

Fuzzy logic is the ability of a controller to self-diagnose and self-correct a system in order to correct various problems. The introduction of low-cost, high-speed controllers with advanced capabilities allows controllers to self-tune control loops without technician assistance. See Figure 26-2. Self-tuning controllers release technicians from tuning control loops. The controller automatically compensates for wear and changing environmental conditions. The result is increased system accuracy and increased energy efficiency. Controllers have evolved to the point where the controller is programmed with basic setpoints and functions and is then allowed to self-correct to the final parameters.

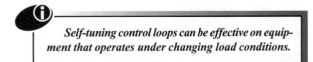

Self-tuning control loops can be effective on equipment that operates under changing load conditions.

SELF-TUNING CONTROL LOOPS

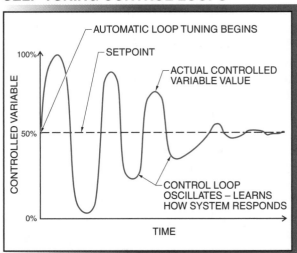

Figure 26-2. Fuzzy logic enables a controller to self-tune control loops without technician assistance to accommodate changing load conditions.

> **ⓘ** *As control system technology evolves, wider implementation of smart (fuzzy) logic systems is possible. These systems may be equipped to communicate with interfaces that have not yet been developed. Future capabilities may be added by upgrading the programming and allowing the hardware to remain unchanged. This increases the need for technicians to update their software skills to remain effective on the job.*

Fuzzy logic may also be used to correct for simple errors in installation or application. For example, a particular VAV terminal box actuator shaft must be turned clockwise to be closed. A VAV terminal box controller may be installed so that the motor turns the actuator shaft counterclockwise to close. A fuzzy logic system may attempt to close the VAV terminal box counterclockwise as programmed. The controller, however, receives signals that the airflow volume rises as it attempts to close the damper. After one or two attempts, the controller changes direction from counterclockwise to clockwise to close the damper. The fuzzy logic controller now receives signals that the airflow volume through the VAV terminal box decreases as the damper is closing. Fuzzy logic systems of today use self-correction only for relatively minor adjustments. See Figure 26-3.

Web-enabled controllers have Ethernet and other connections available for different media types.

Artificial Intelligence

Artificial intelligence is the use of advanced computing power to make decisions about system setup and functionality. Artificial intelligence systems are an extension of fuzzy logic. As computing power becomes more economical, plentiful, and powerful, new software programs are developed that allow advanced control features to be implemented. Controllers that are self-configuring, self-correcting, and self-diagnosing are being developed. System setup time and installation costs are further reduced, with greater customer satisfaction. Artificial intelligence has potential uses in field controllers and in engineering tasks such as mechanical system design and checkout. Artificial intelligence can be used to reduce the number of poorly designed systems that are installed in buildings. The selection of system equipment such as pumps, fans, dampers, and valves is becoming increasingly affected by the application of artificial intelligence.

Intranet and Web Communication

In recent years, the use of the Internet for personal and business use has risen dramatically. Many large companies have internal Internets (intranets) for the specific use of company employees. The widespread use and familiarity of web browsers such as Netscape® and Microsoft® Internet Explorer™ have led to intranet-based and web-enabled building automation system control systems.

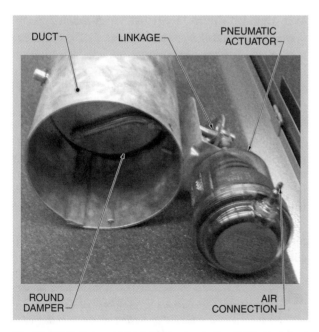

Figure 26-3. Fuzzy logic systems have the ability to detect the changes in operational behavior a variable air volume damper has over time and adapt to the change.

Intranet-Based Control Systems. An intranet-based control system allows technicians to view building functions using a common web browser such as Netscape or Microsoft Internet Explorer. See Figure 26-4. The advantages of using an intranet-based control system instead of a proprietary software system for viewing a building automation system include:

- Widespread availability and low cost of web browsers. Most PCs have manufacturer-supplied web browsers and can be easily adapted to dial into the network of a building.

- Low amount of training required because most employees use PCs and are familiar with using web browsers.

- Easy upgrading because many company applications use the same web browser. When an upgrade is needed, the same web browser can be shared across applications.

INTRANET-BASED SYSTEMS

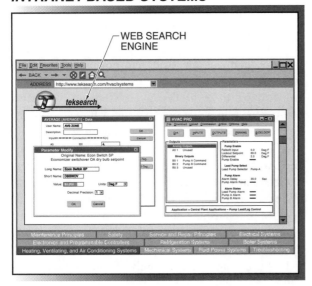

Figure 26-4. An intranet-based control system allows technicians to view building functions using a common web browser.

Intranet-based control systems are implemented by using a PC loaded with building graphics and an intranet software interface for the building automation system. See Figure 26-5. Normally, a large-capacity server PC is used that is loaded with individual web browsers. The upgrading and maintenance of the server PC can be time-consuming.

INTRANET-BASED CONTROL SYSTEMS

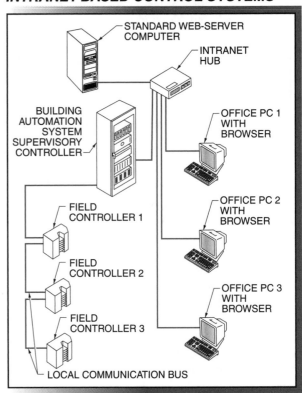

Figure 26-5. An intranet-based control system uses a company's standard web-server computer and web browser to view the building automation system.

Web-Enabled Control Systems. A *web-enabled control system* is a control system consisting of supervisory controllers that have Internet capability. Instead of requiring a server computer that is set up properly, each supervisory controller has all of the required Internet communication hardware and software installed by the manufacturer. See Figure 26-6.

As computing power and memory become less expensive, the reduced costs allow computer technology to be implemented in HVAC systems. For example, the intranet may be used for monitoring building conditions through building automation system controls, obtaining controller files and updates from manufacturer web sites, making purchases, and finding technical information.

> *Web-enabled control systems should be evaluated for the use of a standard operating system, number of concurrent users, upgrade potential, and number, type, and speed of connection media before implementation.*

WEB-ENABLED CONTROL SYSTEMS

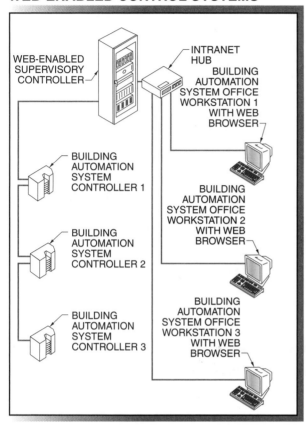

Figure 26-6. A web-enabled system uses a supervisory controller with server capabilities to view the building automation system.

BACnet Protocol

An ongoing concern of customers using building automation system products is that of exclusive communication within proprietary systems. In proprietary systems, each manufacturer has exclusive communication methods and software for particular applications. When a customer attempts to use multiple vendors, the customer finds that the equipment from one vendor cannot communicate with the equipment from another vendor or manufacturer, preventing the systems from sharing data. When building expansions are performed and new equipment is added, the inability to communicate reduces the number of available vendors for the project. Having limited vendors from which to choose prevents competitive pricing and increases overall costs. Also, when vendor equipment is upgraded, the current building automation system equipment may require expensive rework.

The American Society of Heating, Refrigeration, and Air-Conditioning Engineers (ASHRAE) has formed a committee to standardize building automation system industry communication. The ASHRAE committee is the governing organization for many HVAC industry standards, and is composed of representatives from vendors, users, and specifiers. The committee has developed a standard for building automation system controller communication. This standard is referred to as the building automation control network (BACnet). *BACnet* is a communication protocol that uses standard connecting media and enables controllers to be interoperable. The standard connecting media include wires, coaxial cable, unshielded twisted pair cable, and fiber optic cable. *Interoperability* is the ability of devices produced by different manufacturers to communicate and share information.

BACnet operates across multiple levels (layers) of the computer system. For example, the application layer describes the control sequences and how they are defined in the BACnet system. The network layer determines how BACnet will interact across the desired network system. The data link layer is software that determines the appropriate transmission protocol for a particular computer system. The media layer consists of wires, cables, and software protocols to deliver BACnet capability to the network. See Figure 26-7. The controllers, however, are not interchangeable. This means that one controller cannot be substituted for another controller from a different manufacturer. Interchangeability is normally impossible due to different electrical and mounting characteristics of the controllers. The difference between interoperability and interchangeability is key with regard to customer expectations.

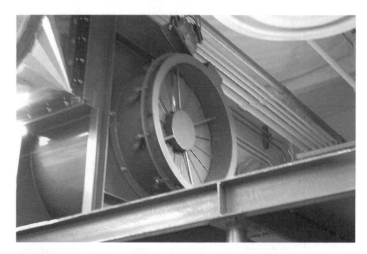

Outputs to devices such as inlet vanes may use BACnet standard definitions.

BACnet Objects. BACnet is commonly used with intranet networks. Intranets are the most commonly used commercial communication systems in building automation systems today. See Figure 26-8. To share data between devices, BACnet determines the points and devices (objects) that have a common look and contain the same information. Analog input objects require specific information such as object name, object type, present value, and status flags. For example, a temperature sensor may have an object name of zone 101, object type of analog input, present value of 76.4, and status flag of normal.

BACNET LAYERS

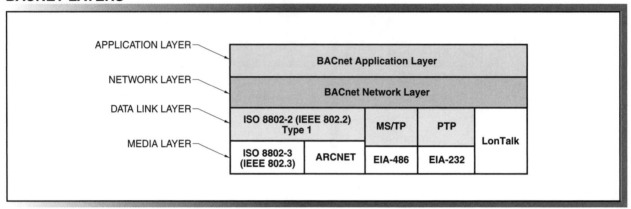

Figure 26-7. BACnet operates across multiple levels (layers) of the computer system and enables controllers to be interoperable.

INTRANET NETWORK CONFIGURATIONS

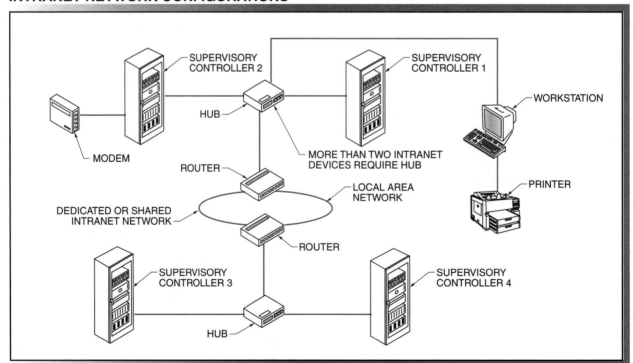

Figure 26-8. Intranet-based systems use hubs and routers and are used to create subnetworks for data transmission.

When evaluating a BACnet system, the BACnet documents should be public and available for examination.

Additional information is desired but not necessarily demanded by the BACnet standard because the BACnet standard is normally viewed as a minimum. The possibility exists that a particular manufacturer has only a few objects that comply with the BACnet standard. When selecting a manufacturer, the actual objects that are BACnet compatible must be clarified.

The BACnet standard is also grouped into controller conformance classes. The conformance classes organize controllers into the type of work expected to be performed. Functions are then assigned to each conformance class. See Figure 26-9. Conformance statements recognize that not all controllers require all of the same functions. For example, a VAV terminal box controller requires different points than a supervisory controller that does share data. The HVAC industry has developed conformance classes that show the degree to which a particular manufacturer's equipment supports the BACnet standard. Conformance statements are public documents that are meant to be shared with customers and other vendors as needed. The BACnet standard is subject to change, is always under review, and is viewed as a work in progress.

BACNET PROTOCOL IMPLEMENTATION CONFORMANCE STATEMENT

Vendor Name: __ACME__
Product Name: __A40__
Product Model Numbers: __MS-A40 1010-0, MS-A40 1310-0, FA-A40 1010-0, FA-A40 1310-0__

PRODUCT DESCRIPTION

Acme A40 Supervisory Controller is designed to manage a small building or campus of buildings. The A40 efficiently supervises the networking of Application Specific Controllers (ASCs) and provides facility management features including weekly scheduling, optimal start, alarm management, and trending.

Facility personnel can review the system status and modify control parameters for the A40 Supervisory Controller and its associated ASCs using a VT100 Terminal or an X3 Workstation.

With the addition of network card, multiple A40s can communicate over an Ethernet peer-to-peer network, providing increased functionality for systems that are more complex.

BACnet Conformance Class Supported			BACnet Functional Groups Supported		
Class 1 ☐	Class 3 ☒	Class 5 ☐	Clock ☒	Files ☐
Class 2 ☐	Class 4 ☐	Class 6 ☐	HHWS ☐	Reinitialize ☒

BACnet Standard Application Services Supported

Application Service	Initiates Requests	Executes Requests	Application Service	Initiates Requests	Executes Requests
CreateObject	☐	☒	UnconfirmedPrivate Transfer	☒	☒
DeleteObject	☐	☒	ReinitializeDevice	☐	☒
ReadProperty	☐	☒	UTCTimesSynchronization	☒	☒
ReadPropertyMultiple	☒	☒	Who-Has	☒	☒
WriteProperty	☐	☒	I-Have	☒	☒
WritePropertyMultiple	☒	☒	Who-Is	☒	☒
ConfirmedPrivateTransfer	☒	☒	I-Am	☒	☒

Figure 26-9. BACnet conformance classes are indicated on a public protocol implementation conformance statement.

Echelon

Echelon is a controller communication system that uses common hardware and software. The Echelon communication system was developed by Echelon, a company based in Palo Alto, California. The driving idea of Echelon is the same as that of BACnet — the desire for interoperability between vendors. In the Echelon communication system, all building automation system vendors must provide devices that meet the Echelon standard for coexistence on the same network.

Echelon Components. Echelon includes hardware and software components. When a vendor manufactures an Echelon-compatible controller, the controller must have an embedded Echelon neuron chip. Echelon also has proprietary communication wiring and processing speeds. An Echelon network is also unique in that each controller communicates directly with other controllers on the network. See Fig 26-10.

ECHELON SYSTEMS

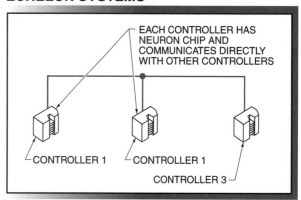

Figure 26-10. In an Echelon system, each controller communicates directly with other controllers by the use of a neuron chip.

Many building automation systems use master/submaster communication systems. For example, communication between controllers in a building automation system is handled by supervisory or master controllers. The Echelon standard demands that tools used to program controllers be supplied as part of the equipment for installation, reducing the need to purchase proprietary programming tools later.

One area in which Echelon and BACnet are similar is that each demands that the controller points have a specific format depending on the classification of the controller, such as universal input/output controllers, VAV terminal box controllers, and rooftop unit controllers. Point formats are contained in information files that are also public property. It is possible for BACnet software to operate on an Echelon hardware system. The preferred method of providing interoperability in the future may be BACnet, Echelon, a combination of the two, or some other technology.

Advanced Communication Applications

In the building automation system industry, improvements in capacity and capability are always being made. The building automation system industry is also influenced by improvements in other industries. For example, radio frequency communication, infrared communication, and personal digital assistants are finding applications in the building automation systems.

Radio Frequency Communication. The need for hard wiring between controllers and devices in a network is a source of expense and problems. When wires are not run and terminated properly, the wires cause intermittent building automation system problems. Also, hard wiring makes quick reconfiguration of a building automation system difficult when tenants move. Hard wiring problems are eliminated and quick reconfiguration of a building automation system is made possible by using radio frequency (RF) communication.

Radio frequency communication is the use of a high-frequency electronic signal to communicate between control system components. This frequency may be very high to avoid interference between it and other building devices. When a room sensor and controller are communicating via radio frequency, the sensor can easily be relocated to a new wall when a room is reconfigured. See Figure 26-11. A drawback of radio frequency technology is that interference occurs between radio frequency devices such as variable speed drives, fluorescent lights, video equipment, and computer monitors. In addition, distance and aiming parameters must be addressed.

> **ℹ** *Many modern building automation system controllers permit use of multiple connections and protocols. Some controllers may have the ability to use a proprietary RS-485 bus or have an Echelon card added if needed for a particular installation. Future software upgradeability should be factored into the job specifications to prevent premature hardware obsolescence.*

RADIO FREQUENCY COMMUNICATION

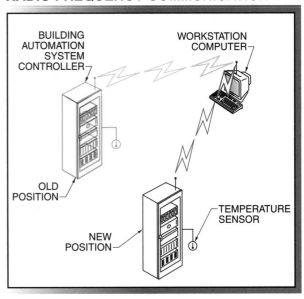

Figure 26-11. Radio frequency communication systems eliminate the need for direct wiring.

Infrared Communication. *Infrared communication* is communication between devices using infrared (IR) radiation. Many modern devices communicate using infrared radiation. See Figure 26-12. Most modern personal computers have infrared communication capability. Infrared communication requires an unobstructed line of sight between devices for proper operation. Infrared communication exists for building automation system applications but still requires development.

INFRARED COMMUNICATION

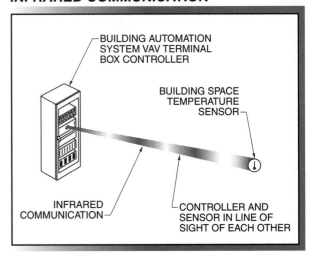

Figure 26-12. Infrared communication requires an unobstructed line of sight between devices for proper operation.

The major drawback of infrared communication is that an unobstructed line of sight is required between the devices. Many VAV terminal box controllers are mounted above drop ceilings, which prevents line-of-sight communication. In addition, if a sensor is moved around a corner, the sensor is no longer capable of communicating with a controller.

Personal Digital Assistants. A *personal digital assistant (PDA)* is a small, hand-held computing device with a small display screen. Personal digital assistant use has increased dramatically in recent years. Personal digital assistants can perform many applications such as word processing, e-mail, spreadsheets, phone number logs, games, etc. The use of personal digital assistants has become so prevalent and the price so reasonable, that manufacturers have developed building automation system interfaces that enable a personal digital assistant to act as a technician interface when connected to the control system. See Figure 26-13.

Personal digital assistants (PDAs) are often loaded with appropriate software and are provided to electrical subcontractors to rapidly check system installation and operation.

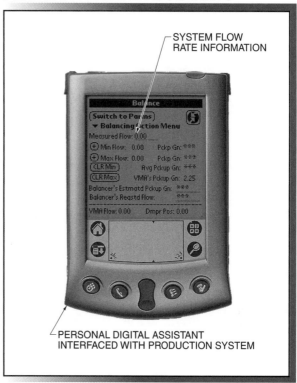

Figure 26-13. Personal digital assistants can be used to communicate with building automation system controllers.

Air Conditioning Contractors of America

OUTDOOR DESIGN TEMPERATURE

State and City	Lat.§	Winter DB*	Summer DB*	Daily Range*	WB*
ALABAMA					
Alexander City	32	18	96	21	79
Auburn	32	18	96	21	79
Birmingham	33	17	96	21	78
Huntsville	34	11	95	23	78
Mobile	30	25	95	18	80
Montgomery	32	22	96	21	79
Talladega	33	18	97	21	79
Tuscaloosa	33	20	98	22	79
ALASKA					
Anchorage	61	−23	71	15	60
Barrow	71	−45	57	12	54
Fairbanks	64	−51	82	24	64
Juneau	58	− 4	74	15	61
Kodiak	57	10	69	10	60
Nome	64	−31	66	10	58
ARIZONA					
Flagstaff	35	− 2	84	31	61
Phoenix	33	31	109	27	76
Tucson	32	28	104	26	72
Yuma	32	36	111	27	79
ARKANSAS					
Fort Smith	35	12	101	24	80
Hot Springs	34	17	101	22	80
Little Rock	34	15	99	22	80
Pine Bluff	34	16	100	22	81
CALIFORNIA					
Bakersfield	35	30	104	32	73
Burbank	34	37	95	25	71
Fresno	36	28	102	34	72
Laguna Beach	33	41	83	18	70
Long Beach	33	41	83	22	70
Los Angeles	33	41	83	15	70
Monterey	36	35	75	20	64
Napa	38	30	100	30	71
Oakland	37	34	85	19	66
Oceanside	33	41	83	13	70
Palm Springs	33	33	112	35	76
Pasadena	34	32	98	29	73
Sacramento	38	30	101	36	72
San Diego	32	42	83	12	71
San Fernando	34	37	95	38	71
San Francisco	37	38	74	14	64
San Jose	37	34	85	26	68
Santa Barbara	34	34	81	24	68
Santa Cruz	36	35	75	28	64
Santa Monica	34	41	83	16	70
Stockton	37	28	100	37	71
COLORADO					
Boulder	40	2	93	27	64

OUTDOOR DESIGN TEMPERATURE

State and City	Lat.§	Winter DB*	Summer DB*	Daily Range*	WB*
Colorado Spgs.	38	− 3	91	30	63
Denver	39	− 5	93	28	64
Pueblo	38	− 7	97	31	67
CONNECTICUT					
Bridgeport	41	6	86	18	75
Hartford	41	3	91	22	77
New Haven	41	3	88	17	76
Waterbury	41	− 4	88	21	75
DELAWARE					
Dover	39	11	92	18	79
Wilmington	39	10	92	20	77
DISTRICT OF COLUMBIA					
Washington	38	14	93	18	78
FLORIDA					
Cape Kennedy	28	35	90	15	80
Daytona Beach	29	32	92	15	80
Fort Lauderdale	26	42	92	15	80
Key West	24	55	90	9	80
Miami	25	44	91	15	79
Orlando	28	35	94	17	79
Pensacola	30	25	94	14	80
St. Petersburg	27	36	92	16	79
Sarasota	27	39	93	17	79
Tallahassee	30	27	94	19	79
Tampa	27	36	92	17	79
GEORGIA					
Athens	33	18	94	21	78
Atlanta	33	17	94	19	77
Augusta	33	20	97	19	80
Griffin	33	18	93	21	78
Macon	32	21	96	22	79
Savannah	32	24	96	20	80
HAWAII					
Honolulu	21	62	87	12	76
Kaneohe Bay	21	65	85	12	76
IDAHO					
Boise	43	3	96	31	68
Idaho Falls	43	−11	89	38	65
Lewiston	46	− 1	96	32	67
Twin Falls	42	− 3	99	34	64
ILLINOIS					
Aurora	41	− 6	93	20	79
Bloomington	40	− 6	92	21	78
Carbondale	37	2	95	21	80
Champaign	40	− 3	95	21	78
Chicago	41	− 9	90	15	79
Galesburg	40	− 7	93	22	78
Joliet	41	− 5	93	20	78
Kankakee	41	− 4	93	21	78
Macomb	40	− 5	95	22	79

* in °F   § in degrees

continued

continued

OUTDOOR DESIGN TEMPERATURE					
State and City	**Lat.§**	**Winter**	**Summer**		
		DB*	**DB***	**Daily Range***	**WB***
Peoria	40	− 8	91	22	78
Rantoul	40	− 4	94	21	78
Rockford	42	− 9	91	24	77
Springfield	39	− 3	94	21	79
Waukegan	42	− 6	92	21	78
INDIANA					
Fort Wayne	41	− 4	92	24	77
Hobart	41	− 4	91	21	77
Indianapolis	39	− 2	92	22	78
Kokomo	40	− 4	91	22	77
Lafayette	40	− 3	94	22	78
Muncie	40	− 3	92	22	76
South Bend	41	− 3	91	22	77
Terre Haute	39	− 2	95	22	79
Valparaiso	41	− 3	93	22	78
IOWA					
Ames	42	−11	93	23	78
Burlington	40	− 7	94	22	78
Cedar Rapids	41	−10	91	23	78
Des Moines	41	−10	94	23	78
Dubuque	42	−12	90	22	77
Iowa City	41	−11	92	22	80
Keokuk	40	− 5	95	22	79
Sioux City	42	−11	95	24	78
KANSAS					
Garden City	37	− 1	99	28	74
Liberal	37	2	99	28	73
Russell	38	0	101	29	78
Topeka	39	0	99	24	79
Wichita	37	3	101	23	77
KENTUCKY					
Bowling Green	35	4	94	21	79
Lexington	38	3	93	22	77
Louisville	38	5	95	23	79
LOUISIANA					
Alexandria	31	23	95	20	80
Baton Rouge	30	25	95	19	80
Lafayette	30	26	95	18	81
Monroe	32	20	99	20	79
New Orleans	29	29	93	16	81
Shreveport	32	20	99	20	79
MAINE					
Augusta	44	− 7	88	22	74
Bangor	44	−11	86	22	73
Caribou	46	−18	84	21	71
Portland	43	− 6	87	22	74
MARYLAND					
Baltimore	39	10	94	21	78
Frederick	39	8	94	22	78
Salisbury	38	12	93	18	79

OUTDOOR DESIGN TEMPERATURE					
State and City	**Lat.§**	**Winter**	**Summer**		
		DB*	**DB***	**Daily Range***	**WB***
MASSACHUSETTS					
Boston	42	6	91	16	75
Clinton	42	− 2	90	17	75
Lawrence	42	− 6	90	22	76
Lowell	42	− 4	91	21	76
New Bedford	41	5	85	19	74
Pittsfield	42	− 8	87	23	73
Worcester	42	0	87	18	73
MICHIGAN					
Battle Creek	42	1	92	23	76
Benton Harbor	42	1	91	20	75
Detroit	42	3	91	20	76
Flint	42	− 4	90	25	76
Grand Rapids	42	1	91	24	75
Holland	42	2	88	22	75
Kalamazoo	42	1	92	23	76
Lansing	42	− 3	90	24	75
Marquette	46	−12	84	18	72
Pontiac	42	0	90	21	76
Port Huron	42	0	90	21	76
Saginaw	43	0	91	23	76
Sault Ste. Marie	46	−12	84	23	72
MINNESOTA					
Alexandria	45	−22	91	24	76
Duluth	46	−21	85	22	72
International Falls	48	−29	85	26	71
Minneapolis	44	−16	92	22	77
Rochester	43	−17	90	24	77
MISSISSIPPI					
Biloxi	30	28	94	16	82
Clarksdale	34	14	96	21	80
Jackson	32	21	97	21	79
Laurel	31	24	96	21	81
Natchez	31	23	96	21	81
Vicksburg	32	22	97	21	81
MISSOURI					
Columbia	38	− 1	97	22	78
Hannibal	39	− 2	96	22	80
Jefferson City	38	2	98	23	78
Kansas City	39	2	99	20	78
Kirksville	40	− 5	96	24	78
Moberly	39	− 2	97	23	78
St. Joseph	39	− 3	96	23	81
St. Louis	38	2	97	21	78
Springfield	37	3	96	23	78
MONTANA					
Billings	45	−15	94	31	67
Butte	45	−24	86	35	60
Great Falls	47	−21	91	28	64
Lewiston	47	−22	90	30	65

* in °F   § in degrees

continued

continued

OUTDOOR DESIGN TEMPERATURE

State and City	Lat.§	Winter DB*	Summer DB*	Daily Range*	WB*
Missoula	46	−13	92	36	65
NEBRASKA					
Columbus	41	− 6	98	25	77
Fremont	41	− 6	98	22	78
Grand Island	40	− 8	97	28	75
Lincoln	40	− 5	99	24	78
Norfolk	41	− 8	97	30	78
North Platte	41	− 8	97	28	74
Omaha	41	− 8	94	22	78
NEVADA					
Carson City	39	4	94	42	63
Las Vegas	36	25	108	30	71
Reno	39	6	96	45	64
NEW HAMPSHIRE					
Claremont	43	− 9	89	24	74
Concord	43	− 8	90	26	74
Manchester	42	− 8	91	24	75
Portsmouth	43	− 2	89	22	75
NEW JERSEY					
Atlantic City	39	10	92	18	78
Long Branch	40	10	93	18	78
Newark	40	10	94	20	77
New Brunswick	40	6	92	19	77
Trenton	40	11	91	19	78
NEW MEXICO					
Albuquerque	35	12	96	27	66
Carlsbad	32	13	103	28	72
Gallup	35	0	90	32	64
Los Alamos	35	5	89	32	62
Santa Fe	35	6	90	28	63
Silver City	32	5	95	30	66
NEW YORK					
Albany	42	− 6	91	23	75
Batavia	43	1	90	22	75
Buffalo	42	2	88	21	74
Geneva	42	− 3	90	22	75
Glens Falls	43	−11	88	23	74
Ithaca	42	− 5	88	24	74
Kingston	41	− 3	91	22	76
Lockport	43	4	89	21	76
New York City	40	11	92	17	76
Niagara Falls	43	4	89	20	76
Rochester	43	1	91	22	75
Syracuse	43	− 3	90	20	75
Utica	43	−12	88	22	75
NORTH CAROLINA					
Charlotte	35	18	95	20	77
Durham	35	16	94	20	78
Greensboro	36	14	93	21	77
Jacksonville	34	20	92	18	80

OUTDOOR DESIGN TEMPERATURE

State and City	Lat.§	Winter DB*	Summer DB*	Daily Range*	WB*
Wilmington	34	23	93	18	81
Winston-Salem	36	16	94	20	76
NORTH DAKOTA					
Bismark	46	−23	95	27	73
Fargo	46	−22	92	25	76
Grand Forks	47	−26	91	25	74
Williston	48	−25	91	25	72
OHIO					
Akron-Canton	40	1	89	21	75
Athens	39	0	95	22	78
Bowling Green	41	− 2	92	23	76
Cambridge	40	1	93	23	78
Cincinnati	39	1	92	21	77
Cleveland	41	1	91	22	76
Columbus	40	0	92	24	77
Dayton	39	− 1	91	20	76
Fremont	41	− 3	90	24	76
Marion	40	0	93	23	77
Newark	40	− 1	94	23	77
Portsmouth	38	5	95	22	78
Toledo	41	− 3	90	25	76
Warren	41	0	89	23	74
OKLAHOMA					
Bartlesville	36	6	101	23	77
Chickasha	35	10	101	24	78
Lawton	34	12	101	24	78
McAlester	34	14	99	23	77
Norman	35	9	99	24	77
Oklahoma City	35	9	100	23	78
Seminole	35	11	99	23	77
Stillwater	36	8	100	24	77
Tulsa	36	8	101	22	79
Woodward	36	6	100	26	78
OREGON					
Albany	44	18	92	31	69
Astoria	46	25	75	16	65
Baker	44	− 1	92	30	65
Eugene	44	17	92	31	69
Grants Pass	42	20	99	33	71
Klamath Falls	42	4	90	36	63
Medford	42	19	98	35	70
Portland	45	17	89	23	69
Salem	44	18	92	31	69
PENNSYLVANIA					
Allentown	40	4	92	22	76
Altoona	40	0	90	23	74
Butler	40	1	90	22	75
Erie	42	4	88	18	75
Harrisburg	40	7	94	21	77
New Castle	41	2	91	23	75

* in °F   § in degrees

continued

continued

OUTDOOR DESIGN TEMPERATURE					
		Winter	Summer		
State and City	Lat.§	DB*	DB*	Daily Range*	WB*
Philadelphia	39	10	93	21	77
Pittsburgh	40	1	89	22	74
Reading	40	9	92	19	76
West Chester	39	9	92	20	77
Williamsport	41	2	92	23	75
RHODE ISLAND					
Newport	41	5	88	16	76
Providence	41	5	89	19	75
SOUTH CAROLINA					
Charleston	32	25	94	13	81
Columbia	33	20	97	22	79
Florence	34	22	94	21	80
Sumter	33	22	95	21	79
SOUTH DAKOTA					
Aberdeen	45	−19	94	27	77
Brookings	44	−17	95	25	77
Huron	44	−18	96	28	77
Rapid City	44	−11	95	28	71
Sioux Falls	43	−15	94	24	76
TENNESSEE					
Athens	35	13	95	22	77
Chattanooga	35	13	96	22	78
Dyersburg	36	10	96	21	81
Knoxville	35	13	94	21	77
Memphis	35	13	98	21	80
Murfreesboro	34	9	97	22	78
Nashville	36	9	97	21	78
TEXAS					
Abilene	32	15	101	22	75
Alice	27	31	100	20	82
Amarillo	35	6	98	26	71
Austin	30	24	100	22	78
Beaumont	29	27	95	19	81
Big Spring	32	16	100	26	74
Brownsville	25	35	94	18	80
Corpus Christi	27	31	95	19	80
Dallas	32	18	102	20	78
El Paso	31	20	100	27	69
Forth Worth	32	17	101	22	78
Galveston	29	31	90	10	81
Houston	29	27	96	18	80
Huntsville	30	22	100	20	78
Laredo	27	32	102	23	78
Lubbock	33	10	98	26	73
Mcallen	26	35	97	21	80
Midland	31	16	100	26	73
Pecos	31	16	100	27	73
San Antonio	29	25	99	19	77
Temple	31	22	100	22	78
Tyler	32	19	99	21	80

OUTDOOR DESIGN TEMPERATURE					
		Winter	Summer		
State and City	Lat.§	DB*	DB*	Daily Range*	WB*
Victoria	28	29	98	18	82
Waco	31	21	101	22	78
Wichita Falls	33	14	103	24	77
UTAH					
Cedar City	37	− 2	93	32	65
Logan	41	− 3	93	33	65
Provo	40	1	98	32	66
Salt Lake City	40	3	97	32	66
VERMONT					
Barre	44	−16	84	23	73
Burlington	44	−12	88	23	74
Rutland	43	−13	87	23	74
VIRGINIA					
Charlottesville	38	14	94	23	77
Fredericksburg	38	10	96	21	78
Harrisonburg	38	12	93	23	75
Lynchburg	37	12	93	21	77
Norfolk	36	20	93	18	79
Petersburg	37	14	95	20	79
Richmond	37	14	95	21	79
Roanoke	37	12	93	23	75
Winchester	39	6	93	21	77
WASHINGTON					
Aberdeen	46	25	80	16	65
Bellingham	48	10	81	19	68
Olympia	46	16	87	32	67
Seattle-Tacoma	47	17	84	22	66
Spokane	47	− 6	93	28	65
Walla Walla	46	0	97	27	69
WEST VIRGINIA					
Charleston	38	7	92	20	76
Clarksburg	39	6	92	21	76
Parkersburg	39	7	93	21	77
Wheeling	40	1	89	21	74
WISCONSIN					
Beloit	42	− 7	92	24	78
Fon Du Lac	43	−12	89	23	76
Green Bay	44	−13	88	23	76
La Crosse	43	−13	91	22	77
Madison	43	−11	91	22	77
Milwaukee	42	− 8	90	21	76
Racine	42	− 6	91	21	77
Sheboygan	43	−10	89	20	77
Wausau	44	−16	91	23	76
WYOMING					
Casper	42	−11	92	31	63
Cheyenne	41	− 9	89	30	63
Laramie	41	−14	84	28	61
Rock Springs	41	− 9	86	32	59
Sheridan	44	−14	94	32	66

* in °F   § in degrees   •*Air Conditioning Contractors of America*

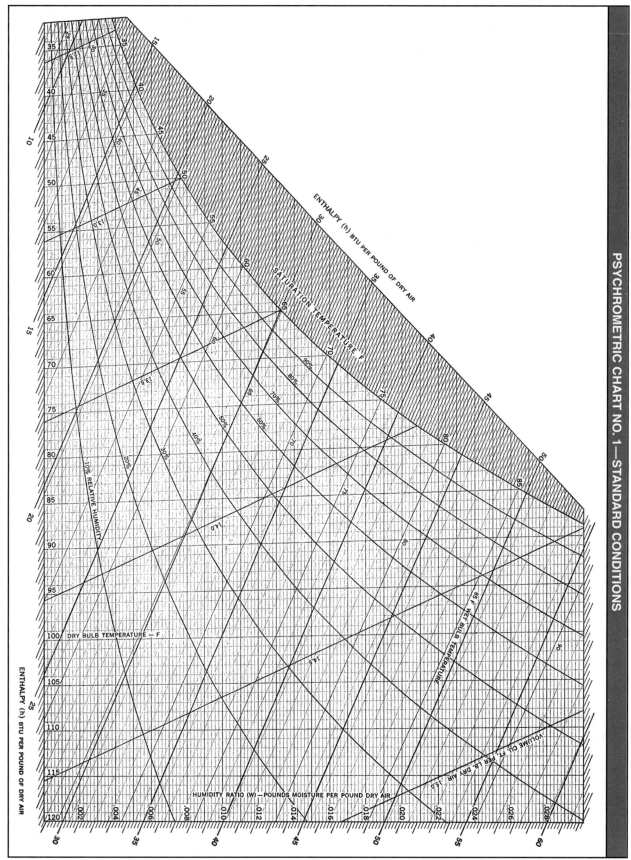

FLOW COEFFICIENTS (C_v) FOR WATER VALVE SIZING*

C_v	Differential Pressure†							C_v	Differential Pressure†						
	3	5	7	9	12	15	20		3	5	7	9	12	15	20
0.2	0.3	0.4	0.5	.06	0.7	0.8	0.9	26	45	58	69	78	90	101	118
0.4	0.7	0.9	1.1	1.2	1.4	1.5	1.8	28	48	63	74	84	97	108	125
0.6	1.05	1.35	1.6	1.8	2.05	2.3	2.7	30	52	67	79	90	104	116	135
0.8	1.3	1.7	2.1	2.4	2.7	3	3.5	32	55	72	85	96	111	124	144
1.0	1.7	2.2	2.6	3	3.5	3.9	4.5	34	59	76	90	102	118	132	152
1.2	2	2.6	3.1	3.6	4.1	4.6	5.25	36	62.5	80	95	108	125	140	161
1.4	2.4	3.1	3.7	4.2	4.8	5.4	6.1	38	66	85	101	114	132	147	170
1.6	2.8	3.6	4.2	4.8	5.5	6.2	7.1	40	69	89	106	120	139	155	180
1.8	3.1	4	4.7	5.4	6.2	6.9	8.1	45	78	101	119	135	156	—	—
2.0	3.4	4.4	5.2	6	6.9	7.7	9	50	87	112	132	150	173	—	—
2.2	3.8	4.9	5.8	6.6	7.6	8.5	9.75	55	95	125	145	165	190	—	—
2.4	4.1	5.3	6.2	7.1	8.2	9.25	10.8	60	104	134	159	180	208	—	—
2.6	4.5	4.5	5.8	6.8	7.8	9	11.9	65	113	145	172	195	225	—	—
2.8	4.8	6.3	7.4	8.4	10	11	12.5	70	121	157	185	210	242	—	—
3.0	5.2	6.7	7.9	9	10	12	13.5	80	139	179	212	240	277	—	—
3.2	5.5	7.1	8.4	9.6	11	12	14.2	90	156	201	238	270	312	—	—
3.4	5.9	7.6	9	10	12	13	15.2	100	175	225	265	300	350	—	—
3.6	6.2	8	9.5	11	12.5	14	16	120	208	268	318	360	416	—	—
3.8	6.5	8.4	10	11.5	13	15	17	140	242	313	370	420	485	—	—
4.0	6.9	9	11	12	14	15	18	160	280	350	425	475	560	—	—
4.5	7.7	10	12	14	16	17	20	180	313	400	475	540	625	—	—
5.0	8.6	11	13	15	17	19	22.5	200	346	447	529	600	693	—	—
5.5	9.5	12	15	17	19	21	24.2	220	381	492	582	660	762	—	—
6.0	10	13	16	18	21	23	27	240	416	537	635	720	831	—	—
6.5	11	14	17	19	23	25	29	260	450	580	680	775	900	—	—
7.0	12	16	19	21	24	27	31.9	280	485	626	741	840	970	—	—
7.5	13	17	20	23	26	29	34	300	520	671	794	900	1039	—	—
8.0	14	18	21	24	28	31	36	320	550	720	840	950	1100	—	—
8.5	14.5	19	22.5	25.9	29	33	38	340	590	755	900	1000	1175	—	—
9.0	15.5	20	24	27	31	35	40	360	625	800	950	1100	1250	—	—
9.5	16.5	21	25	28.5	33	36.5	42.5	380	660	850	1000	1150	1350	—	—
10	17	22	26	30	35	39	45	400	700	890	1075	1200	1400	—	—
11	19	25	29	33	38	43	48	450	780	1000	1200	1375	1580	—	—
12	21	27	32	36	42	46	54	500	866	1118	1323	1500	1732	—	—
13	23	29	34	39	45	50	58	550	950	1230	1450	1650	1900	—	—
14	24	31	37	42	48	54	62.5	600	1039	1342	1587	1800	2078	—	—
15	26	34	40	45	52	58	68	650	1125	1475	1700	1950	2250	—	—
16	28	36	42	48	55	62	72	700	1212	1565	1852	2100	2425	—	—
17	29	38	45	51	59	66	75	750	1300	1675	2000	2250	2600	—	—
18	31	40	48	54	62	70	80	800	1400	1800	2250	2400	2800	—	—
19	33	42	50	57	66	74	85	850	1475	1900	2400	2575	2950	—	—
20	34	44	52	60	69	77	89	900	1559	2012	2381	2700	3117	—	—
22	38	49	58	66	76	85	98	950	1645	2124	2514	2850	3291	—	—
24	42	54	64	72	83	93	108	1000	1750	2250	2650	3000	3450	—	—

* water capacity in gpm

† in psi

FLOW COEFFICIENTS (C_V) FOR STEAM VALVE SIZING*												
	Inlet Pressure†											
C_V	2		5		10			15	25	35	50	75	100
	Differential Pressure†											
	1	2	2	5	2	5	10	13	18	22	29	39	52
0.2	2.4	3.3	3.6	5.5	4.1	6.3	8.4	10	14	18	23	32	44
0.4	4.8	6.7	7.3	11	8.2	13	17	21	28	35	46	63	85
0.6	7.2	10	11	17	12	19	25	31	42	53	68	95	121
0.9	11	15	16	25	19	28	38	47	63	79	103	142	190
1	12	17	18	28	21	32	42	52	70	88	114	155	202
1.2	14	20	21	33	25	38	50	62	84	105	136	179	242
1.4	17	23	25	39	29	44	59	73	98	123	159	221	283
1.6	19	26	29	44	33	50	67	83	112	140	181	252	323
1.8	22	30	33	50	37	57	75	94	126	158	205	284	363
2	24	33	36	55	41	63	84	105	140	175	228	316	404
2.2	26	37	40	61	45	69	92	115	154	193	251	347	444
2.4	28	40	43	66	49	75	100	125	168	210	273	379	485
2.6	31	43	47	72	54	82	109	136	182	228	296	411	525
2.8	34	47	51	78	58	88	117	146	196	245	319	444	565
3	36	51	55	83	62	95	126	157	210	263	342	474	606
3.2	38	55	58	89	66	101	134	167	224	280	364	505	646
3.4	41	58	62	94	70	107	139	178	238	296	387	537	686
3.6	43	62	66	100	74	113	151	188	252	315	410	568	727
3.8	46	66	69	105	78	120	159	199	266	333	433	600	767
4	48	69	73	110	82	126	168	209	280	350	456	632	808
4.5	54	73	82	125	93	142	189	235	315	394	513	711	909
5	60	82	91	139	103	158	210	262	350	438	570	790	1010
5.5	66	91	100	152	113	173	230	288	384	481	626	868	1110
6	72	100	109	166	124	189	251	314	419	525	683	947	1211
6.5	78	109	118	180	134	205	272	340	454	569	740	1026	1312
7	84	118	127	194	144	221	293	366	489	613	797	1105	1450
7.5	90	127	137	208	155	236	314	392	524	656	854	1184	1514
8	96	137	146	222	165	252	335	418	559	700	911	1263	1615
8.5	102	146	155	235	175	268	356	444	594	744	965	1342	1709
9	108	155	164	249	185	284	377	471	629	788	1025	1421	1803
9.5	114	164	173	263	195	299	398	497	644	831	1090	1500	1900
10	120	173	182	277	206	315	419	523	699	875	1139	1579	2019
11	132	182	200	300	226	346	461	575	768	962	1252	1736	2221
12	144	200	218	332	247	378	503	628	839	1050	1367	1895	2600
13	156	218	237	360	268	410	545	680	909	1100	1481	2052	2625
14	168	237	255	388	288	441	587	732	979	1225	1595	2211	2827
15	180	251	273	416	309	473	629	785	1049	1313	1709	2369	3029
16	192	267	291	443	330	504	670	837	1118	1400	1822	2526	3230

* steam capacity in lb/hr

† in psi

continued

	FLOW COEFFICIENTS (C$_v$) FOR STEAM VALVE SIZING*												
	Inlet Pressure†												
C$_v$	2		5		10			15	25	35	50	75	100
	Differential Pressure†												
	1	2	2	5	2	5	10	13	18	22	29	39	52
17	204	284	309	471	350	536	712	889	1188	1488	1936	2684	3432
18	216	301	328	499	371	567	754	941	1258	1575	2050	2842	3634
19	228	317	346	526	391	598	796	994	1328	1665	2164	3000	3836
20	240	334	364	554	412	630	838	1046	1398	1750	2278	3158	4038
22	264	367	400	609	453	693	899	1150	1538	1925	2506	3474	4422
24	288	400	436	664	494	756	1006	1252	1678	2100	2734	3790	4846
26	312	434	473	720	536	819	1089	1360	1817	2275	2961	4105	5249
28	336	468	510	776	577	882	1173	1464	1979	2450	3189	4421	5653
30	360	501	546	831	618	945	1257	1569	2097	2625	3417	4737	6057
32	384	534	582	886	659	1008	1341	1674	2237	2800	3645	5053	6461
34	408	567	619	942	700	1071	1425	1778	2377	2975	3873	5368	6865
36	432	601	655	997	742	1134	1508	1883	2516	3150	4100	5684	7268
38	456	635	692	1053	783	1197	1592	1987	2625	3325	4328	6000	7672
40	480	668	728	1108	824	1260	1676	2092	2900	3600	4785	6631	8479
45	540	751	819	1246	927	1417	1885	2363	3225	4000	5240	7262	9286
50	600	835	910	1385	1030	1575	2095	2615	3495	4375	5695	7895	10095
55	660	919	1001	1524	1133	1733	2305	2877	3845	4812	6265	8685	11104
60	720	1002	1092	1662	1236	1890	2514	3138	4194	5250	6834	9474	12114
65	780	1086	1183	1801	1339	2048	2724	3400	4544	5688	7404	10264	13124
70	840	1169	1274	1939	1442	2205	2933	3661	4893	6125	8354	11580	14806
80	960	1335	1456	2215	1648	2520	3351	4184	5591	7000	9304	12896	16488
90	1080	1503	1638	2493	1854	2835	3771	4707	6291	7875	10251	14211	18171
100	1200	1669	1820	2770	2063	3150	4190	5230	6990	8750	12339	17106	21872
120	1440	2004	2184	3324	2472	3780	5028	6276	8388	10500	14427	20001	25573
140	1680	2338	2548	3878	2884	4410	5866	8322	9786	12250	16515	22896	29274
160	1920	2672	2912	4432	3296	5040	6704	9034	11184	14000	18603	25791	32975
180	2160	3006	3276	4986	3708	5670	7542	9746	12582	15750	20691	28686	36676
200	2400	3340	3640	5540	4120	6300	8380	10460	13980	17500	22780	31580	40380
220	2640	3674	4004	6094	4532	6930	9128	11511	15378	19250	25058	34738	44418
240	2880	4008	4368	6648	4944	7560	10056	12552	16776	21000	27366	37896	48608
260	3120	4342	4732	7202	5356	8190	10894	13598	18174	22750	29614	41054	52794
280	3360	4677	5096	7757	5768	8820	11732	14644	19572	24500	32057	44447	56982
300	3600	5010	5460	8310	6180	9450	12570	15690	20970	26250	34500	47840	61170
320	3840	5344	5824	8864	6592	10080	13408	16736	22370	28000	37707	52284	66853
340	4080	5678	6488	9418	7004	10710	14246	17782	23770	29750	40914	56728	72536
360	4320	6012	6552	9972	7416	11340	15084	18828	25170	31500	44121	61172	78219
380	4560	6346	6916	10536	7828	11970	15922	19874	26570	33250	47328	65616	83902
400	4800	6680	7280	11080	8240	12600	16760	20920	27970	35000	50535	70060	89585
450	5400	7515	8190	12465	9270	14175	18855	23535	31465	39375	53742	74504	95268
500	6000	8350	9100	13850	10300	15750	20950	26150	34950	43750	56950	78950	10000
550	6594	9176	10000	15221	11319	17309	23024	28738	38410	48081	59000	84000	10500
600	7188	10002	10900	16592	12338	18868	25098	31326	41870	52417	64000	88000	11000

* steam capacity in lb/hr

† in psi

PNEUMATIC RELAY MODEL NUMBER COMPARISON

Relay Type	Honeywell	Johnson	Kreuter	Pneuline	Siemens (Powers)
Diverting/Switching	RP670A1001	V-6135-1	RCC-1009	RPP100	243-0001
High-Pressure Selector	RP470A1003	C-5226-3	RCC-1008	RPP400	243-0018
Low-Pressure Selector	RP970A1008	C-5226-3	RCC-1006	RPP500	243-0020
High/Low-Pressure Selector	RP913A1008	C-2220	RCC-1111	RPP900	243-0019
Booster/Repeater Ratio 1:1	RP970A1008	R-2080-1	RCC-1516	RPP700	
Reversing	RP972A1006	R-3030-1	RCC-1504	RPP300	243-0009
Minimum Position	SP970A1005	C-5230-2	—	—	
Averaging	RP973A1013	C-2040-1	RCC-1514	RPP800	243-0011

RECEIVER CONTROLLER MODEL NUMBER COMPARISON

Type	CPA	Barber-Colman	Honeywell	Johnson	Robertshaw	Siemens (Powers)
Single Input	with	RKS-2001	RP920A1025	T-5800-1	2341-512	195-0011
	without	RKS-1001	RP920A1033			
Dual Input	with	RKS-4002	RP920B1023	T-5800-3		195-0003
	without	RKS-3002	RP920B1031			

PNEUMATIC THERMOSTAT MODEL NUMBER COMPARISON

Type	Action	Barber-Colman	Honeywell	Johnson	Robertshaw	Siemens (Powers)
Single Temperature	Direct-Acting	TKR-1001	TP970A2004	T-4002-201	2212-118	192-202
Two-Pipe	Reverse-Acting	TKR-1101	TP970B2002	T-4002-202	2212-119	192-203
Day/Night	Direct-Acting	TK-1301	TP971A2003	T-4506-201	2216-126	192-204
	Reverse-Acting	TK-1381	TP971B2001	T-4506-209	2214-122	192-205
Summer/Winter	Direct-Acting (lo), Reverse-Acting (hi)	TK-1741	—	T-4756-201	2218-133	192-209
	Reverse-Acting (lo), Direct-Acting (hi)	TK-1731	TP972A2002	T-4756-205	2218-132	192-208
Covers						
Blank		Included	14001984-500	T-4000-2138	21-928	192-256
With Exposed Setpoint and Thermometer			14002132-101	T-4000-2142	21-933	192-266
Universal Conversion Kits		AT-536	14002573-001	T-4000-605	22-022	192-300

TRANSMITTER TEMPERATURE/PRESSURE RELATIONSHIPS

Output Pressure*	Temperature Range†															
	−25 125	−20 80	30 80	50 90	0 100	50 100	−40 120	20 120	25 125	35 135	40 140	50 150	−40 160	0 200	40 240	80 240
3.0	−25	−20	30	50	0	50	−40	20	25	35	40	50	−40	0	40	80
3.24	−22	−18	31	50.8	2	51	−36.8	22	27	37	42	52	−36	4	44	83.2
3.48	−19	−16	32	51.6	4	52	−33.6	24	29	39	44	54	−32	8	48	86.4
3.72	−16	−14	33	52.4	6	53	−30.4	26	31	41	46	56	−28	12	52	89.6
3.96	−13	−12	34	53.2	8	54	−27.2	28	33	43	48	58	−24	16	56	92.8
4.2	−10	−10	35	54	10	55	−24	30	35	45	50	60	−20	20	60	96
4.44	−7	−8	36	54.8	12	56	−20.8	32	37	47	52	62	−16	24	64	99.2
4.68	−4	−6	37	55.6	14	57	−17.6	34	39	49	54	64	−12	28	68	102.4
4.92	−1	−4	38	56.4	16	58	−14.4	36	41	51	56	66	−8	32	72	105.6
5.16	2	−2	39	57.2	18	59	−11.2	38	43	53	58	68	−4	36	76	108.8
5.4	5	0	40	58	20	60	−8	40	45	55	60	70	0	40	80	112
5.64	8	2	41	58.8	22	61	−4.8	42	47	57	62	72	4	44	84	115.2
5.88	11	4	42	59.6	24	62	−1.6	44	49	59	64	74	8	48	88	118.4
6.12	14	6	43	60.4	26	63	1.6	46	50	61	66	76	12	52	92	121.6
6.36	17	8	44	61.2	28	64	4.8	48	51	63	68	78	16	56	96	124.8
6.6	20	10	45	62	30	65	8	50	53	65	70	80	20	60	100	128
6.84	23	12	46	62.8	32	66	11.2	52	55	67	72	82	24	64	104	131.2
7.08	26	14	47	63.6	34	67	14.4	54	57	69	74	84	28	68	108	134.4
7.32	29	16	48	64.4	36	68	17.6	56	59	71	76	86	32	72	112	137.6
7.56	32	18	49	65.2	38	69	20.8	58	61	73	78	88	36	76	116	140.8
7.8	35	20	50	66	40	70	24	60	63	75	80	90	40	80	120	144
8.04	38	22	51	66.8	42	71	27.2	62	65	77	82	92	44	84	124	147.2
8.28	41	24	52	67.6	44	72	30.4	64	67	79	84	94	48	88	128	150.4
8.52	44	26	53	68.4	46	73	33.6	66	69	81	86	96	52	92	132	153.6
8.76	47	28	54	69.2	48	74	36.8	68	71	83	88	98	56	96	136	156.8
9.0	50	30	55	70	50	75	40	70	73	85	90	100	60	100	140	160
9.24	53	32	56	70.8	52	76	43.2	72	75	87	92	102	64	104	144	163.2
9.48	56	34	57	71.6	54	77	46.4	74	77	89	94	104	68	108	148	166.4
9.72	59	36	58	72.4	56	78	49.6	76	79	91	96	106	72	112	152	169.6
9.96	62	38	59	73.2	58	79	52.8	78	81	93	98	108	76	116	156	172.8
10.2	65	40	60	74	60	80	56	80	83	95	100	110	80	120	160	176
10.44	68	42	61	74.8	62	81	59.2	82	85	97	102	112	84	124	164	179.2
10.68	71	44	62	75.6	64	82	62.4	84	87	99	104	114	88	128	168	182.4
10.92	74	46	63	76.4	66	83	65.6	86	89	101	106	116	92	132	172	185.6
11.16	77	48	64	77.2	68	84	68.8	88	91	103	108	118	96	136	176	188.8
11.4	80	50	65	78	70	85	72	90	93	105	110	120	100	140	180	192
11.64	83	52	66	78.8	72	86	75.2	92	95	107	112	122	104	144	184	195.2
11.88	86	54	67	79.6	74	87	78.4	94	97	109	114	124	108	148	188	198.4
12.12	89	56	68	80.4	76	88	81.6	96	99	111	116	126	112	152	192	201.6
12.36	92	58	69	81.2	78	89	84.8	98	101	113	118	128	116	156	196	204.8
12.6	95	60	70	82	80	90	88	100	103	115	120	130	120	160	200	208
12.84	98	62	71	82.8	82	91	91.2	102	105	117	122	132	124	164	204	211.2
13.08	101	64	72	83.6	84	92	94.4	104	107	119	124	134	128	168	208	214.4
13.32	104	66	73	84.4	86	93	97.6	106	109	121	126	136	132	172	212	217.6
13.56	107	68	74	85.2	88	94	100.8	108	111	123	128	138	136	176	216	220.8
13.8	110	70	75	86	90	95	104	110	113	125	130	140	140	180	220	224
14.04	113	72	76	86.8	92	96	107.2	112	115	127	132	142	144	184	224	227.2
14.28	116	74	77	87.6	94	97	110.4	114	117	129	134	144	148	188	228	230.4
14.52	119	76	78	88.4	96	98	113.6	116	119	131	136	146	152	192	232	233.6
14.76	122	78	79	89.2	98	99	116.8	118	123	133	138	148	156	196	236	236.8
15.0	125	80	80	90	100	100	120	120	125	135	140	150	160	200	240	240

* in psi

† in °F

RTD ELECTRONIC SENSOR TEMPERATURE/RESISTANCE RELATIONSHIPS

Temperature		Resistance*			Temperature		Resistance*		
°F	°C	Nickel	Platinum	Silicon	°F	°C	Nickel	Platinum	Silicon
–50	–46	674	821	—	110	43	1121	1168	1186
–40	–40	699	843	602	120	49	1152	1190	1234
–30	–34	725	865	633	130	54	1184	1211	1283
–20	–29	751	887	665	140	60	1216	1232	1333
–10	–23	777	909	698	150	66	1248	1254	1385
0	–18	803	930	732	160	71	1281	1275	1437
10	–12	830	952	768	170	77	1314	1296	1491
20	–7	858	994	804	180	82	1348	1317	1546
30	–1	885	996	842	190	88	1382	1339	1602
40	4	914	1017	881	200	93	1417	1360	1659
50	10	942	1039	921	210	99	1452	1381	1718
60	16	971	1061	962	220	104	1487	1402	—
70	21	1000	1082	1005	230	110	1524	1423	—
80	27	1030	1104	1048	240	116	1560	1444	—
90	32	1060	1125	1093	250	121	1597	1465	—
100	38	1090	1147	1139					

* in Ω

HVAC GENERAL GUIDELINES

When solving for. . .	General Guideline	When solving for. . .	General Guideline
Absorbers	18 lb steam/t	1φ electric motors	
Air movement	6 air changes/hr, 1 cfm/sq ft	universal	1.07 kW/hp
Boiler horsepower	33,479 Btu/hr, 9.8 kW	repulsion	2.25 kW/hp
		capacitor	2.97 kW/hp
Chilled water	2.4 gpm/t*	split-phase	1.86 kW/hp
Chiller input		Fan energy	1000–1500 cfm/hp
centrifugal	.54 kW/t	Heat transmission	
screw	.7 kW/t	general building	.15–.5 Btu/sq ft/°F
reciprocating	.8 kW/t	glass	1 Btu/sq ft/°F
		infiltration	.5–1.5 air changes/hr
Chiller size	300–400 sq ft/t	masonry wall	.15–.3 Btu/sq ft/°F
Condenser water	3 gpm/t*	Lighting	2–4 W/sq ft
Cooling towers		Occupant density	100–150 sq ft/person
small (under 100 t)	1.5–5 kW/t	Occupant load	450 Btu/person/hr
medium (100 t–1000 t)	4–10 kW/t	Reheat systems	55°F supply air, 30°F–40°F reheat coil temperature rise
large (1000 t–10,000 t)	9–14 kW/t	Ton	12,000 Btu/hr
Dual-duct system	55°F cold deck, 70°F–105°F hot deck†	VAV systems	55°F cooling, 10% box leakage flow, 40%–50% minimum fan volume
3φ electric motors	.88 kW/hp	Ventilation rate	15 cfm/person

* with 10°F temperature rise
† w/outside air reset schedule

REFRIGERANT PROPERTIES

Refrigerant No.	Refrigerant Name	Chemical Formula	Molecular Mass	Normal Boiling Point*	Freezing Point*	Critical Temperature*	Critical Pressure§	Critical Volume#
704	Helium	He	4.0026	−452.1	None	−450.3	33.21	.2311
702	Hydrogen (normal)	H₂	2.0159	−423.0	−434.5	−399.9	190.8	.5320
702	Hydrogen (para)	H₂	2.0159	−423.2	−434.8	−400.3	187.5	.5097
720	Neon	Ne	20.183	−410.9	−415.5	−379.7	493.1	.03316
728	Nitrogen	N₂	28.013	−320.4	−346.0	−232.4	492.9	.05092
729	Air	. . .	28.97	−317.8	. . .	−221.3	547.4	.04883
740	Argon	A	39.948	−302.6	−308.7	−188.1	710.4	.02990
732	Oxygen	O₂	31.9988	−297.3	−361.8	−181.1	736.9	.0375
50	Methane	CH₄	16.04	−258.7	−296	−116.5	673.1	.099
14	Tetrafluoromethane	CF₄	88.01	−198.3	−299	− 50.2	543	.0256
1150	Ethylene	C₂H₄	28.05	−154.7	−272	48.8	742.2	.070
170	Ethane	C₂H₆	30.07	−127.85	−297	90.0	709.8	.0830
744	Nitrous Oxide	N₂O	44.02	−129.1	−152	97.7	1048	.0355
23	Trifluoromethane	CHF₃	70.02	−115.7	−247	78.1	701.4	.0311
13	Chlorotrifluoromethane	CClF₃	104.47	−114.6	−294	83.9	561	.0277
744	Carbon Dioxide	CO₂	44.01	−109.2	− 69.9	87.9	1070.0	.0342
13	Bromotrifluoromethane	CBrF₃	148.93	− 71.95	−270	152.6	575	.0215
1270	Propylene	C₃H₆	42.09	− 53.86	−301	197.2	670.3	.0720
290	Propane	C₃H₈	44.10	− 43.73	−305.8	206.3	617.4	.0728
22	Chlorodifluoromethane	CHClF₂	86.48	− 41.36	−256	204.8	721.9	.0305
115	Chloropentafluoroethane	CClF₂CF₃	154.48	− 38.4	−159	175.9	457.6	.0261
717	Ammonia	NH₃	17.03	− 28.0	−107.9	271.4	1657	.068
12	Dichlorodifluoromethane	CCl₂F₂	120.93	− 21.62	−252	233.6	596.9	.0287
134	Tetrafluoroethane	CF₃CH₂F	102.03	− 15.08	−141.9	214.0	589.8	.029
152	Difluoroethane	CH₃CHF₂	66.05	− 13.0	−178.6	236.3	652	.0439
40	Methyl Chloride	CH₃Cl	50.49	− 11.6	−144	289.6	968.7	.0454
600	Isobutane	C₄H₁₀	58.13	10.89	−255.5	275.0	529.1	.0725
764	Sulfur Dioxide	SO₂	64.07	14.0	−103.9	315.5	1143	.0306
142	Chlorodifluoroethane	CH₃CClF₂	100.5	14.4	−204	278.8	598	.0368
630	Methyl Amine	CH₃NH₂	31.06	19.9	−134.5	314.4	1082	
318	Octafluorocyclobutane	C₄F₈	200.04	21.5	− 42.5	239.6	403.6	.0258
600	Butane	C₄H₁₀	58.13	31.1	−217.3	305.6	550.7	.0702
114	Dichlorotetrafluoroethane	CClF₂CClF₂	170.94	38.8	−137	294.3	473	.0275
21	Dichlorofluoromethane	CHCl₂F	102.92	47.8	−211	353.3	750	.0307
160	Ethyl Chloride	C₂H₅Cl	64.52	54.32	−216.9	369.0	764.4	.0485
631	Ethyl Amine	C₂H₅NH₂	45.08	61.88	−113	361.4	815.6	
11	Trichlorofluoromethane	CCl₃F	137.38	74.87	−168	388.4	639.5	.0289
611	Methyl Formate	C₂H₄O₂	60.05	89.22	−146	417.2	870	.0459
610	Ethyl Ether	C₄H₁₀O	74.12	94.3	−177.3	381.2	523	.0607
216	Dichlorohexafluoropropan	C₃Cl₂F₆	220.93	96.24	−193.7	356.0	399.5	.0279
30	Methylene Chloride	CH₂Cl₂	84.93	104.4	−142	458.6	882	
113	Trichlorotrifluoroethane	CCl₂FCClF₂	187.39	117.63	− 31	417.4	498.9	.0278
1130	Dichloroethylene	CHCl=CHCl	96.95	118	− 58	470	795	
1120	Trichloroethylene	CHCl=CCl₂	131.39	189.0	− 99	520	728	
718	Water	H₂O	18.02	212	32	705.6	3208	.0501

* in °F
§ in psia
# in cu ft/lb

•*ASHRAE 1989 Handbook — Fundamentals*

RECEIVER CONTROLLER PORT CROSS REFERENCE

Manufacturer	Model Number	Branch (output)	Main	Primary	Secondary (reset)	Control Point Adjustment
Johnson Controls	T-5800 T-9000	O O	S S	CV II	M III	SP I
Siemens (Powers)	RC-195	C	S	1 DA 2 RA	3	2 DA 1 RA
Honeywell	RP-908 RP-920	B 2	M 1	1 3	2 5	CPA 9
Robertshaw	2341	B	M	1	3	2

THROTTLING RANGE/TRANSMITTER SPAN/PROPORTIONAL BAND RELATIONSHIPS

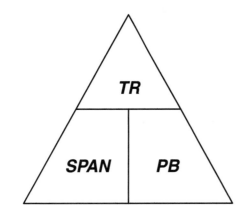

TR = THROTTLING RANGE (in ˚F, %rh, or in. wc)
$SPAN$ = TRANSMITTER SPAN
PB = PROPORTIONAL BAND (in %)

Example A:
A receiver controller is connected to a pneumatic mixed air transmitter with a 0˚F–100˚F span. The receiver controller has a 5% proportional band. What is the throttling range?

$TR = SPAN \times PB$
$TR = 100˚F \times 5\%$
$TR = \mathbf{5˚F}$

Example B:
A hot water system has a throttling range of 10˚F and a proportional band setpoint of 10%. What is the transmitter span?

$SPAN = \dfrac{TR}{PB}$

$SPAN = \dfrac{10˚F}{10\%}$

$SPAN = \mathbf{100˚F}$

PROPERTIES OF SATURATED STEAM						
Gauge Pressure*	Absolute Pressure†	Temperature‡	Sensible Heat§	Latent Heat§	Total Heat§	Specific Volume Steam**
32	46.7	276.7	245.9	927.6	1173.5	9.08
34	48.7	279.4	248.5	925.8	1174.3	8.73
36	50.7	281.9	251.1	924.0	1175.1	8.40
38	52.7	284.4	253.7	922.1	1175.8	8.11
40	54.7	286.7	256.1	920.4	1176.5	7.83
42	56.7	289.0	258.5	918.6	1177.1	7.57
44	58.7	291.3	260.8	917.0	1177.8	7.33
46	60.7	293.5	263.0	915.4	1178.4	7.10
48	62.7	295.6	265.2	913.8	1179.0	6.89
50	64.7	297.7	267.4	912.2	1179.6	6.68
52	66.7	299.7	269.4	910.7	1180.1	6.50
54	68.7	301.7	271.5	909.2	1180.7	6.32
56	70.7	303.6	273.5	907.8	1181.3	6.16
58	72.7	305.5	275.3	906.5	1181.8	6.00
60	74.7	307.4	277.1	905.3	1182.4	5.84
62	76.7	309.2	279.0	904.0	1183.0	5.70
64	78.7	310.9	280.9	902.6	1183.5	5.56
66	80.7	312.7	282.8	901.2	1184.0	5.43
68	82.7	314.3	284.5	900.0	1184.5	5.31
70	84.7	316.0	286.2	898.8	1185.0	5.19
72	86.7	317.7	288.0	897.5	1185.5	5.08
74	88.7	319.3	289.4	896.5	1185.9	4.97
76	90.7	320.9	291.2	895.1	1186.3	4.87
78	92.7	322.4	292.9	893.9	1186.8	4.77
80	94.7	323.9	294.5	892.7	1187.2	4.67
82	96.7	325.5	296.1	891.5	1187.6	4.58
84	98.7	326.9	297.6	890.3	1187.9	4.49
86	100.7	328.4	299.1	889.2	1188.3	4.41
88	102.7	329.9	300.6	888.1	1188.7	4.33
90	104.7	331.2	302.1	887.0	1189.1	4.25
92	106.7	332.6	303.5	885.8	1189.3	4.17
94	108.7	333.9	304.9	884.8	1189.7	4.10
96	110.7	335.3	306.3	883.7	1190.0	4.03
98	112.7	336.6	307.7	882.6	1190.3	3.96
100	114.7	337.9	309.0	881.6	1190.6	3.90
102	116.7	339.2	310.3	880.6	1190.9	3.83
104	118.7	340.5	311.6	879.6	1191.2	3.77
106	120.7	341.7	313.0	878.5	1191.5	3.71
108	122.7	343.0	314.3	877.5	1191.8	3.65
110	124.7	344.2	315.5	876.5	1192.0	3.60
112	126.7	345.4	316.8	875.5	1192.3	3.54
114	128.7	346.5	318.0	874.5	1192.5	3.49

* in psig

† in psia

‡ in °F

§ in Btu/lb

** in cu ft/lb

continued

PROPERTIES OF SATURATED STEAM						
Gauge Pressure*	Absolute Pressure†	Temperature‡	Sensible Heat§	Latent Heat§	Total Heat§	Specific Volume Steam**
116	130.7	347.7	319.3	873.5	1192.8	3.44
118	132.7	348.9	320.5	872.5	1193.0	3.39
120	134.7	350.1	321.8	871.5	1193.3	3.34
125	139.7	352.8	324.7	869.3	1194.0	3.23
130	144.7	355.6	327.6	866.9	1194.5	3.12
135	149.7	358.3	330.6	864.5	1195.1	3.02
140	154.7	360.9	333.2	862.5	1195.7	2.93
145	159.7	363.5	335.9	860.3	1196.2	2.84
150	164.7	365.9	338.6	858.0	1196.6	2.76
155	169.7	368.3	341.1	856.0	1197.1	2.68
160	174.7	370.7	343.6	853.9	1197.5	2.61
165	179.7	372.9	346.1	851.8	1197.9	2.54
170	184.7	375.2	348.5	849.8	1198.3	2.48
175	189.7	377.5	350.9	847.9	1198.8	2.41
180	194.7	379.6	353.2	845.9	1199.1	2.35
185	199.7	381.6	355.4	844.1	1199.5	2.30
190	204.7	383.7	357.6	842.2	1199.8	2.24
195	209.7	385.7	359.9	840.2	1200.1	2.18
200	214.7	387.7	362.0	838.4	1200.4	2.14
210	224.7	391.7	366.2	834.8	1201.0	2.04
220	234.7	395.5	370.3	831.2	1201.5	1.96
230	244.7	399.1	374.2	827.8	1202.0	1.88
240	254.7	402.7	378.0	824.5	1202.5	1.81
250	264.7	406.1	381.7	821.2	1202.9	1.74
260	274.7	409.3	385.3	817.9	1203.2	1.68
270	284.7	412.5	388.8	814.8	1203.6	1.62
280	294.7	415.8	392.3	811.6	1203.9	1.57
290	304.7	418.8	395.7	808.5	1204.2	1.52
300	314.7	421.7	398.9	805.5	1204.4	1.47
310	324.7	424.7	402.1	802.6	1204.7	1.43
320	334.7	427.5	405.2	799.7	1204.9	1.39
330	344.7	430.3	408.3	796.7	1205.0	1.35
340	354.7	433.0	411.3	793.8	1205.1	1.31
350	364.7	435.7	414.3	791.0	1205.3	1.27
360	374.7	438.3	417.2	788.2	1205.4	1.24
370	384.7	440.8	420.0	785.4	1205.4	1.21
380	394.7	443.3	422.8	782.7	1205.5	1.18
390	404.7	445.7	425.6	779.9	1205.5	1.15
400	414.7	448.1	428.2	777.4	1205.6	1.12
420	434.7	452.8	433.4	772.2	1205.6	1.07
440	454.7	457.3	438.5	767.1	1205.6	1.02
460	474.7	461.7	443.4	762.1	1205.5	.98

* in psig
† in psia
‡ in °F
§ in Btu/lb
** in cu ft/lb

continued

PROPERTIES OF SATURATED STEAM

Gauge Pressure*	Absolute Pressure†	Temperature‡	Sensible Heat§	Latent Heat§	Total Heat§	Specific Volume Steam**
480	494.7	465.9	448.3	757.1	1205.4	.94
500	514.7	470.0	453.0	752.3	1205.3	.902
520	534.7	474.0	457.6	747.5	1205.1	.868
540	554.7	477.8	462.0	742.8	1204.8	.835
560	574.7	481.6	466.4	738.1	1204.5	.805
580	594.7	485.2	470.7	733.5	1204.2	.776
600	614.7	488.8	474.8	729.1	1203.9	.750
620	634.7	492.3	479.0	724.5	1203.5	.726
640	654.7	495.7	483.0	720.1	1203.1	.703
660	674.7	499.0	486.9	715.8	1202.7	.681
680	694.7	502.2	490.7	711.5	1202.2	.660
700	714.7	505.4	494.4	707.4	1201.8	.641
720	734.7	508.5	498.2	703.1	1201.3	.623
740	754.7	511.5	501.9	698.9	1200.8	.605
760	774.7	514.5	505.5	694.7	1200.2	.588
780	794.7	517.5	509.0	690.7	1199.7	.572
800	814.7	520.3	512.5	686.6	1199.1	.557

Spirax Sarco, Inc.

* in psig
† in psia
‡ in °F
§ in Btu/lb
** in cu ft/lb

CONVERSION FACTORS

1 atmosphere (atm) (standard) = 14.7 psi
1 atmosphere (standard) = 29.92 in Hg
1 horsepower = 746 watts (W)
1 horsepower = 33,000 ft-lb/minute
1 British thermal unit = 778 ft-lb
1 horsepower-hour (HP-hour) = 2545 Btu
1 therm-hour = 100,000 Btu/hour
1 boiler horsepower = 33,475 Btu/hour
1 boiler horsepower = 34.5 lb of steam per hour at 212°F
1 kilowatt (kW) = 3413 Btu/hour
1 kilowatt-hour (kWh) = 3413 Btu
1 kilowatt = 1000 W
1 kilowatt = 1.341 HP
1 horsepower = .746 kW
1 cubic foot = 7.48 gal.
1 gallon = 231 cu in.
1 cubic foot of fresh water = 62.5 lb
1 gallon of fresh water = 8.33 lb
1 cubic foot of salt water = 64 lb
1 long ton of fresh water = 36 cu ft
1 long ton of salt water = 35 cu ft
1 foot of head of water = .434 psi
1 inch of head of mercury = .491 psi
1 barrel (oil) = 42 gal.

ENERGY LOSS FROM SCALE DEPOSITS IN BOILERS

Scale Thickness*	Extra Fuel Cost†
1/32	8.5
1/25	9.3
1/20	11.1
1/16	12.4
1/8	25.0
1/4	40.0
3/8	55.0
1/2	70.0

* in in.
† in %

HVAC COMMON EQUATIONS

Latent Air Conditioning

Btu/hr = 60 min/hr × .075 lb/ft^3 × cfm × ΔH
Btu/hr = 4.5 × cfm × ΔH

Sensible Air Conditioning

Btu/hr = cfm × 60 min/hr × .075 lb/ft^3 × .24 Btu/lb × ΔT
Btu/hr = 1.08 × cfm × ΔT

Water Heating/Cooling

Btu/hr = gpm × 60 min/hr × 8.33 lb/gal. × 1 Btu/lb/°F × ΔT

HVAC SYMBOLS

EQUIPMENT SYMBOLS	DUCTWORK	HEATING PIPING
EXPOSED RADIATOR	DUCT (1ST FIGURE, WIDTH; 2ND FIGURE, DEPTH) 12 X 20	HIGH-PRESSURE STEAM —— HPS ——
RECESSED RADIATOR	DIRECTION OF FLOW	MEDIUM-PRESSURE STEAM —— MPS ——
FLUSH ENCLOSED RADIATOR	FLEXIBLE CONNECTION	LOW-PRESSURE STEAM —— LPS ——
PROJECTING ENCLOSED RADIATOR	DUCTWORK WITH ACOUSTICAL LINING	HIGH-PRESSURE RETURN —— HPR ——
UNIT HEATER (PROPELLER) – PLAN	FIRE DAMPER WITH ACCESS DOOR FD AD	MEDIUM-PRESSURE RETURN —— MPR ——
		LOW-PRESSURE RETURN —— LPR ——
UNIT HEATER (CENTRIFUGAL) – PLAN	MANUAL VOLUME DAMPER — VD	BOILER BLOW OFF —— BD ——
UNIT VENTILATOR – PLAN	AUTOMATIC VOLUME DAMPER	CONDENSATE OR VACUUM PUMP DISCHARGE —— VPD ——
STEAM	EXHAUST, RETURN OR OUTSIDE AIR DUCT – SECTION 20 X 12	FEEDWATER PUMP DISCHARGE —— PPD ——
		MAKEUP WATER —— MU ——
DUPLEX STRAINER	SUPPLY DUCT – SECTION 20 X 12	AIR RELIEF LINE —— V ——
PRESSURE-REDUCING VALVE	CEILING DIFFUSER SUPPLY OUTLET 20" DIA CD 1000 CFM	FUEL OIL SUCTION —— FOS ——
		FUEL OIL RETURN —— FOR ——
AIR LINE VALVE	CEILING DIFFUSER SUPPLY OUTLET 20 X 12 CD 700 CFM	FUEL OIL VENT —— FOV ——
		COMPRESSED AIR —— A ——
STRAINER	LINEAR DIFFUSER 96 X 6-LD 400 CFM	HOT WATER HEATING SUPPLY —— HW ——
THERMOMETER	FLOOR REGISTER 20 X 12 FR 700 CFM	HOT WATER HEATING RETURN —— HWR ——
		AIR CONDITIONING PIPING
PRESSURE GAUGE AND COCK	TURNING VANES	REFRIGERANT LIQUID —— RL ——
RELIEF VALVE		REFRIGERANT DISCHARGE —— RD ——
		REFRIGERANT SUCTION —— RS ——
AUTOMATIC 3-WAY VALVE	FAN AND MOTOR WITH BELT GUARD	CONDENSER WATER SUPPLY —— CWS ——
		CONDENSER WATER RETURN —— CWR ——
AUTOMATIC 2-WAY VALVE		CHILLED WATER SUPPLY —— CHWS ——
		CHILLED WATER RETURN —— CHWR ——
	LOUVER OPENING 20 X 12-L 700 CFM	MAKEUP WATER —— MU ——
SOLENOID VALVE		HUMIDIFICATION LINE —— H ——
		DRAIN —— D ——

REFRIGERATION SYMBOLS

GAUGE		PRESSURE SWITCH		DRYER	
SIGHT GLASS		HAND EXPANSION VALVE		FILTER AND STRAINER	
HIGH SIDE FLOAT VALVE		AUTOMATIC EXPANSION VALVE		COMBINATION STRAINER AND DRYER	
LOW SIDE FLOAT VALVE		THERMOSTATIC EXPANSION VALVE		EVAPORATIVE CONDENSOR	
IMMERSION COOLING UNIT		CONSTANT PRESSURE VALVE, SUCTION		HEAT EXCHANGER	
COOLING TOWER		THERMAL BULB		AIR-COOLED CONDENSING UNIT	
NATURAL CONVECTION, FINNED TYPE EVAPORATOR		SCALE TRAP			
FORCED CONVECTION EVAPORATOR		SELF-CONTAINED THERMOSTAT		WATER-COOLED CONDENSING UNIT	

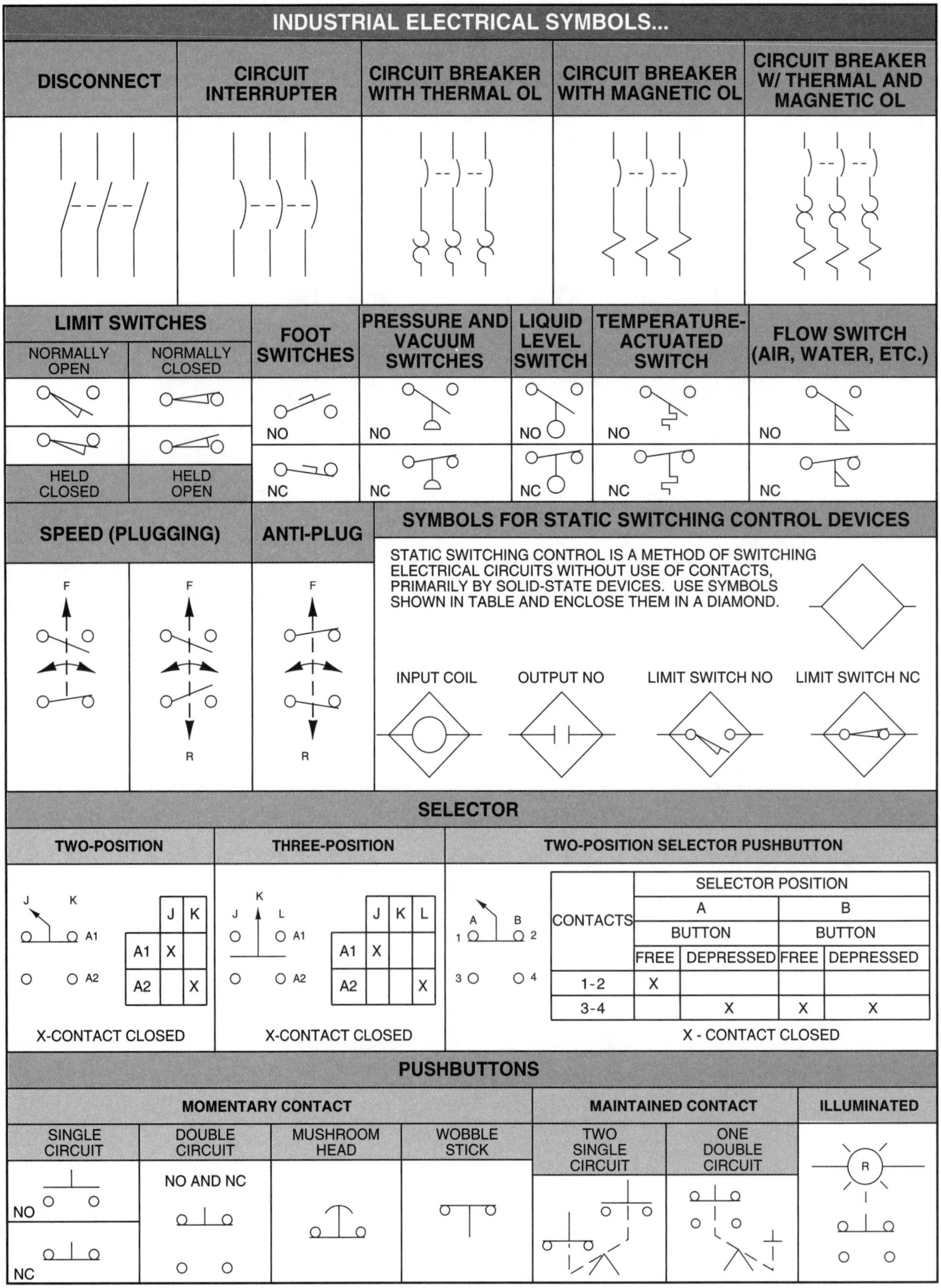

INDUSTRIAL ELECTRICAL SYMBOLS...

DISCONNECT	CIRCUIT INTERRUPTER	CIRCUIT BREAKER WITH THERMAL OL	CIRCUIT BREAKER WITH MAGNETIC OL	CIRCUIT BREAKER W/ THERMAL AND MAGNETIC OL

LIMIT SWITCHES		FOOT SWITCHES	PRESSURE AND VACUUM SWITCHES	LIQUID LEVEL SWITCH	TEMPERATURE-ACTUATED SWITCH	FLOW SWITCH (AIR, WATER, ETC.)
NORMALLY OPEN	NORMALLY CLOSED	NO	NO	NO	NO	NO
HELD CLOSED	HELD OPEN	NC	NC	NC	NC	NC

SPEED (PLUGGING)	ANTI-PLUG	SYMBOLS FOR STATIC SWITCHING CONTROL DEVICES

STATIC SWITCHING CONTROL IS A METHOD OF SWITCHING ELECTRICAL CIRCUITS WITHOUT USE OF CONTACTS, PRIMARILY BY SOLID-STATE DEVICES. USE SYMBOLS SHOWN IN TABLE AND ENCLOSE THEM IN A DIAMOND.

INPUT COIL   OUTPUT NO   LIMIT SWITCH NO   LIMIT SWITCH NC

SELECTOR

TWO-POSITION	THREE-POSITION	TWO-POSITION SELECTOR PUSHBUTTON

TWO-POSITION:

	J	K
A1	X	
A2		X

X-CONTACT CLOSED

THREE-POSITION:

	J	K	L
A1	X		
A2			X

X-CONTACT CLOSED

TWO-POSITION SELECTOR PUSHBUTTON:

CONTACTS	SELECTOR POSITION			
	A		B	
	BUTTON		BUTTON	
	FREE	DEPRESSED	FREE	DEPRESSED
1-2	X			
3-4		X	X	X

X - CONTACT CLOSED

PUSHBUTTONS

MOMENTARY CONTACT				MAINTAINED CONTACT		ILLUMINATED
SINGLE CIRCUIT	DOUBLE CIRCUIT	MUSHROOM HEAD	WOBBLE STICK	TWO SINGLE CIRCUIT	ONE DOUBLE CIRCUIT	
NO	NO AND NC					
NC						

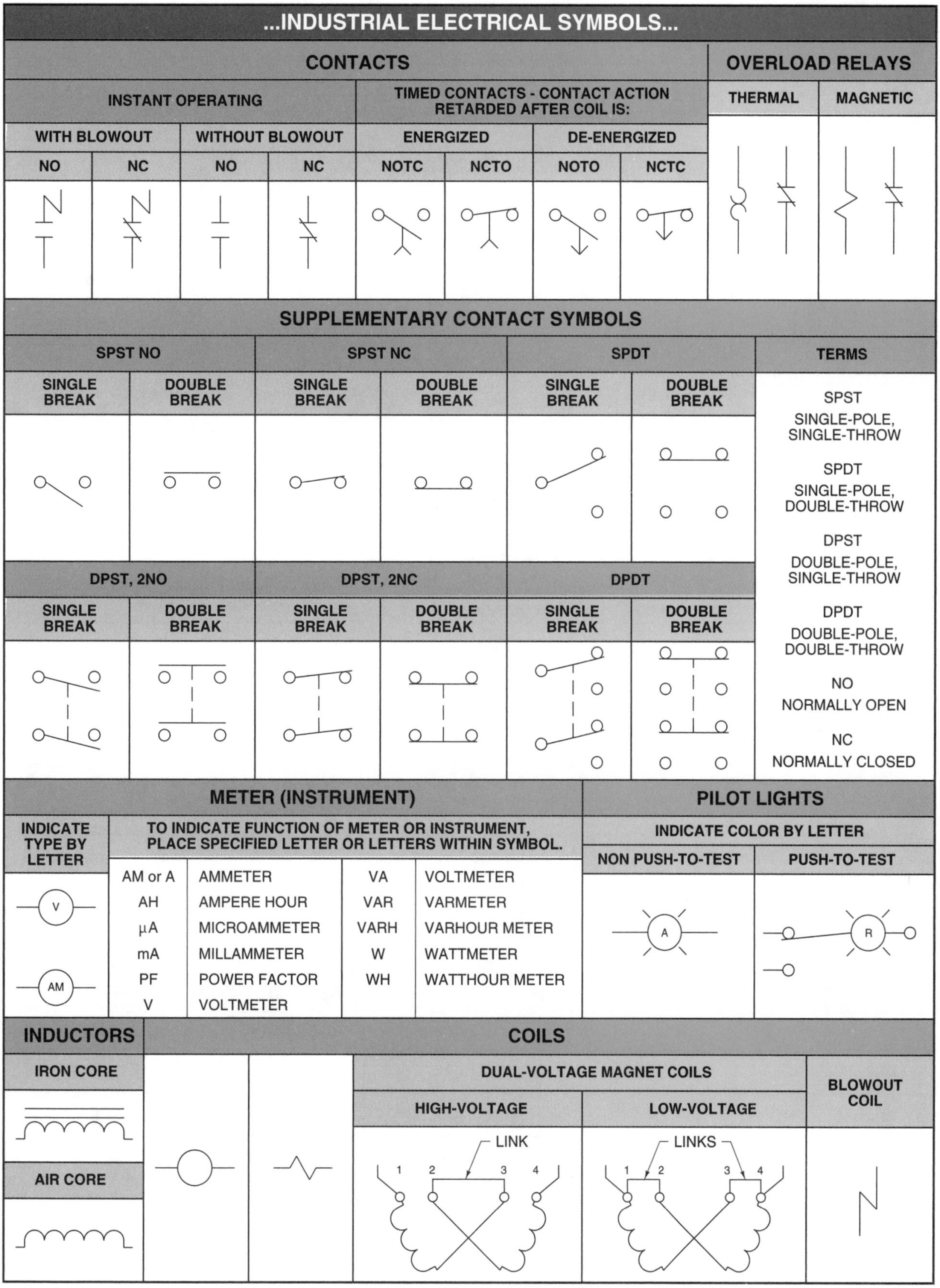

...INDUSTRIAL ELECTRICAL SYMBOLS...

CONTACTS

INSTANT OPERATING				TIMED CONTACTS - CONTACT ACTION RETARDED AFTER COIL IS:			
WITH BLOWOUT		WITHOUT BLOWOUT		ENERGIZED		DE-ENERGIZED	
NO	NC	NO	NC	NOTC	NCTO	NOTO	NCTC

OVERLOAD RELAYS

THERMAL	MAGNETIC

SUPPLEMENTARY CONTACT SYMBOLS

SPST NO		SPST NC		SPDT		TERMS
SINGLE BREAK	DOUBLE BREAK	SINGLE BREAK	DOUBLE BREAK	SINGLE BREAK	DOUBLE BREAK	SPST SINGLE-POLE, SINGLE-THROW

DPST, 2NO		DPST, 2NC		DPDT		
SINGLE BREAK	DOUBLE BREAK	SINGLE BREAK	DOUBLE BREAK	SINGLE BREAK	DOUBLE BREAK	

TERMS

SPST
SINGLE-POLE,
SINGLE-THROW

SPDT
SINGLE-POLE,
DOUBLE-THROW

DPST
DOUBLE-POLE,
SINGLE-THROW

DPDT
DOUBLE-POLE,
DOUBLE-THROW

NO
NORMALLY OPEN

NC
NORMALLY CLOSED

METER (INSTRUMENT)

INDICATE TYPE BY LETTER	TO INDICATE FUNCTION OF METER OR INSTRUMENT, PLACE SPECIFIED LETTER OR LETTERS WITHIN SYMBOL.			
	AM or A	AMMETER	VA	VOLTMETER
	AH	AMPERE HOUR	VAR	VARMETER
	µA	MICROAMMETER	VARH	VARHOUR METER
	mA	MILLAMMETER	W	WATTMETER
	PF	POWER FACTOR	WH	WATTHOUR METER
	V	VOLTMETER		

PILOT LIGHTS

INDICATE COLOR BY LETTER	
NON PUSH-TO-TEST	PUSH-TO-TEST

INDUCTORS

IRON CORE

AIR CORE

COILS

DUAL-VOLTAGE MAGNET COILS		BLOWOUT COIL
HIGH-VOLTAGE	LOW-VOLTAGE	

LINK

LINKS

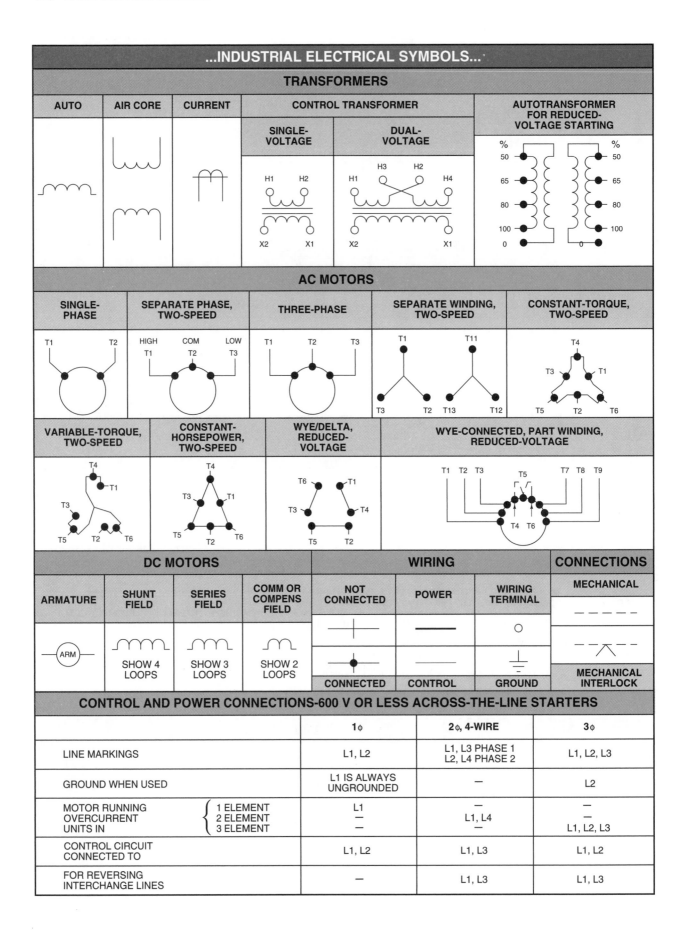

...INDUSTRIAL ELECTRICAL SYMBOLS...

TRANSFORMERS

AUTO	AIR CORE	CURRENT	CONTROL TRANSFORMER		AUTOTRANSFORMER FOR REDUCED-VOLTAGE STARTING
			SINGLE-VOLTAGE	DUAL-VOLTAGE	

AC MOTORS

SINGLE-PHASE	SEPARATE PHASE, TWO-SPEED	THREE-PHASE	SEPARATE WINDING, TWO-SPEED	CONSTANT-TORQUE, TWO-SPEED

VARIABLE-TORQUE, TWO-SPEED	CONSTANT-HORSEPOWER, TWO-SPEED	WYE/DELTA, REDUCED-VOLTAGE	WYE-CONNECTED, PART WINDING, REDUCED-VOLTAGE

DC MOTORS / WIRING / CONNECTIONS

ARMATURE	SHUNT FIELD	SERIES FIELD	COMM OR COMPENS FIELD	NOT CONNECTED	POWER	WIRING TERMINAL	MECHANICAL
	SHOW 4 LOOPS	SHOW 3 LOOPS	SHOW 2 LOOPS	CONNECTED	CONTROL	GROUND	MECHANICAL INTERLOCK

CONTROL AND POWER CONNECTIONS-600 V OR LESS ACROSS-THE-LINE STARTERS

		1φ	2φ, 4-WIRE	3φ
LINE MARKINGS		L1, L2	L1, L3 PHASE 1 L2, L4 PHASE 2	L1, L2, L3
GROUND WHEN USED		L1 IS ALWAYS UNGROUNDED	—	L2
MOTOR RUNNING OVERCURRENT UNITS IN	1 ELEMENT	L1	—	—
	2 ELEMENT	—	L1, L4	—
	3 ELEMENT	—	—	L1, L2, L3
CONTROL CIRCUIT CONNECTED TO		L1, L2	L1, L3	L1, L2
FOR REVERSING INTERCHANGE LINES		—	L1, L3	L1, L3

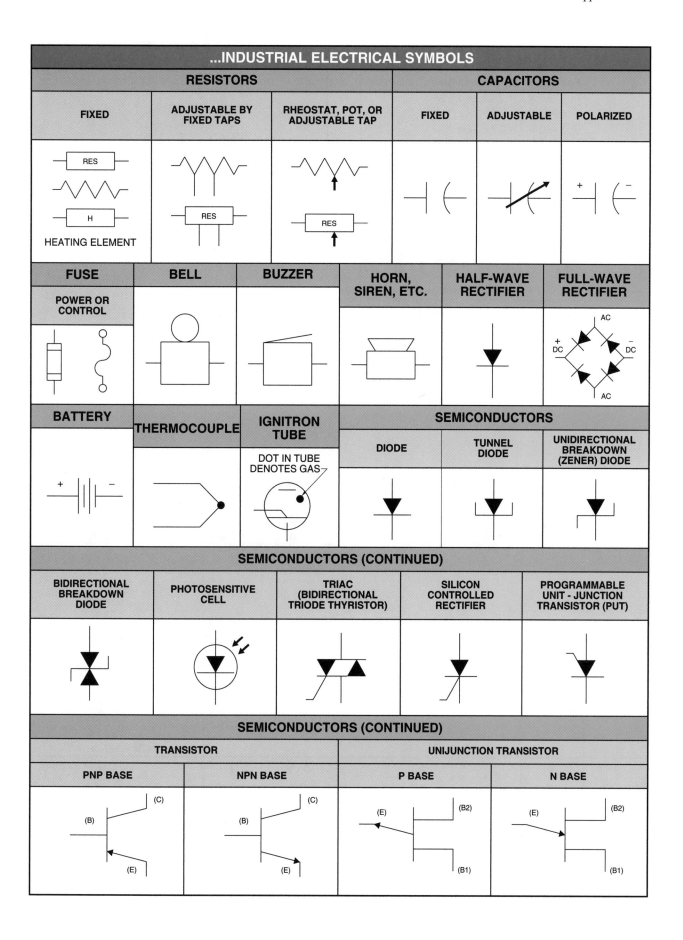

THREE-PHASE VOLTAGE VALUES		
For 208 V	1.732, use 360	
For 230 V	1.732, use 398	
For 240 V	1.732, use 416	
For 440 V	1.732, use 762	
For 460 V	1.732, use 797	
For 480 V	1.732, use 831	

POWER FORMULA ABBREVIATIONS AND SYMBOLS	
P = Watts	V = Volts
I = Amps	VA = Volt Amps
A = Amps	ϕ = Phase
R = Ohms	$\sqrt{}$ = Square Root
E = Volts	

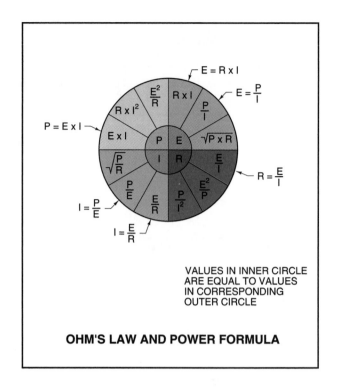

VALUES IN INNER CIRCLE ARE EQUAL TO VALUES IN CORRESPONDING OUTER CIRCLE

OHM'S LAW AND POWER FORMULA

POWER FORMULAS – 1ϕ, 3ϕ					
Phase	To Find	Use Formula	Example		
			Given	Find	Solution
1ϕ	I	$I = \dfrac{VA}{V}$	32,000 VA, 240 V	I	$I = \dfrac{VA}{V}$ $I = \dfrac{32,000\,VA}{240\,V}$ **I = 133 A**
1ϕ	VA	$VA = I \times V$	100 A, 240 V	VA	$VA = I \times V$ $VA = 100\,A \times 240\,V$ **VA = 24,000 VA**
1ϕ	V	$V = \dfrac{VA}{I}$	42,000 VA, 350 A	V	$V = \dfrac{VA}{I}$ $V = \dfrac{42,000\,VA}{350\,A}$ **V = 120 V**
3ϕ	I	$I = \dfrac{VA}{V \times \sqrt{3}}$	72,000 VA, 208 V	I	$I = \dfrac{VA}{V \times 3}$ $I = \dfrac{72,000\,VA}{360\,V}$ **I = 200 A**
3ϕ	VA	$VA = I \times V \times \sqrt{3}$	2 A, 240 V	VA	$VA = I \times V \times \sqrt{3}$ $VA = 2 \times 416$ **VA = 832 VA**

AC/DC FORMULAS				
To Find	**DC**	**AC**		
		1ϕ, 115 or 220 V	**1ϕ, 208, 230, or 240 V**	**3ϕ – All Voltages**
I, HP known	$\dfrac{HP \times 746}{E \times E_{ff}}$	$\dfrac{HP \times 746}{E \times E_{ff} \times PF}$	$\dfrac{HP \times 746}{E \times E_{ff} \times PF}$	$\dfrac{HP \times 746}{1.73 \times E \times E_{ff} \times PF}$
I, kW known	$\dfrac{kW \times 1000}{E}$	$\dfrac{kW \times 1000}{E \times PF}$	$\dfrac{kW \times 1000}{E \times PF}$	$\dfrac{kW \times 1000}{1.73 \times E \times PF}$
I, kVA known		$\dfrac{kVA \times 1000}{E}$	$\dfrac{kVA \times 1000}{E}$	$\dfrac{kVA \times 1000}{1.763 \times E}$
kW	$\dfrac{I \times E}{1000}$	$\dfrac{I \times E \times PF}{1000}$	$\dfrac{I \times E \times PF}{1000}$	$\dfrac{I \times E \times 1.73 \times PF}{1000}$
kVA		$\dfrac{I \times E}{1000}$	$\dfrac{I \times E}{1000}$	$\dfrac{I \times E \times 1.73}{1000}$
HP (output)	$\dfrac{I \times E \times E_{ff}}{746}$	$\dfrac{I \times E \times E_{ff} \times PF}{746}$	$\dfrac{I \times E \times E_{ff} \times PF}{746}$	$\dfrac{I \times E \times 1.73 \times E_{ff} \times PF}{746}$

E_{ff} = efficiency

HORSEPOWER FORMULAS				
To Find	**Use Formula**	**Example**		
		Given	**Find**	**Solution**
HP	$HP = \dfrac{I \times E \times E_{ff}}{746}$	240 V, 20 A, 85% E_{ff}	HP	$HP = \dfrac{I \times E \times E_{ff}}{746}$ $HP \quad \dfrac{20 \text{ A} \times 240 \text{ V} \times 85\%}{746}$ $HP = \textbf{5.5}$
I	$I = \dfrac{HP \times 746}{E \times E_{ff} \times PF}$	10 HP, 240 V, 90% E_{ff}, 88% PF	I	$I = \dfrac{HP \times 746}{E \times E_{ff} \times PF}$ $I \quad \dfrac{10 \text{ HP} \times 746}{240 \text{ V} \times 90\% \times 88\%}$ $I = \textbf{39 A}$

VOLTAGE DROP FORMULAS – 1ϕ, 3ϕ					
Phase	**To Find**	**Use Formula**	**Example**		
			Given	**Find**	**Solution**
1ϕ	VD	$VD \quad \dfrac{2 \times R \times L \times I}{1000}$	240 V, 40 A, 60 L, .764 R	VD	$VD \quad \dfrac{2 \times R \times L \times I}{1000}$ $VD \quad \dfrac{2 \times .764 \times 60 \times 40}{1000}$ $VD = \textbf{3.67 V}$
3	VD	$VD \quad \dfrac{2 \times R \times L \times I}{1000} \times .866$	208 V, 110 A, 75 L, .194 R, .866 multiplier	VD	$VD \quad \dfrac{2 \times R \times L \times I}{1000} \times .866$ $VD \quad \dfrac{2 \times .194 \times 75 \times 110}{1000} \times .866$ $VD = \textbf{2.77 V}$

$*\dfrac{\sqrt{3}}{2} = .866$

ELECTRICAL/ELECTRONIC ABBREVIATIONS/ACRONYMS

Abbr/ Acronym	Meaning	Abbr/ Acronym	Meaning	Abbr/ Acronym	Meaning
A	Ammeter; Ampere; Anode; Armature	FU	Fuse	PNP	Positive-Negative-Positive
AC	Alternating Current	FWD	Forward	POS	Positive
AC/DC	Alternating Current; Direct Current	G	Gate; Giga; Green; Conductance	POT.	Potentiometer
A/D	Analog to Digital	GEN	Generator	P-P	Peak-to-Peak
AF	Audio Frequency	GRD	Ground	PRI	Primary Switch
AFC	Automatic Frequency Control	GY	Gray	PS	Pressure Switch
Ag	Silver	H	Henry; High Side of Transformer; Magnetic Flux	PSI	Pounds Per Square Inch
ALM	Alarm			PUT	Pull-Up Torque
AM	Ammeter; Amplitude Modulation	HF	High Frequency	Q	Transistor
AM/FM	Amplitude Modulation; Frequency Modulation	HP	Horsepower	R	Radius; Red; Resistance; Reverse
		Hz	Hertz	RAM	Random-Access Memory
ARM.	Armature	I	Current	RC	Resistance-Capacitance
Au	Gold	IC	Integrated Circuit	RCL	Resistance-Inductance-Capacitance
AU	Automatic	INT	Intermediate; Interrupt	REC	Rectifier
AVC	Automatic Volume Control	INTLK	Interlock	RES	Resistor
AWG	American Wire Gauge	IOL	Instantaneous Overload	REV	Reverse
BAT.	Battery (electric)	IR	Infrared	RF	Radio Frequency
BCD	Binary Coded Decimal	ITB	Inverse Time Breaker	RH	Rheostat
BJT	Bipolar Junction Transistor	ITCB	Instantaneous Trip Circuit Breaker	rms	Root Mean Square
BK	Black	JB	Junction Box	ROM	Read-Only Memory
BL	Blue	JFET	Junction Field-Effect Transistor	rpm	Revolutions Per Minute
BR	Brake Relay; Brown	K	Kilo; Cathode	RPS	Revolutions Per Second
C	Celsius; Capacitance; Capacitor	L	Line; Load; Coil; Inductance	S	Series; Slow; South; Switch
CAP.	Capacitor	LB-FT	Pounds Per Foot	SCR	Silicon Controlled Rectifier
CB	Circuit Breaker; Citizen's Band	LB-IN.	Pounds Per Inch	SEC	Secondary
CC	Common-Collector Configuration	LC	Inductance-Capacitance	SF	Service Factor
CCW	Counterclockwise	LCD	Liquid Crystal Display	1 PH; 1φ	Single-Phase
CE	Common-Emitter Configuration	LCR	Inductance-Capacitance-Resistance	SOC	Socket
CEMF	Counter Electromotive Force	LED	Light Emitting Diode	SOL	Solenoid
CKT	Circuit	LRC	Locked Rotor Current	SP	Single-Pole
CONT	Continuous; Control	LS	Limit Switch	SPDT	Single-Pole, Double-Throw
CPS	Cycles Per Second	LT	Lamp	SPST	Single-Pole, Single-Throw
CPU	Central Processing Unit	M	Motor; Motor Starter; Motor Starter Contacts	SS	Selector Switch
CR	Control Relay			SSW	Safety Switch
CRM	Control Relay Master	MAX.	Maximum	SW	Switch
CT	Current Transformer	MB	Magnetic Brake	T	Tera; Terminal; Torque; Transformer
CW	Clockwise	MCS	Motor Circuit Switch	TB	Terminal Board
D	Diameter; Diode; Down	MEM	Memory	3 PH; 3φ	Three-Phase
D/A	Digital to Analog	MED	Medium	TD	Time Delay
DB	Dynamic Braking Contactor; Relay	MIN	Minimum	TDF	Time Delay Fuse
DC	Direct Current	MN	Manual	TEMP	Temperature
DIO	Diode	MOS	Metal-Oxide Semiconductor	THS	Thermostat Switch
DISC.	Disconnect Switch	MOSFET	Metal-Oxide Semiconductor Field-Effect Transistor	TR	Time Delay Relay
DMM	Digital Multimeter			TTL	Transistor-Transistor Logic
DP	Double-Pole	MTR	Motor	U	Up
DPDT	Double-Pole, Double-Throw	N; NEG	North; Negative	UCL	Unclamp
DPST	Double-Pole, Single-Throw	NC	Normally Closed	UHF	Ultrahigh Frequency
DS	Drum Switch	NEUT	Neutral	UJT	Unijunction Transistor
DT	Double-Throw	NO	Normally Open	UV	Ultraviolet; Undervoltage
DVM	Digital Voltmeter	NPN	Negative-Positive-Negative	V	Violet; Volt
EMF	Electromotive Force	NTDF	Nontime-Delay Fuse	VA	Volt Amp
F	Fahrenheit; Fast; Field; Forward; Fuse	O	Orange	VAC	Volts Alternating Current
FET	Field-Effect Transistor	OCPD	Overcurrent Protection Device	VDC	Volts Direct Current
FF	Flip-Flop	OHM	Ohmmeter	VHF	Very High Frequency
FLC	Full-Load Current	OL	Overload Relay	VLF	Very Low Frequency
FLS	Flow Switch	OZ/IN.	Ounces Per Inch	VOM	Volt-Ohm-Milliammeter
FLT	Full-Load Torque	P	Peak; Positive; Power; Power Consumed	W	Watt; White
FM	Frequency Modulation	PB	Pushbutton	w/	With
FREQ	Frequency	PCB	Printed Circuit Board	X	Low Side of Transformer
FS	Float Switch	PH; φ	Phase	Y	Yellow
FTS	Foot Switch	PLS	Plugging Switch	Z	Impedance

ENGLISH SYSTEM

LENGTH

Unit	Abbr	Equivalents
mile	mi	5280′, 320 rd, 1760 yd
rod	rd	5.50 yd, 16.5′
yard	yd	3′, 36″
foot	ft *or* ′	12″, .333 yd
inch	in. *or* ″	.083′ .028 yd

AREA

$A = l \times w$

Unit	Abbr	Equivalents
square mile	sq mi *or* mi^2	640 A, 102,400 sq rd
acre	A	4840 sq yd, 43,560 sq ft
square rod	sq rd *or* rd^2	30.25 sq yd, .00625 A
square yard	sq yd *or* yd^2	1296 sq in., 9 sq ft
square foot	sq ft *or* ft^2	144 sq in., .111 sq yd
square inch	sq in. *or* in^2	.0069 sq ft, .00077 sq yd

VOLUME

$V = l \times w \times t$

Unit	Abbr	Equivalents
cubic yard	cu yd *or* yd^3	27 cu ft, 46,656 cu in.
cubic foot	cu ft *or* ft^3	1728 cu in., .0370 cu yd
cubic inch	cu in. *or* in^3	.00058 cu ft, .000021 cu yd

CAPACITY

WATER, FUEL, ETC.

	Unit	Abbr	Equivalents
U.S. liquid measure	gallon	gal.	4 qt (231 cu in.)
	quart	qt	2 pt (57.75 cu in.)
	pint	pt	4 gi (28.875 cu in.)
	gill	gi	4 fl oz (7.219 cu in.)
	fluidounce	fl oz	8 fl dr (1.805 cu in.)
	fluidram	fl dr	60 min (.226 cu in.)
	minim	min	1/6 fl dr (.003760 cu in.)

VEGETABLES, GRAIN, ETC.

	Unit	Abbr	Equivalents
U.S. dry measure	bushel	bu	4 pk (2150.42 cu in.)
	peck	pk	8 qt (537.605 cu in.)
	quart	qt	2 pt (67.201 cu in.)
	pint	pt	1/2 qt (33.600 cu in.)

DRUGS

	Unit	Abbr	Equivalents
British imperial liquid and dry measure	bushel	bu	4 pk (2219.36 cu in.)
	peck	pk	2 gal. (554.84 cu in.)
	gallon	gal.	4 qt (277.420 cu in.)
	quart	qt	2 pt (69.355 cu in.)
	pint	pt	4 gi (34.678 cu in.)
	gill	gi	5 fl oz (8.669 cu in.)
	fluidounce	fl oz	8 fl dr (1.7339 cu in.)
	fluidram	fl dr	60 min (.216734 cu in.)
	minim	min	1/60 fl dr (.003612 cu in.)

MASS AND WEIGHT

COAL, GRAIN, ETC.

	Unit	Abbr	Equivalents
avoirdupois	ton		2000 lb
	short ton	t	2000 lb
	long ton		2240 lb
	pound	lb *or* #	16 oz, 7000 gr
	ounce	oz	16 dr, 437.5 gr
	dram	dr	27.344 gr, .0625 oz
	grain	gr	.037 dr, .002286 oz

GOLD, SILVER, ETC.

	Unit	Abbr	Equivalents
troy	pound	lb	12 oz, 240 dwt, 5760 gr
	ounce	oz	20 dwt, 480 gr
	pennyweight	dwt *or* pwt	24 gr, .05 oz
	grain	gr	.042 dwt, .002083 oz

DRUGS

	Unit	Abbr	Equivalents
apothecaries'	pound	lb ap	12 oz, 5760 gr
	ounce	oz ap	8 dr ap, 480 gr
	dram	dr ap	3 s ap, 60 gr
	scruple	s ap	20 gr, .333 dr ap
	grain	gr	.05 s, .002083 oz, .0166 dr ap

METRIC SYSTEM

LENGTH	Unit	Abbreviation	Number of Base Units
	kilometer	km	1000
	hectometer	hm	100
	dekameter	dam	10
	meter*	m	1
	decimeter	dm	.1
	centimeter	cm	.01
	millimeter	mm	.001
AREA $A = l \times w$	square kilometer	sq km or km^2	1,000,000
	hectare	ha	10,000
	are	a	100
	square centimeter	sq cm or cm^2	.0001
VOLUME $V = l \times w \times t$	cubic centimeter	cu cm, cm^3, or cc	.000001
	cubic decimeter	dm^3	.001
	cubic meter*	m^3	1
CAPACITY WATER, FUEL, ETC. VEGETABLES, GRAIN, ETC. DRUGS	kiloliter	kl	1000
	hectoliter	hl	100
	dekaliter	dal	10
	liter*	l	1
	cubic decimeter	dm^3	1
	deciliter	dl	.10
	centiliter	cl	.01
	milliliter	ml	.001
MASS AND WEIGHT COAL, GRAIN, ETC. GOLD, SILVER, ETC. DRUGS	metric ton	t	1,000,000
	kilogram	kg	1000
	hectogram	hg	100
	dekagram	dag	10
	gram*	g	1
	decigram	dg	.10
	centigram	cg	.01
	milligram	mg	.001

* base units

INCH — MILLIMETER EQUIVALENTS*

Inches		MM	Inches		MM	Inches		MM
	.00004	.001		.11811	3		.550	13.970
	.00039	.01	1/8	.1250	3.175		.55118	14
	.00079	.02		.13780	3.5	9/16	.56250	14.2875
	.001	.025	9/64	.14063	3.5719		.57087	14.5
	.00118	.03		.150	3.810	37/64	.57813	14.6844
	.00157	.04	5/32	.15625	3.9688		.59055	15
	.00197	.05		.15748	4	19/32	.59375	15.0812
	.002	.051	11/64	.17188	4.3656		.600	15.24
	.00236	.06		.1750	4.445	39/64	.60938	15.4781
	.00276	.07		.17717	4.5		.61024	15.5
	.003	.0762	3/16	.18750	4.7625	5/8	.6250	15.875
	.00315	.08		.19685	5		.62992	16
	.00354	.09		.20	5.08	41/64	.64063	16.2719
	.00394	.1	13/64	.20313	5.1594		.64961	16.5
	.004	.1016		.21654	5.5		.650	16.51
	.005	.1270	7/32	.21875	5.5562	21/32	.65625	16.6688
	.006	.1524		.2250	5.715		.66929	17
	.007	.1778	15/64	.23438	5.9531	43/64	.67188	17.0656
	.00787	.2		.23622	6	11/16	.68750	17.4625
	.008	.2032	1/4	.250	6.35		.68898	17.5
	.009	.2286					.700	17.78
	.00984	.25		.25591	6.5	45/64	.70313	17.8594
	.01	.254	17/64	.26563	6.7469		.70866	18
	.01181	.3		.275	6.985	23/32	.71875	18.2562
1/64	.01563	.3969		.27559	7		.72835	18.5
	.01575	.4	9/32	.28125	7.1438	47/64	.73438	18.6531
	.01969	.5		.29528	7.5		.74803	19
	.02	.508	19/64	.29688	7.5406	3/4	.750	19.050
	.02362	.6		.30	7.62			
	.025	.635	5/16	.3125	7.9375	49/64	.76563	19.4469
	.02756	.7		.31496	8		.76772	19.5
	.0295	.75	21/64	.32813	8.3344	25/32	.78125	19.8438
	.03	.762		.33465	8.5		.78740	20
1/32	.03125	.7938	11/32	.34375	8.7375	51/64	.79688	20.2406
	.0315	.8		.350	8.89		.800	20.320
	.03543	.9		.35433	9		.80709	20.5
	.03937	1	23/64	.35938	9.1281	13/16	.81250	20.6375
	.04	1.016		.37402	9.5		.82677	21
3/64	.04687	1.191	3/8	.375	9.525	53/64	.82813	21.0344
	.04724	1.2	25/64	.39063	9.9219	27/32	.84375	21.4312
	.05	1.27		.39370	10		.84646	21.5
	.05512	1.4		.400	10.16		.850	21.590
	.05906	1.5	13/32	.40625	10.3188	55/64	.85938	21.8281
	.06	1.524		.41339	10.5		.86614	22
1/16	.06250	1.5875	27/64	.42188	10.7156	7/8	.875	22.225
	.06299	1.6		.43307	11		.88583	22.5
	.06693	1.7	7/16	.43750	11.1125	57/64	.89063	22.6219
	.07	1.778		.450	11.430		.900	22.860
	.07087	1.8		.45276	11.5		.90551	23
	.075	1.905	29/64	.45313	11.5094	29/32	.90625	23.0188
5/64	.07813	1.9844	15/32	.46875	11.9062	59/64	.92188	23.4156
	.07874	2		.47244	12		.92520	23.5
	.08	2.032	31/64	.48438	12.3031	15/16	.93750	23.8125
	.08661	2.2		.49213	12.5		.94488	24
	.09	2.286	1/2	.50	12.7		.950	24.130
	.09055	2.3				61/64	.95313	24.2094
3/32	.09375	2.3812		.51181	13		.96457	24.5
	0.9843	2.5	33/64	.51563	13.0969	31/32	.96875	24.6062
	.1	2.54	17/32	.53125	13.4938		.98425	25
	.10236	2.6		.53150	13.5	63/64	.98438	25.0031
7/64	.10937	2.7781	35/64	.54688	13.8906	1	1.0000	25.4

AREA – PLANE FIGURES

$A = l \times w$

where
A = area
l = length
w = width

SQUARE OR RECTANGLE

$A = \frac{1}{2} \times b \times h$

where
A = area
b = base
h = height

RIGHT TRIANGLE

$A = \pi r^2$

where
A = area
π = 3.1416
r = radius

CIRCLE

VOLUME – SOLID FIGURES

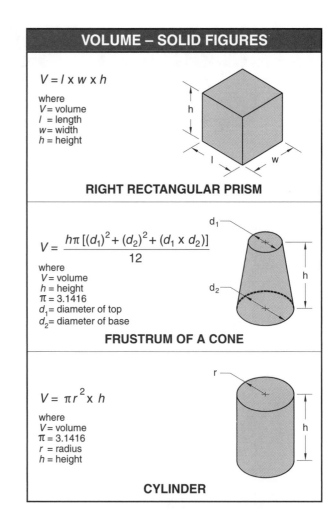

$V = l \times w \times h$

where
V = volume
l = length
w = width
h = height

RIGHT RECTANGULAR PRISM

$V = \dfrac{h\pi\,[(d_1)^2 + (d_2)^2 + (d_1 \times d_2)]}{12}$

where
V = volume
h = height
π = 3.1416
d_1 = diameter of top
d_2 = diameter of base

FRUSTRUM OF A CONE

$V = \pi r^2 \times h$

where
V = volume
π = 3.1416
r = radius
h = height

CYLINDER

DECIMAL EQUIVALENTS OF AN INCH

Fraction	Decimal	Fraction	Decimal	Fraction	Decimal	Fraction	Decimal
1/64	0.015625	17/64	0.265625	33/64	0.515625	49/64	0.765625
1/32	0.03125	9/32	0.28125	17/32	0.53125	25/32	0.78125
3/64	0.046875	19/64	0.296875	35/64	0.546875	51/64	0.796875
1/16	0.0625	5/16	0.3125	9/16	0.5625	13/16	0.8125
5/64	0.078125	21/64	0.328125	37/64	0.578125	53/64	0.828125
3/32	0.09375	11/32	0.34375	19/32	0.59375	27/32	0.84375
7/64	0.109375	23/64	0.359375	39/64	0.609375	55/64	0.859375
1/8	0.125	3/8	0.375	5/8	0.625	7/8	0.875
9/64	0.140625	25/64	0.390625	41/64	0.640625	57/64	0.890625
5/32	0.15625	13/32	0.40625	21/32	0.65625	29/32	0.90625
11/64	0.171875	27/64	0.421875	43/64	0.671875	59/64	0.921875
3/16	0.1875	7/16	0.4375	11/16	0.6875	15/16	0.9375
13/64	0.203125	29/64	0.453125	45/64	0.703125	61/64	0.953125
7/32	0.21875	15/32	0.46875	23/32	0.71875	31/32	0.96875
15/64	0.234375	31/64	0.484375	47/64	0.734375	63/64	0.984375
1/4	0.250	1/2	0.500	3/4	0.750	1	1.000

CHEMICAL ELEMENTS

Name	Symbol	Atomic Weight[a]	Atomic Number	Name	Symbol	Atomic Weight[a]	Atomic Number
Actinium	Ac	[227]	89	Neon	Ne	20.183	10
Aluminum	Al	26.9815	13	Neptunium	Np	[237]	93
Americium	Am	[243]	95	Nickel	Ni	58.71	28
Antimony	Sb	121.75	51	Niobium	Nb	92.906	41
Argon	Ar	39.948	18	Nitrogen	N	14.0067	7
Arsenic	As	74.9216	33	Nobelium	No	[255]	102
Astatine	At	[210]	85	Osmium	Os	190.2	76
Barium	Ba	137.34	56	Oxygen	O	15.9994	8
Berkelium	Bk	[247]	97	Palladium	Pd	106.4	46
Beryllium	Be	9.0122	4	Phosphorus	P	30.9738	15
Bismuth	Bi	208.980	83	Platinum	Pt	195.09	78
Boron	B	10.811	5	Plutonium	Pu	[244]	94
Bromine	Br	79.909	35	Polonium	Po	[210]	84
Cadmium	Cd	112.40	48	Potassium	K	39.102	19
Calcium	Ca	40.08	20	Praseodymium	Pr	140.907	59
Californium	Cf	[251]	98	Promethium	Pm	[145]	61
Carbon	C	12.01115	6	Protactinium	Pa	[231]	91
Cerium	Ce	140.12	58	Radium	Ra	[226]	88
Cesium	Cs	132.905	55	Radon	Rn	[222]	86
Chlorine	Cl	35.453	17	Rhenium	Re	186.2	75
Chromium	Cr	51.996	24	Rhodium	Rh	102.905	45
Cobalt	Co	58.9332	27	Rubidium	Rb	85.47	37
Copper	Cu	63.54	29	Ruthenium	Ru	101.07	44
Curium	Cm	[247]	96	Samarium	Sm	150.35	62
Dysprosium	Dy	162.50	66	Scandium	Sc	44.956	21
Einsteinium	Es	[254]	99	Selenium	Se	78.96	34
Erbium	Er	167.26	68	Silicon	Si	28.086	14
Europium	Eu	151.96	63	Silver	Ag	107.870	47
Fermium	Fm	[257]	100	Sodium	Na	22.9898	11
Fluorine	F	18.9984	9	Strontium	Sr	87.62	38
Francium	Fr	[223]	87	Sulfur	S	32.064	16
Gadolinium	Gd	157.25	64	Tantalum	Ta	180.948	73
Gallium	Ga	69.72	31	Technetium	Tc	[97]	43
Germanium	Ge	72.59	32	Tellurium	Te	127.60	52
Gold	Au	196.967	79	Terbium	Tb	158.924	65
Hafnium	Hf	178.49	72	Thallium	Tl	204.37	81
Helium	He	4.0026	2	Thorium	Th	232.038	90
Holmium	Ho	164.930	67	Thulium	Tm	168.934	69
Hydrogen	H	1.00797	1	Tin	Sn	118.69	50
Indium	In	114.82	49	Titanium	Ti	47.90	22
Iodine	I	126.9044	53	Tungsten	W	183.85	74
Iridium	Ir	192.2	77	Unnilennium	Une	[266]	109
Iron	Fe	55.847	26	Unnilhexium	Unh	[263]	106
Krypton	Kr	83.80	36	Unniloctium	Uno	[265]	108
Lanthanum	La	138.91	57	Unnilpentium	Unp	[262]	105
Lawrencium	Lr	[256]	103	Unnilquadium	Unq	[261]	104
Lead	Pb	207.19	82	Unnilseptium	Uns	[262]	107
Lithium	Li	6.939	3	Uranium	U	238.03	92
Lutetium	Lu	174.97	71	Vanadium	V	50.942	23
Magnesium	Mg	24.312	12	Xenon	Xe	131.30	54
Manganese	Mn	54.9380	25	Ytterbium	Yb	173.04	70
Mendelevium	Md	[258]	101	Yttrium	Y	88.905	39
Mercury	Hg	200.59	80	Zinc	Zn	65.37	30
Molybdenum	Mo	95.94	42	Zirconium	Zr	91.22	40
Neodymium	Nd	144.24	60				

[a] a number in brackets indicates the mass number of the most stable isotope

absolute pressure: Pressure above a perfect vacuum.

absorbent: A fluid that has a strong attraction for another fluid.

absorber: An absorption refrigeration system component in which refrigerant is absorbed by the absorbent.

absorption refrigeration system: A nonmechanical refrigeration system that uses a fluid with the ability to absorb a vapor when it is cool and release a vapor when heated.

accumulator (counter): A device that records the number of occurrences of a signal.

actuator: A device that accepts a signal from a controller and causes a proportional mechanical motion to occur.

adapter card (network interface card): A card (circuit board) installed in a network component to allow it to communicate with the network.

adaptive control algorithm: A control algorithm that automatically adjusts its response time based on environmental conditions.

adaptive start control: A control method that adjusts (learns) its control settings based on the condition of a building.

adsorption: The adhesion of a gas or liquid to the surface of a porous material.

air compressor: A component that takes air from the atmosphere and compresses it to increase its pressure.

air conditioning: The process of cooling the air in building spaces to provide a comfortable temperature.

air conditioning system: A system that produces a refrigeration effect and distributes the cool air or water to building spaces.

air handling unit: A device consisting of a fan, ductwork, filters, dampers, heating coils, cooling coils, humidifiers, dehumidifiers, sensors, and controls to condition and distribute air throughout a building.

air handling unit (AHU) controller: A controller that contains inputs and outputs required to operate large central-station air handling units.

air line filter: A device that consists of a plastic housing containing a centrifugal deflector plate and a small filtration element.

air-to-air heat pump: A heat pump that uses air as the heat source and heat sink.

air velocity: The speed at which air moves from one point to another.

alarm : A notification of improper temperature or other conditions existing in a building.

alarm differential: The amount of change required in a variable for the alarm to return to normal after it has been in alarm status.

alarm printer: A printer used with a building automation system to produce hard copies of alarms (indications of improper system operation), preventive maintenance messages, and data trends. An alarm printer may be connected to a desktop PC or directly to a network communication module.

algorithm: A mathematical equation used by a building automation system controller to determine a desired setpoint.

alternating current (AC): Current that reverses its direction at regular intervals.

alternator: A device that operates one compressor during one pumping cycle and the other compressor during the next pumping cycle.

analog control system: A control system that uses a variable signal.

analog input: A device that senses a variable such as temperature, pressure, or humidity and causes a proportional electrical signal change at the building automation system controller.

analog output: A device that produces a continuous signal between two values.

anticipator: A device that turns heating or cooling equipment ON or OFF before it normally would.

application-specific controller (ASC): A controller designed to control only one type of HVAC system.

Arcnet: An early network architecture that uses token ring data transmission.

artificial intelligence: The use of advanced computing power to make decisions about system setup and functionality.

atmospheric air: A mixture of dry air, moisture, and particles.

atmospheric pressure: The pressure due to the weight of the Earth's atmosphere pressing against an object on the Earth's surface.

authority: The relationship of the primary variable change to the secondary variable change, expressed as a percentage.

automated control system: A control system that uses digital solid-state components.

automatic drain: A device that opens and closes automatically at a predetermined interval to drain moisture from the receiver.

automatic drain valve: A device that is normally piped to the lowest part of the receiver and opens based on differential pressure or moisture level buildup.

auxiliary device: A device used in a control system that produces a desired function when actuated by the output signal from a controller.

averaging control: A direct digital control feature that calculates an average value from all selected inputs.

averaging element temperature transmitter: A pneumatic transmitter that uses a long tube filled with a liquid or gas to sense duct temperature.

averaging relay: A relay that calculates the mathematical average of two input pressures and sends this pressure to a controlled device.

BACnet™: A communication protocol that uses standard connecting media and enables controllers to be interoperable.

baseband: A data transmission method where data is sent without frequency modulation.

baseplate: A flat piece of metal to which the thermostat components are mounted.

biasing (ratio) relay: A relay used to perform a mathematical function on an input signal. Biasing relays are used to add, subtract, multiply, or divide.

bipolar junction transistor (BJT): A transistor that controls the flow of current through the emitter (E) and collector (C) with a properly biased base (B).

bimetallic element: A sensing device that consists of two different metals joined together.

bleedport: An orifice that allows a small volume of air to be expelled to the atmosphere.

bleed-type thermostat: A thermostat that changes the air pressure to a valve or damper actuator by changing the amount of air expelled to the atmosphere.

booster (capacity) relay: A device that increases the air volume available to a damper or valve while maintaining the air pressure at a 1:1 ratio.

branch line pressure: Pressure in the air line that is piped from the thermostat to the controlled device.

bridge rectifier: A circuit containing four diodes that permits both halves of the input AC sine wave to pass.

British thermal unit (Btu): The amount of heat energy required to raise the temperature of 1 lb of water 1°F.

building automation system: A system that uses microprocessors (computer chips) to control the energy-using devices in a building.

building automation system input: A device that senses and sends building condition information to a controller.

building automation system output: A device that changes the state of a controlled device in response to a command from a building automation system controller.

building survey: An inventory of the energy-consuming equipment in a commercial building.

bulb-type temperature transmitter: A pneumatic transmitter that uses a capillary tube and bulb filled with a liquid or gas to sense temperature.

bulk polymer sensor: A humidity sensor that consists of a polymer saturated with a salt compound.

calibration simulator kit: A portable precision pressure-regulating kit with precise adjustment capabilities and gauges.

capacitive sensor: A humidity sensor that uses a thin film of hygroscopic element to alter the capacitance of a circuit.

central-direct digital control system: A control system in which all decisions are made in one location and which provides closed loop control.

central supervisory control system: A control system in which the decision-making equipment is located in one place and the system enables/disables local (primary) controllers.

centrifugal compressor: A compressor that uses centrifugal force to move air.

change of state: The process that occurs when enough heat is added to or removed from a substance to change it from one physical state to another, such as from ice to water or water to steam.

changeover relay: A relay that causes the operation of the thermostat to change between two or more modes such as day/night.

circuit breaker: An overcurrent protection device with a mechanism that automatically opens the circuit when an overload condition or short circuit occurs.

circulation: The movement of air.

closed loop control: Control in which feedback occurs between the controller, sensor, and controlled device.

collision: A data transmission overlap.

combustion: The chemical reaction that occurs when oxygen reacts with the hydrogen and carbon present in a fuel at ignition temperature.

comfort: A condition that occurs when people cannot sense a difference between themselves and the surrounding air.

commercial HVAC system: A heating, ventilating, and air conditioning system used in office buildings, strip malls, stores, restaurants, and other commercial buildings.

commissioning: The checkout procedure used to start up a new building automation system.

compressor: A mechanical device that compresses refrigerant or other fluid.

concentrator: A network switchboard that allows a number of nodes to communicate with each other.

condensation: The formation of liquid (condensate) as moisture or other vapor cools below its dew point.

condenser: A heat exchanger that removes heat from high-pressure refrigerant vapor.

conduction: Heat transfer that occurs when molecules in a material are heated and the heat is passed from molecule to molecule through the material.

conductor: A material that has little resistance and permits electrons to move through it easily.

constant-volume air handling unit: An air handling unit that moves a constant volume of air.

contactor: A control device that uses a small control current to energize or de-energize the load connected to it.

continuity: The presence of a complete path for current flow.

control agent: The fluid that flows through controlled devices to produce the heating or cooling effect in the system or building spaces.

control drawing: A drawing of a mechanical system that illustrates actual controls and piping between devices.

controlled device: The object that regulates the flow of fluid in a system to provide a heating, air conditioning, or ventilation effect.

controller: A device that receives a signal from a sensor, compares it to a setpoint value, and sends an appropriate output signal to a controlled device.

control loop: The arrangement of a controller, sensor, and controlled device in a system.

control point: The actual value that a control system experiences at any given time.

control strategy: A building automation system software method used to control the energy-using equipment in a building.

control system: An arrangement of a sensor, controller, and controlled device to maintain a specific controlled variable value in a building space, pipe, or duct.

convection: Heat transfer that occurs when currents circulate between warm and cool regions of a fluid.

cooling coil: A finned heat exchanger that removes heat from the air flowing over it.

cooling tower: An evaporative water cooler that uses natural evaporation to cool water.

crossover cable: A cable used for connections from node to node.

current: The amount of electrons flowing through a conductor.

current-sensing relay: A device which surrounds a wire and detects the electromagnetic field due to electricity passing through the wire.

daisy chain network configuration: A configuration in which multiple controllers are connected in series.

damper: An adjustable metal blade or set of blades used to control the flow of air.

database: The completed programming information of a controller.

data logging: The recording of information such as temperature and equipment ON/OFF status at regular time intervals.

data trending: The use of past building equipment performance data to determine future system needs.

day/auto lever: A lever that is used to override a thermostat to the day temperature during the night mode.

day/night (occupied/unoccupied) thermostat: A thermostat that has two setpoints, one setpoint for day (occupied time period) and one setpoint for night (unoccupied time period).

deadband: The range between two temperatures in which no heating or cooling takes place.

deadband pneumatic thermostat: A thermostat that allows no heating or cooling to take place between two temperatures.

dead short: A short circuit that opens the overcurrent protection device as soon as the circuit is energized or when the section of the circuit containing the short is energized.

defrost cycle: A mechanical procedure that consists of reversing refrigerant flow in a system to melt frost or ice that builds up on the evaporator coil.

defrost timer: A timer used to initiate a defrost cycle.

dehumidification: The process of removing moisture from air.

dehumidifier: A device that removes moisture from air by causing moisture to condense.

delay on break timer: A timer that begins its operation when a circuit is de-energized after equipment is turned off.

delay on make timer: A timer that begins its timed operation when power is applied to a circuit.

derivative control algorithm: A control algorithm that determines the instantaneous rate of change of a variable.

desiccant dehumidification: The process in which air contacts a chemical substance (desiccant) that adsorbs moisture.

desiccant drier: A device that removes moisture by adsorption.

dew point: The temperature of air below which moisture begins to condense from the air.

dew point temperature: The dry bulb temperature of the air at which the moisture in the air condenses and falls out as dew, rain, sleet, ice, or snow.

diaphragm: A flexible device that transmits the force of incoming air pressure to the piston cup and then to the spring and shaft assembly of an actuator.

differential pressure switch: A digital input device that switches open or closed because of the difference between two pressures.

digital (binary) input: A device that produces an ON or OFF signal.

digital multimeter: A test tool used to measure two or more electrical values.

digital output: A device that accepts an ON or OFF signal.

dimensional change hygrometer: A hygrometer that operates on the principle that some materials absorb moisture and change size and shape depending on the amount of moisture in the air.

DIN rail: A flat mounting rail that is attached to a control panel.

diode: A semiconductor device that allows current to flow in one direction only.

direct current (DC): Current that flows in one direction only.

direct digital control (DDC) strategy: A control strategy in which a building automation system performs closed loop temperature, humidity, or pressure control.

direct digital control system: A control system in which the building automation system controller is wired directly to controlled devices and can turn them ON or OFF, or start a motion.

direct expansion cooling: Cooling produced by the vaporization of refrigerant in a closed system.

direct readjustment (summer reset): A schedule in which the primary variable increases as the secondary variable increases, and decreases as the secondary variable decreases.

discomfort: The condition that occurs when people can sense a difference between themselves and the surrounding air.

disconnect: A switch that isolates electrical circuits from their voltage source to allow safe access for maintenance or repair.

distributed direct digital control system: A control system that has multiple central processing units at the controller level.

diverting (bypass) valve: A three-way valve that has one inlet and two outlets.

docking station: A base station for a notebook computer that includes a power supply, expansion slots, and monitor and keyboard connectors.

doping: The addition of a material to a base element to alter the crystal structure of the element.

downloading: The process of sending a controller database from a personal computer to a controller.

dry air: The elements that make up atmospheric air with the moisture and particles removed.

dual-duct air handling unit: An air handling unit that has hot and cold air ducts connected to mixing boxes at each building space.

dual-input (reset) receiver controller: A receiver controller in which the change of one variable, commonly outside air temperature, causes the setpoint of the controller to automatically change (reset) to match the changing condition.

duct pressure transmitter: A device mounted in a duct that senses the static pressure due to air movement.

ductwork: The distribution system for a forced-air heating or cooling system.

dumb terminal: A display monitor and keyboard, with no processing capabilities.

duplex air compressor: An air compressor that consists of two air compressors and two electric motors on one common receiver.

duty cycle: The percentage of time a load or circuit is ON compared to the time a load or circuit is ON and OFF (total cycle time).

dynamic compressor: A compressor that adds kinetic energy to accelerate air and convert velocity energy to pressure energy with a diffuser.

Echelon: A controller communication system that uses common hardware and software.

economizer: A system that uses outside air to provide cooling for a building space.

economizer cycle: An economizer cycle is an HVAC system cycle in which building spaces are cooled using only outside air.

economizer lockout control: A direct digital control feature in which the economizer damper function is locked out or discontinued.

EEPROM: An integrated-circuit memory chip that has an internal switch to permit the user to erase the contents and write new contents by means of electrical signals.

electrical circuit: The interconnection of conductor(s) and electrical elements through which current is designed to flow.

electrical consumption (usage): The total amount of electricity used during a billing period.

electrical control system: A control system that uses electricity (24 VAC or higher) to operate the devices in the system.

electrical demand: The highest amount of electricity used during a specific period of time.

electrical demand supervisory control: A control strategy designed to reduce the electrical demand component of an electric bill.

electrical impedance hygrometer: A hygrometer based on the principle that the electrical conductivity of a material changes as the amount of moisture in the air changes.

electrical interface diagram: A drawing showing the interconnection between the pneumatic components and electrical equipment in a system.

electric baseboard heater: A device that uses electric resistance heating elements and is located along the base (bottom) of outside walls of a building.

electric control system: A control system in which the power supply is line voltage (120 VAC or 220 VAC) or low voltage (24 VAC) from a step-down transformer that is wired into the building power supply.

electric heating element: A device that consists of wire coils that become hot when energized.

electric heating system: A heating system that consists of electric resistance heating elements and a fan that circulates air across the heating elements.

electricity: The energy released by the flow of electrons in a conductor (wire).

electric motor drive: An electronic device that controls the direction, speed, and torque of an electric motor.

electric/pneumatic (EP) switch: A device that enables an electric control system to interface with a pneumatic control system.

electromechanical control system: A control system that uses electricity (24 VAC or higher) in combination with a mechanism such as a pivot, mechanical bellows, or other device.

electron: A particle that has a negative electrical charge of one unit.

electronic control system: A control system in which the power supply is 24 VDC or less.

electrostatic filter: A device that cleans air by passing the air through electrically charged plates and collector cells.

emergency heat: Heat provided if the outside air temperature drops below a set temperature or if a heat pump fails.

energy cost index (ECI): The amount spent on energy in a commercial building divided by the square feet in the building.

energy utilization index (EUI): The amount of heat energy (in Btu) used in a commercial building divided by the number of square feet in the building.

enthalpy (h): The total heat contained in a material.

equal percentage valve: A valve that provides incremental flow at light loads and large flow capabilities as the valve strokes (opens) farther.

estimation control: A control method that uses the latest building temperature data to estimate the actual start time to heat or cool a building before occupancy.

Ethernet: A local area network (LAN) architecture than can connect up to 1024 nodes and supports data transfer rates of 10 megabits per second (Mbps).

Ethernet software: The program that enables nodes to transmit and receive information on a network.

evaporation: The process that occurs when a liquid changes to a vapor by absorbing heat.

executive summary: A one-page summary of the significant parts of a proposal.

external restriction: A restriction in which a restrictor is placed outside the receiver controller.

fan: A device with rotating blades or vanes that move air.

feedback: The measurement of the results of a controller action by a sensor or switch.

field-effect transistor (FET): A transistor that controls the flow of current through a drain (D) and source (S) with a properly biased gate (G).

field interface device (FID): An electronic device that follows commands sent to it from the CPU of a central-direct digital control system.

filter: A porous material that removes particles from a moving fluid.

filtration: The process of removing particulate matter from air that circulates through an air distribution system.

first law of thermodynamics: The law of conservation of energy, which states that energy cannot be created or destroyed but may be changed from one form to another.

flow switch: A switch that contains a paddle that moves when contacted by air or water flow.

fluidic controller: A controller that uses vector analysis to arrive at an output.

footcandle: The amount of light produced by a lamp (lumens) divided by the area that is illuminated.

force-balance design: A design in which the controller output is determined by the relationship of mechanical pressures.

fuel oil: A petroleum-based product made from crude oil.

fuel recovery rate: The amount of money a utility is permitted to charge to reflect the constantly changing cost of energy.

full-wave rectifier: A circuit containing two diodes and a center-tapped transformer that permits both halves of the input AC sine wave to pass.

fuse: An overcurrent protection device with a fusible link that melts and opens the circuit when an overload condition or short circuit occurs.

fuzzy logic: The ability of a controller to self-diagnose and self-correct a system in order to correct various problems.

gain: The mathematical relationship between the controller output pressure change and the transmitter pressure change that causes it.

gauge pressure: The pressure above atmospheric pressure that is used to express pressures inside a closed system.

generator: An absorption refrigeration system component that vaporizes and separates the refrigerant from the absorbent.

global data: Data needed by all controllers in a network.

ground loop: A circuit that has more than one point connected to earth ground, with a voltage potential difference between the two ground points high enough to produce a circulating current in the ground system.

half-wave rectifier: A circuit containing one diode that allows only half of the input AC sine wave to pass.

hardware: The physical parts that make up a device.

heat: The measurement of energy contained in a substance and identified by a temperature difference or a change of state.

heat exchanger: A device that transfers heat from one substance to another substance without allowing the substances to mix.

heating anticipator: A small heater that causes an HVAC unit to stop heating before normal and avoid overshooting.

heating coil: A finned heat exchanger that adds heat to the air flowing over it.

heating system: A system that increases the temperature of a building space.

heating value: The amount of British thermal units (Btu) per pound or gallon of fuel.

heat pump: A direct expansion refrigeration system that contains devices and controls that reverse the flow of refrigerant.

heat recovery device: A heat exchanger that transfers heat between a medium at two different temperatures.

heat transfer: The movement of heat from one material to another.

hertz (Hz): The international unit of frequency equal to one cycle per second.

high/low signal select: A direct digital control feature in which a building automation system selects among the highest or lowest values from multiple inputs.

high-priority load: A load that is important to the operation of a building and is shed last when demand goes up.

high signal selection relay: A signal selection relay that selects the higher of two input pressures and outputs the higher pressure to a controlled device.

high voltage: Voltage over 600 VAC.

hot water heating system: A heating system that uses hot water as a medium to heat building spaces.

hub: A concentrator that manages the communication between components on a local area network.

human-computer interface: A central computer that enables the building staff to view the operation of the building automation system.

humidification: The process of adding moisture to air.

humidifier: A device that adds moisture to air by causing water to evaporate into the air.

humidistat: A device that senses the humidity level in the air.

humidity: The amount of moisture present in the air.

humidity sensor: A device which senses the amount of moisture in the air and sends a signal to a controller.

hygrometer: Any instrument used for measuring humidity.

hygroscopic element: A device that changes its characteristics as the humidity changes.

instantaneous demand: Electrical demand calculated on a moment-by-moment basis.

insulator: A material that has a high resistance and resists the flow of electrons.

integral control algorithm: A control algorithm that eliminates any offset after a certain length of time.

integrated circuit: An electronic device in which all components (transistors, diodes, and resistors) are contained in a single package or chip.

integrated demand: Electrical demand calculated as an average over a time interval.

integration: A function that calculates the amount of difference between the setpoint and control point (offset) over time.

interface: A device that allows two different types of components, voltage levels, voltage types, or systems to be interconnected.

internal restriction: A restriction in which a receiver controller has an internally mounted restrictor.

Internet Protocol (IP) addressing: A method of assigning addresses to nodes on a network.

interoperability: The ability of devices produced by different manufacturers to communicate and share information.

ignition temperature: The intensity of heat required to start the chemical reaction.

incremental output: A digital output device used to position a bidirectional electric motor.

indoor air quality (IAQ): A designation of the contaminants present in the air.

induction air handling unit: An air handling unit that maintains a constant 55°F air temperature and delivers the air to the building spaces at a high duct pressure.

inert-element temperature transmitter: A pneumatic transmitter that is used to sense outside air temperatures in through-the-wall applications.

infrared communication: Communication between devices using infrared (IR) radiation.

jumper: A conductor used to connect pins on a controller or device.

keypad display: A controller-mounted device that consists of a small number of keys and a small display.

latching relay: A relay that requires a short pulse to energize the relay and turn ON the load.

latent heat: Heat identified by a change of state and no temperature change.

lead/lag control: The alternation of operation between two or more similar pieces of equipment.

lead/lag switch: A pressure switch that determines which compressor is the primary (lead) compressor and which compressor is the backup (lag) compressor.

life safety supervisory control: A control strategy for life safety issues such as fire prevention, detection, and suppression.

light-emitting diode: A diode designed to produce light when forward biased.

light level switch: A device that indicates if a light level is above or below a setpoint.

light-troffer application: An application in which the thermostat is located in a return air slot that is integral to a light fixture.

limit thermostat: A thermostat that maintains a temperature above or below an adjustable setpoint.

linear valve: A valve in which the flow through the valve is equal to the amount of valve stroke.

line voltage: Voltage at 120 VAC up to 4160 VAC. Control systems at 120 VAC are the most common line-voltage control systems.

liquid chiller: A system that uses a liquid (normally water) to cool building spaces.

local area network (LAN): A communication network that spans a relatively small area.

lockout: The use of locks, chains, or other physical restraints to positively prevent the operation of equipment.

low-priority load: A load that is shed first when a high electrical demand period occurs.

low temperature cutout control: A device that protects against damage due to a low temperature condition.

low voltage: Voltage at 30 VAC or less (commonly 24 VAC).

makeup air: Air that is used to replace air that is lost to exhaust.

manual drain: A device that is opened and closed by a person to drain moisture from a receiver.

master/submaster thermostat: A pneumatic thermostat that has a third air connection piped to either a manual air regulator or an outside air transmitter.

micron: A unit of measure equal to .000039″.

minimum-position relay: A relay that prevents outside air dampers from completely closing.

minimum ventilation air percentage: The minimum amount of outside air that must be mixed with return air before the air is allowed to enter a building space.

mixed air: The combination of return air and outside air.

mixed network: A network in which new Ethernet network components are added to early Arcnet network components.

mixing box: A sheet metal box with inlets for hot and cold air.

mixing valve: A three-way valve that has two inlets and one outlet.

moist air: The mixture of dry air and moisture.

motor starter: An electrically operated switch (contactor) that includes motor overload protection.

multidrop network configuration: A configuration in which controllers are connected to the network that runs throughout a building through spliced drops for each controller.

multizone air handling unit: An air handling unit that is designed to provide heating, ventilation, and air conditioning for more than one building zone or area.

natural gas: A colorless, odorless fossil fuel.

network: Interconnected equipment used for sending and receiving information.

network address: A unique number assigned to each building automation system controller on an RS-485 communications network.

network communication module (NCM): A controller that coordinates communication from controller to controller on a network and provides a location for operator interface.

network topology: The map (description) of a network configuration.

neutron: A particle that has no electrical charge.

node: A device, such as a computer or a printer, that has a unique address and is attached to a network.

normally closed (NC): A valve that does not allow fluid to flow when the valve is in its normal position.

normally open (NO): A valve that allows fluid to flow when the valve is in its normal position.

N-type material: Material created by doping a region of a crystal with atoms from an element that has more electrons in its outer shell than the crystal.

nucleus: The heavy, dense center of an atom.

offset: The difference between a control point and a setpoint.

Ohm's law: A physical law which expresses the relationship between voltage, current, and resistance in a circuit.

oil carryover: Lubricating oil that leaks by the piston rings and is carried into a compressed air system.

oil removal filter (separator): A device that removes oil droplets from a pneumatic system by forcing compressed air to change direction quickly.

one-pipe (low-volume) device: A device that uses a small amount of the compressed air supply (restricted main air).

ON/OFF (digital) control: Control in which a controller produces only a 0% or 100% output signal.

open loop control: Control in which no feedback occurs between the controller, sensor, and controlled device.

operator interface: A device that allows an individual to access and respond to building automation system information.

opposed blade damper: A damper in which adjacent blades move in opposite directions from one another.

optical fiber: Glass fiber used to transmit information using light.

optimum start: A supervisory control strategy in which the HVAC load is turned ON as late as possible to obtain the proper building space temperature at the beginning of building occupancy.

optimum stop: A supervisory control strategy in which the HVAC load is turned OFF as early as possible to maintain the proper building space temperature at the end of building occupancy.

optoisolation: A communication method in which controllers use photonic (light-sensitive) components to prevent communication problems.

outside air: Air brought into a building space from outside the building.

outside air economizer: A unit that uses outside air to cool building spaces.

overcurrent protection: A device that shuts OFF the power supply when current flow is excessive.

overload: A device that prevents overcurrent in an electrical control system.

overshooting: The increasing of a controlled variable above the controller setpoint.

packing: A bulk deformable material or one or more mating deformable elements reshaped by manually adjustable compression.

pager (beeper): A small portable electronic device that vibrates, emits a beeping signal, or displays a text message when the individual carrying it is paged.

parallel blade damper: A damper in which adjacent blades are parallel and move in the same direction with one another.

parallel connection: A connection that has two or more components connected so that there is more than one path for current flow.

parallel device: A device that transmits multiple bits of information (0s or 1s) simultaneously.

partial short: A short circuit of only a section or several sections of a machine.

parts list: A reference list that indicates part description acronyms and actual manufacturer part names and numbers.

passive dehumidification: The process of removing moisture from air by using the existing cooling coils of a system.

patch cable: A cable used for connections from a node to a concentrator.

peripheral device: A device that is connected to a personal computer or building automation system controller to perform a specific function.

permanent holiday: A holiday that remains on the same date each year.

personal digital assistant (PDA): A small hand-held computing device with a small display screen.

photodarlington transistor: A transistor that consists of a phototransistor and a standard NPN transistor in a single package.

photodiode: An electronic device that changes resistance or switches ON when exposed to light.

phototransistor: An NPN transistor that has a large, thin base region that is switched ON when exposed to a light source.

physical layer: The cables and other network devices, such as hubs and switches, that make up a network.

physiological function: A natural physical or chemical function of an organism.

pilot bleed (two-pipe) thermostat: A thermostat that uses air volume amplified by a relay to control the temperature in a building space or area.

ping: An echo message and its reply sent by one network device to detect the presence of another device.

piston cup: A device that transmits the force generated by the air pressure against the diaphragm to the spring and shaft assembly of an actuator.

pitot tube: A device that senses static pressure and total pressure in a duct.

pneumatic control diagram: A pictorial and written representation of pneumatic controls and related equipment.

pneumatic control system: A control system in which compressed air is used to provide power for the control system.

pneumatic/electric (PE) switch: A device that allows an air pressure signal to energize or de-energize an electrical device such as a fan, pump, compressor, or electric heating element.

pneumatic/electronic transducer (PET): A device that converts an air pressure input signal to an electronic output signal.

pneumatic humidistat: A controller that uses compressed air to open or close a device which maintains a certain humidity level inside a duct or area.

pneumatic humidity transmitter: A device that measures the amount of moisture in the air compared to the amount of moisture the air could hold if it were saturated.

pneumatic positioner (pilot positioner): An auxiliary device mounted to a damper or valve actuator that ensures that the damper or actuator moves to a given extension.

pneumatic pressure switch: A controller that maintains a constant air pressure in a duct or area.

pneumatic thermostat: A thermostat that uses changes in compressed air to control the temperature in individual rooms inside a commercial building.

pneumatic transmitter: A device that senses temperature, pressure, or humidity and sends a proportional (3 psig to 15 psig) signal to a receiver controller.

point mapping: The process of adding the individual input and output points to the database of the human-computer interface.

polarity: The positive (+) or negative (−) state of an object.

portable operator terminal (POT): A small, lightweight, hand-held device that allows access to basic building automation system functions from various controllers throughout a building.

positive-displacement compressor: A compressor that compresses a fixed quantity of air with each cycle.

power factor (PF): The ratio of true power used in an AC circuit to apparent power delivered to the circuit.

predictive maintenance: The monitoring of wear conditions and equipment characteristics and comparing them to a predetermined tolerance to predict possible malfunctions or failures.

pressure: The force created by a substance per unit of area.

pressure regulator: A valve that restricts and/or blocks downstream air flow.

pressure sensor: A device which measures the pressure in a duct, pipe, or room and sends a signal to a controller.

pressure switch: An electric switch operated by pressure that acts on a diaphragm or bellows element.

pressure test: A test that determines the time required for an air compressor to reach the pressure switch shut-off pressure.

pressure transmitter: A device used to sense the pressure due to air flow in a duct or water flow through a pipe.

preventive maintenance: Scheduled inspection and work (lubrication, adjustment, cleaning) required to maintain equipment in peak operating condition.

preventive maintenance reporting: The generating of forms to notify maintenance personnel of routine maintenance procedures.

programming: The creation of software.

proportional band: The number of units of a controlled variable that causes the actuator to move through its entire spring range.

proportional (analog) control: Control in which the controlled device is positioned in direct response to the amount of offset in the system.

proportional control algorithm: A control algorithm that positions the controlled device in direct response to the amount of offset in a building automation system.

proportional/integral/derivative (PID) controller: A direct digital control system that uses proportional, integral, and derivative algorithms.

proton: A particle that has a positive electrical charge of one unit.

pseudopoint (data point): A point that exists only in software and is not a hard-wired point.

psychrometric chart: A graph that defines the properties of the air at various conditions.

psychrometrics: The scientific study of the properties of air and the relationships between them.

P-type material: Material created by doping a region of a crystal with atoms from an element that has fewer electrons in its outer shell than the crystal.

pulse width modulation (PWM): A control technique in which a sequence of short pulses is used to position an actuator.

quick-opening valve: A valve in which flow increases rapidly as soon as the valve is opened.

radiant heat panel: A device that uses an electric resistance heating element embedded in a ceiling panel.

radiation: Heat transfer in the form of radiant energy (electromagnetic waves).

radio frequency communication: The use of a high-frequency electronic signal to communicate between control system components.

ratchet clause: Documentation that permits an electric utility to charge for demand based on the highest amount of electricity used in a 12-month period, not the amount actually measured.

readjustment: The relationship between the primary and secondary (reset) variables from the schedule.

receiver controller: A device which accepts one or more input signals from pneumatic transmitters and produces an output signal based on its calibration.

reciprocating compressor: A compressor that uses reciprocating pistons to compress air.

rectifier: A device that changes AC voltage into DC voltage.

refrigerant: Fluid used for transferring heat (energy) in a refrigeration system.

refrigerated air drier: A device that uses refrigeration to lower the temperature of compressed air.

relative humidity: The amount of moisture in the air compared to the amount of moisture that it could hold if it were saturated.

relay: A device that uses a low voltage to switch a high voltage.

remote bulb controller: A device that consists of a controller mounted to a duct or pipe that is connected by a capillary tube to a bulb that is inserted into the duct or pipe.

remote control point adjustment: The ability to adjust the controller setpoint from a remote location.

reset control: A direct digital control feature in which a primary setpoint is reset automatically as another value (reset variable) changes.

reset schedule: A chart that describes the setpoint changes in a pneumatic control system.

resistance temperature detector (RTD): A resistor made of a material for which the electrical resistance is a known function of the temperature.

restored load: A shed load that has been turned ON by an electrical demand supervisory control strategy.

restrictor: A fixed orifice that meters air flow through a port and allows fine output pressure adjustments and precise circuit control.

retrofit: The process of upgrading a building automation system by replacing obsolete or worn parts, components, and controls with new or modern ones.

reverse readjustment (winter reset): A schedule in which the primary variable increases as the secondary variable decreases and decreases as the secondary variable increases.

reversing relay: A relay designed to invert the output signal relative to the input signal.

RJ-45 connector: A connector that contains eight pins instead of four as in standard phone connectors.

rod-and-tube temperature transmitter: A pneumatic transmitter that uses a high-quality metal rod with precision expansion and contraction characteristics as the sensing element.

room temperature transmitter: A transmitter used in applications that require a receiver controller to measure and control the temperature in an area.

rotary compressor: A compressor that uses a rotating motion to compress air. Rotary compressors may be screw or vane compressors.

rotating priority load shedding: An electrical demand supervisory control strategy in which the order of loads to be shed is changed with each high electrical demand condition.

round blade damper: A damper having a round blade. Round blade dampers are used in systems consisting of round ductwork.

router: A concentrator that manages the communication between different networks.

RS-485: A multipoint communication standard that incorporates low-impedance drivers and receivers providing high tolerance to noise.

run time test: A test that measures the percentage of time that a compressor runs to enable it to maintain a supply of compressed air to the control system.

safety relief valve: A device that prevents excessive pressure from building up by venting air to the atmosphere.

screw compressor: A compressor that contains a pair of screw-like rotors that interlock as they rotate.

second law of thermodynamics: A law of physics that states that heat always flows from a material at a high temperature to a material at a low temperature.

semiconductor: A material in which electrical conductivity is between that of a conductor (high conductivity) and that of an insulator (low conductivity).

sensible heat: Heat measured with a thermometer or sensed by a person. Sensible heat does not involve a change of state.

sensor: A device that measures a controlled variable such as temperature, pressure, or humidity and sends a signal to a controller.

sequence chart: A chart that shows the numerical relationship between the different values in a pneumatic system.

serial device: A device that transmits one bit of information (0 or 1) at a time.

serial information: Information that is transmitted sequentially one bit (0 or 1) at a time.

series connection: A connection that has two or more components connected so that there is one path for current flow.

series/parallel connection: A combination of series- and parallel-connected components.

setback: The unoccupied heating setpoint.

setpoint: The desired value to be maintained by a system.

setpoint schedule: A description of the amount a reset variable resets the primary setpoint.

setup: The unoccupied cooling setpoint.

shed load: An electric load that has been turned OFF by an electrical demand supervisory control strategy.

shed table: A table that prioritizes the order in which electrical loads are turned OFF.

short circuit: Current that leaves the normal current-carrying path by going around the load and back to the power source or to ground.

signal selection relay: A multiple-input device that selects the higher or lower of two pneumatic signal levels.

silicon-controlled rectifier: A thyristor that is capable of switching direct current.

single-input receiver controller: A receiver controller that is designed to be connected to only one transmitter and to maintain only one temperature, pressure, or humidity setpoint.

single-temperature-setpoint thermostat: A pilot bleed thermostat that has one setpoint year-round for a building space or area.

single-zone air handling unit: An air handling unit that provides heating, ventilation, and air conditioning to only one building zone or area.

software: The program that enables a controller to function.

solar energy: Energy transmitted from the sun by radiation.

specific heat: The ability of a material to hold heat.

specific humidity: A measurement of the exact amount or weight of the moisture in the air.

spring range: Difference in pressure at which an actuator shaft moves and stops.

spring range shift: The process by which the nominal spring range changes to the actual spring range.

spring range overlap: A condition in which actuators with different spring ranges interfere with each other.

stagnant air: Air that contains an excess of impurities and lacks the oxygen required for comfort.

stale indoor air: Air that contains odors or contaminants.

star network configuration: A configuration in which communication network cables radiate out to remote controllers located within a building.

starts per hour test: A test that records the number of times an air compressor starts per hour under a standard load.

steam heating system: A heating system that uses steam as a medium to heat building spaces.

step-down transformer: A transformer in which the secondary coil (low-voltage side) has more turns of wire than the primary coil (high-voltage side).

step-up transformer: A transformer in which the secondary coil (high-voltage side) has more turns of wire than the primary coil (low-voltage side).

strain gauge: A spring or thin piece of metal that measures movement of a sensing element.

submasking: An addressing scheme that filters messages and determines if a message is to be passed to local nodes on a subnetwork or if it is to be sent on to the main network through the router.

subnetwork: A local area network that is made up of components that communicate frequently.

supervisory control strategy: A programmable software method used to control the energy-consuming functions of a commercial building.

switch: A concentrator that manages the communication between networks or parts of networks that operate at different data transmission speeds.

switching relay: A device that switches air flow from one circuit to another.

tagout: The process of attaching a danger tag to the source of power to indicate that the equipment may not be operated until the tag is removed.

temperature: The measurement of the intensity of the heat of a substance.

temperature sensor: A device which measures the temperature in a duct, pipe, or room and sends a signal to a controller.

temperature stratification: The variation of air temperature in a building space that occurs when warm air rises to the ceiling and cold air drops to the floor.

terminal reheat air handling unit: An air handling unit that delivers air at a constant 55°F temperature to building spaces.

therm: The quantity of gas required to produce 100,000 Btu.

thermal recovery coefficient: The ratio of an indoor temperature change and the length of time it takes to obtain that temperature change.

thermistor: A resistor made of semiconductor material in which electrical resistance varies based on changes in temperature.

thermodynamics: The science of thermal energy (heat) and how it transforms to and from other forms of energy.

thermostat: A temperature-actuated switch.

thicknet: Thick coaxial cable classified as 10base5.

thinnet: Thin coaxial cable classified as 10base2.

throttling range: The number of units of a controlled variable that causes the actuator to move through its entire spring range.

thyristor: A solid-state switching device that switches current ON by a quick pulse of control current.

tier: The classification of a customer depending on the type of service and the amount of electricity used.

time-based supervisory control: A control strategy in which the time of day is used to determine the desired operation of a load.

timed override: A time-based supervisory control strategy in which the occupants can change a zone from an unoccupied to occupied mode for temporary occupancy.

timed override initiator: A device that, when closed, sends a signal to a controller which indicates that a timed override period is to begin.

token ring: A data transmission method in which a device obtains a token (ticket) that allows it to transmit on a network while the other nodes listen.

ton of cooling: The amount of heat required to melt a ton of ice (2000 lb) over a 24-hour period.

transducer: A device that changes one type of proportional control signal into another.

transformer: A device that steps up or steps down alternating current.

transient holiday: A holiday that changes its date each year.

transistor: A three-terminal semiconductor device that controls current according to the amount of voltage applied to the base.

transmitter: A device that sends an air pressure signal to a receiver controller.

transmitter range: The temperatures between which a transmitter is capable of sensing.

transmitter sensitivity: The output pressure change that occurs per unit of measured variable change.

transmitter span: The difference between the minimum and maximum sensing capability of a transmitter (number of units between the endpoints of the transmitter range).

triac: A solid-state switching device used to switch alternating current.

tuning: The downloading of the proper response times into the controller and checking the response of the control systems.

two-pipe (high-volume) device: A device that uses the full volume of compressed air available.

two-way valve: A valve that has two pipe connections.

undershooting: Decreasing of a controlled variable below the controller setpoint.

unitary controller: A controller designed for basic zone control using a standard wall-mount temperature sensor.

unit ventilator: A small air handling unit mounted on the outside wall of each room in a building.

universal input-output controller (UIOC): A controller designed to control most HVAC equipment.

unshielded twisted pair (UTP): A pair of wires that are twisted around each other with no electromagnetic shielding.

vacuum tube: A device that switches or amplifies electronic signals.

valve: A device that controls the flow of fluids in an HVAC system.

valve flow characteristics: The relationship between the valve stroke and flow through the valve.

valve shut-off rating: The maximum fluid pressure against which a valve can completely close.

valve stem: A valve component that consists of a metal shaft, normally made of stainless steel, that transmits the force of the actuator to the valve plug.

valve turn down ratio: The relationship between the maximum flow and the minimum controllable flow through the valve.

vane compressor: A positive-displacement compressor that has multiple vanes located in an offset rotor.

variable air volume air handling unit: An air handling unit that moves a variable volume of air.

variable air volume (VAV) terminal box: A device that controls the air flow to a building space, matching the building space requirements for comfort.

variable air volume terminal box controller: A controller that modulates the damper inside a variable air volume (VAV) terminal box to maintain a specific building space temperature.

vector analysis: The calculation of outputs based on the direction and strength of forces.

ventilation: The process of introducing fresh air into a building.

viscosity: The ability of a liquid to resist flow.

voice phone interface: An automated phone service in which a voice on a computer chip prompts the caller to press various numbers for different functions when dialing in on a standard touch-tone phone.

vortex damper: A pie-shaped damper located at the inlet of a centrifugal fan.

water chiller: A device that cools water.

water-to-air heat pump: A heat pump that uses water as the heat source and heat sink.

web-enabled control system: A control system consisting of supervisory controllers that have Internet capability.

weight: The force with which a body is pulled downward by gravity.

wet bulb temperature: The temperature of the air taking into account the amount of humidity in the air.

wide area network (WAN): A communication network that spans a relatively large area.

winter/summer thermostat: A pilot bleed thermostat that changes the setpoint and action of the thermostat from the winter (heating) to the summer (cooling) mode.

written sequence of operation: A written description of the operation of a control system.

zener diode: A diode designed to operate in a reverse-biased mode without being damaged.

HVAC *Control Systems*   Index

A

absolute pressure, 13
absorbent, 40
absorber, 40
absorption refrigeration system, 40
access levels and codes, 232
accumulators, *240*
actuator, *83*
  color codes, *86*
  components, 84
  spring range, 86
    shift, 86, *87*
    testing, *86*
AC voltage, 180–181, *180*
adapter card, 261
adaptive control algorithm, 273–*274*
adaptive start control, 281
adsorption, 71
air
  circulated, 3
  intake
    filter, 69
    outside, 68
  mixed, 20
  quality, 3
    indoor, 21, 56
  return, 4
  stagnant, 4
  stale indoor, 20
  velocity, 4
air compressor, 65
  performance tests, 77
  preventive maintenance, 80–81, *80*
  replacement parts, *81*
  second, 78
air compressor station, 59, *66*
air conditioning, 39
air conditioning system, 39
air filter replacement, 81
air handling unit, 24–*25,* 31–32
  control of VAV systems, 275
  controllers, 219–220

  induction, *32*
  multizone, 169
  single-zone, 167–169, *168*
  variable air volume, *32*
air line filter, 72
air lines, 76
  installation of, 76
alarm
  differential, 300
  monitoring, 299–300
  parameter, 322–323
  printer, 226
  reporting, 301
algorithms, 272–274
alternate retrofit strategies, 298
alternate scheduling, 280
alternating, 179
alternators, *79*
aluminum end cap, *84*
aluminum piston cup, *85*
American Society for Testing and Materials (ASTM), 37
analog control system, 61
analog input, *233*
  resistance value testing, 257
analog output, 242
anticipator, 53
appendices and documentation, 314
application-specific controllers, 218–219
Arcnet, 265
artificial intelligence, 327
as-built, 165
ASHRAE psychrometric chart, *14*
atmospheric air, 8
atmospheric pressure, 13
atomic structure, 177
authority/ratio calculation, *141*
automated control system, *62*
  applications, 63
automatic control system operation, 54
automatic drain, 70, *71*
  valve, 70
automation system installation, 290

Using the HVAC Control Systems CD-ROM

Before removing the CD-ROM from the protective sleeve, please note that the book cannot be returned for refund or credit if the CD-ROM sleeve seal is broken.

System Requirements

The *HVAC Control Systems* CD-ROM is designed to work best on a computer meeting the following hardware/software requirements:

- Intel® Pentium® processor or equivalent
- Microsoft® Windows® 95, 98, 98 SE, Me, NT®, 2000, or XP® operating system
- 64 MB of free available system RAM (128 MB recommended)

- 90 MB of available disk space
- 800 × 600 16-bit (thousands of colors) color display or better
- Sound output capability and speakers
- CD-ROM drive

Opening Files

Insert the CD-ROM into the computer CD-ROM drive. Within a few seconds, the home screen will be displayed allowing access to all features of the CD-ROM. Information about the usage of the CD-ROM can be accessed by clicking on USING THIS CD-ROM. The Chapter Quick Quizzes™, Illustrated Glossary, Media Clips, and Reference Material can be accessed by clicking on the appropriate button on the home screen. Clicking on the American Tech logo accesses the American Tech web site (www.go2atp.com) for information on related educational products. Unauthorized reproduction of the material on this CD-ROM is strictly prohibited.

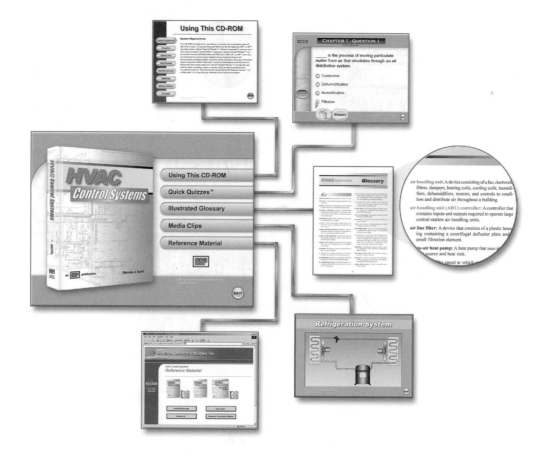